GREEN CHEMISTRY AND SUSTAINABLE TECHNOLOGY

Biological, Pharmaceutical, and Macromolecular Systems

GREEN CHEMISTRY AND SUSTAINABLE TECHNOLOGY

Biological, Pharmaceutical, and Macromolecular Systems

Edited by

Satish A. Dake, PhD

Ravindra S. Shinde, PhD

Suresh C. Ameta, PhD

A. K. Haghi, PhD

Apple Academic Press Inc.
4164 Lakeshore Road
Burlington ON L7L 1A4
Canada

Apple Academic Press, Inc.
1265 Goldenrod Circle NE
Palm Bay, Florida 32905
USA

First issued in paperback 2021

Exclusive worldwide distribution by CRC Press, a member of Taylor & Francis Group
No claim to original U.S. Government works

ISBN 13: 978-1-77463-495-0 (pbk)
ISBN 13: 978-1-77188-862-2 (hbk)

Library and Archives Canada Cataloguing in Publication

Title: Green chemistry and sustainable technology : biological, pharmaceutical, and macromolecular systems / edited by Satish A. Dake, PhD, Ravindra S. Shinde, PhD, Suresh C. Ameta, PhD, A. K. Haghi, PhD.
Names: Dake, Satish A., editor. | Shinde, Ravindra S., 1979- editor. | Ameta, Suresh C., editor. | Haghi, A. K., editor.
Description: Includes bibliographical references and index.
Identifiers: Canadiana (print) 20200224263 | Canadiana (ebook) 20200224328 | ISBN 9781771888622 (hardcover) | ISBN 9780367808310 (ebook)
Subjects: LCSH: Green chemistry. | LCSH: Technology—Environmental aspects.
Classification: LCC TP155.2.E58 G74 2020 | DDC 660.028/6—dc23

CIP data on file with US Library of Congress

Apple Academic Press also publishes its books in a variety of electronic formats. Some content that appears in print may not be available in electronic format. For information about Apple Academic Press products, visit our website at **www.appleacademicpress.com** and the CRC Press website at **www.crcpress.com**

About the Editors

Satish A. Dake, PhD

Department of Basic Science and Humanities, Deogiri Institute of Engineering and Management Studies, Aurangabad, Maharashtra, India

Satish A. Dake, PhD, is working as an Assistant Professor in Engineering Chemistry at the Department *of Basic Science and Humanities,* Deogiri Institute of Engineering and Management Studies, Aurangabad, Maharashtra, India. He has published 27 research papers in national and international journals and four book chapters. His interests are in the fields of green chemistry and synthesis of bioactive heterocyclic compounds such as antibacterial and *in vitro* antiproliferative, anti-inflammatory, antibacterial, anticancer, and antimicrobial activity. He has worked as an Associate Editor and Member of the journal *Anti-Infective Agents* and the *Journal of Basic and Applied Sciences*. He is a life member of the Indian Science Congress Association (ISCA).

Ravindra S. Shinde, PhD

Assistant Professor, Department of Chemistry and Industrial Chemistry, Dayanand Science College, Latur, Maharashtra, India

Ravindra S. Shinde, PhD, is presently serving as Assistant Professor, Department of Chemistry at Dayanand Science College, Latur, in Maharashtra, India. He has around 14 years of teaching experience at the BSc and MSc levels. Having more than 22 research publications to his credit in journals of national and international repute, he is also the author of many undergraduate- and postgraduate-level books. He has published two books with Apple Academic Press: *Green Chemistry and Sustainable Technology* and *Modern Green Chemistry and Heterocyclic Compounds,* and one with Lulu Press: *Practical Chemistry.* He has also written chapters in books published by several other international publishers. Dr. Shinde has delivered lectures and chaired sessions at national conferences and is a reviewer of a number of international journals. In addition, he has completed minor research projects from different funding agencies. He is a university-approved and recognized postgraduate teacher of chemistry. He has around one decade of administrative experience as N.S.S. Programme Officer, member of the University Exam Committee, coordinator of UGC-sponsored NET/SET coaching cell, and One Teacher One Skill Committee of the college. He was also a Brand Ambassador of the online NPTEL Examination run by I.I.T. and I.I.Sc. He was awarded a number of prestigious awards during his careers, such as from the Vidyabushan Puraskarby Indian NET/SET Association (2010) and a National Teacher Award (2019). He also has working experience in S. R. T. M. University, Nanded, in various departments, NSS District Area Co-coordinator, an internal and external examiner of MSC and BSC chemistry examination and member of the university skill development committee. His areas of research are heterocyclic chemistry, medicinal chemistry, multicomponent reactions, nanocatalysis, organic synthesis, click chemistry, green chemistry, synthetic organic chemistry, and nanomaterials science. His fields of interest are medicinal chemistry and drug design.

Suresh C. Ameta, PhD

Dean, Faculty of Science, PAHER University, Udaipur, India

Suresh C. Ameta, PhD, is currently Dean, Faculty of Science at PAHER University, Udaipur, India. He has served as Professor and Head of the Department of Chemistry at North Gujarat University Patan and at M. L. Sukhadia University, Udaipur, and as Head of the Department of Polymer Science. He also served as Dean of Postgraduate Studies. Prof. Ameta has held the position of President of the Indian Chemical Society, Kolkata, and is now a life-long Vice President. He was awarded a number of prestigious awards during his career, such as national prizes twice for writing chemistry books in Hindi. He also received the Prof. M. N. Desai Award (2004), Prof. W. U. Malik Award (2008), the National Teacher Award (2011), the Prof. G. V. Bakore Award (2007), a Life-Time Achievement Award by the Indian Chemical Society (2011) as well as the Indian Council of Chemist (2015), etc. He has successfully guided 81 PhD students. Having more than 350 research publications to his credit in journals of national and international repute, he is also the author of many undergraduate- and postgraduate-level books. He has published three books with Apple Academic Press: *Chemical Applications of Symmetry and Group Theory; Microwave-Assisted Organic Synthesis*; and *Green Chemistry: Fundamentals and Applications*; and two with Taylor and Francis: *Solar Energy Conversion and Storage* and *Photocatalysis.* He has also written chapters in books published by several other international publishers. Prof. Ameta has delivered lectures and chaired sessions at national conferences and is a reviewer of a number of international journals. In addition, he has completed five major research projects from different funding agencies, such as DST, UGC, CSIR, and Ministry of Energy, Govt. of India.

A. K. Haghi, PhD

Editor-in-Chief, International Journal of Chemoinformatics and Chemical Engineering and Polymers Research Journal;
Member, Canadian Research and Development Center of Sciences and Cultures (CRDCSC), Montreal, Quebec, Canada

A. K. Haghi, PhD, is the author and editor of 165 books, as well as 1000 published papers in various journals and conference proceedings. Dr. Haghi has received several grants, consulted for a number of major corporations, and is a frequent speaker to national and international audiences. Since 1983, he served as professor at several universities. He served as Editor-in-Chief of the *International Journal of Chemoinformatics* and *Chemical Engineering* and *Polymers Research Journal* and is on the editorial boards of many International journals. He is also a member of the Canadian Research and Development Center of Sciences and Cultures (CRDCSC), Montreal, Quebec, Canada. He holds a BSc in urban and environmental engineering from the University of North Carolina (USA), an MSc in mechanical engineering from North Carolina A&T State University (USA), a DEA in applied mechanics, acoustics, and materials from the Université de Technologie de Compiègne (France), and a PhD in engineering sciences from Université de Franche-Comté (France).

Contents

Contributors

Rakshit Ameta
Department of Chemistry, J. R. N. Rajasthan Vidyapeeth (Deemed to be University), Udaipur–313001, Rajasthan, India

Suresh C. Ameta
Department of Chemistry, PAHER University, Udaipur–313003, Rajasthan, India,
E-mail: ameta_sc@yahoo.com

J. Aneli
Institute of Macromolecular Chemistry and Polymeric Materials, Iv. Javakhishvili Tbilisi State University, I. Chavchavadze Ave., 13, Tbilisi 0179, Georgia

Akshada A. Bakliwal
Department of Pharmaceutics, Sandip Institute of Pharmaceutical Sciences, Mahiravani, Nashik, Maharashtra, India

Jayesh Bhatt
Department of Chemistry, PAHER University, Udaipur–313003, Rajasthan, India

Eszter Bögi
CEM, Institute of Experimental Pharmacology and Toxicology, Slovak Academy of Sciences, Dúbravská Cesta 9, SK–84104 Bratislava, Slovakia

Vikas V. Borgaonkar
Department of Chemistry, Shri Siddheshwar Mahavidyalaya, Majalgaon–431131, Maharashtra, India

W. Brostow
Laboratory of Advanced Polymers and Optimized Materials (LAPOM) Department of Materials Science and Engineering and Department of Physics, University of North Texas,
3940 North Elm Street, Denton, TX 76207, USA, E-mail: wkbrostow@gmail.com

Satish A. Dake
Department of Engineering Chemistry, Deogiri Institute of Engineering and Management Studies, Station Road, Aurangabad–431005, Maharashtra, India, E-mail: satish_dake57@yahoo.com

Shweta S. Gedam
Department of Pharmaceutics, Sandip Institute of Pharmaceutical Sciences, Mahiravani, Nashik, Maharashtra, India

N. Jalagonia
Institute of Macromolecular Chemistry and Polymeric Materials, Iv. Javakhishvili Tbilisi State University, I. Chavchavadze Ave., 13, Tbilisi 0179, Georgia

Monika Jangid
Department of Chemistry, PAHER University, Udaipur–313003, Rajasthan, India

Onkar K. Jogdand
Department of Environmental Science, Deogiri College, Station Road, Aurangabad–431005, Maharashtra, India, E-mail: onkar.jogdand@gmail.com

Poonam Khullar
Department of Chemistry, B.B.K. D.A.V. College for Women, Amritsar–143005, Punjab, India

Seema Kothari
PAHER University, Udaipur–313003, Rajasthan, India

E. Markarashvili
Department of Chemistry, Iv. Javakhishvili Tbilisi State University, I. Chavchavadze Ave., 1, Tbilisi 0179, Georgia, Institute of Macromolecular Chemistry and Polymeric Materials, IV. Javakhishvili Tbilisi State University, I. Chavchavadze Ave., 13, Tbilisi 0179, Georgia

Sunil R. Mirgane
Department of Chemistry, J.E.S. College, Jalna–431203 Maharashtra, India

O. Mukbaniani
Department of Chemistry, Iv. Javakhishvili Tbilisi State University, I. Chavchavadze Ave., 1, Tbilisi 0179, Georgia, E-mail: omar.mukbaniani@tsu.ge

Sukanchan Palit
43, Judges Bagan, Post-Office-Haridevpur, Kolkata–700082, India, Tel.: 0091-8958728093, E-mails: sukanchan68@gmail.com, sukanchan92@gmail.com, sukanchanp@rediffmail.com

Ajay M. Patil
Department of Chemistry, Pratishthan Mahavidyalaya, Paithan–431107, Maharashtra, India

Avinash Kumar Rai
PAHER University, Udaipur–313003, Rajasthan, India

B. R. Sharma
Department of Physics, Pratishthan Mahavidyalaya, Paithan–431107, Maharashtra, India

Ravindra S. Shinde
Department of Chemistry and Industrial Chemistry, Dayanand Science College, Latur–413512, Maharashtra, India

Nana V. Shitole
Department of Chemistry, Shri Shivaji College, Basmat Road, Parbhani–431401, Maharashtra, India, E-mail: nvshitole@gmail.com

Ladislav Šoltés
CEM, Institute of Experimental Pharmacology and Toxicology, Slovak Academy of Sciences, Dúbravská Cesta 9, SK–84104 Bratislava, Slovakia

Ivana Šušaníková
University of Comenius, Pharmaceutical Faculty, Odbojárov 65/10, SK–83104 Bratislava, Slovakia

Swati G. Talele
Department of Pharmaceutics, Sandip Institute of Pharmaceutical Sciences, Mahiravani, Nashik, Maharashtra, India, E-mail: swatitalele77@gmail.com

T. Tatrishvili
Department of Chemistry, Iv. Javakhishvili Tbilisi State University, I. Chavchavadze Ave., 1, Tbilisi 0179, Georgia, Institute of Macromolecular Chemistry and Polymeric Materials, IV. Javakhishvili Tbilisi State University, I. Chavchavadze Ave., 13, Tbilisi 0179, Georgia

Dominika Topoľská
IQVIA, Vajnorská 100/B, SK – 83104 Bratislava, Slovakia

Katarína Valachová
CEM, Institute of Experimental Pharmacology and Toxicology, Slovak Academy of Sciences, Dúbravská Cesta 9, SK – 84104 Bratislava, Slovakia, E-mail: katarina.valachova@savba.sk

Abbreviations

[Bdmim]Cl	1-butyl-2,3-dimethylimidazolium chloride
[Bmim][Cl]	1-butyl-3-methylimidazolium chloride
[Bmp]Cl	1-butyl-4-methylpyridinium chloride
[Emim]Cl	1-ethyl-3-methylimidazolium chloride
2-HEAA	2-hydroxy ethylammonium acetate
2-HEAF	2-hydroxyethylammonium formate
Ac%	acetyl content
AE	Afsin-Elbistan
AIL	acidic ionic liquid
BBB	blood-brain barrier
BD	bulk density
BHET	bis-(β-hydroxyethyleneterephthalate)
BMVECs	brain microvascular endothelial cells
BSu	butylene succinate
C	carbon
CAILD	computer-aided ionic liquid design
CDA-g-PLAs	cellulose diacetate-graft-poly(lactic acid)s
CDFAs	carbonyl difatty amides
CFCs	chlorofluorocarbons
CH_4	methane
CIs	corrosion inhibitors
CMC/GO	carboxymethyl cellulose/graphene oxide
CMPS	chloromethylated polystyrene
CNF	cellulose nanofiber
CNs	cellulose nanocrystals
CNTs	carbon nanotubes
CO_2	carbon dioxide
COSMO-RS	conductor like screening model for real solvents
DG	diglycerides
DGEBA	diglycidyl ether of bisphenol A
DHA	docosahexaenoic acid
DILs	dicationic ionic liquids
DMA	dynamic mechanical analysis
DMEM	Dulbecco's modified eagle's medium

DMH	dimethylhydantoin
DMSO	dimethylsulfoxide
DP	degree of polymerization
DS	degree of substitution
DSC	differential scanning calorimetry
DTG	differential TG thermograms
EA	ethylene adipate
EDG	electron-donating group
EDS	extractive desulfurization
EM	elastic modulus
EMS	enhanced microwave synthesis
ENR	epoxy novolac
EPA	Environmental Protection Agency
EPO	epoxidized palm oil
EPSPS	enolpyruvoylshikimate 3-phosphate synthase
$EtNH_3$	ethyl ammonium nitrate
FAME	fatty acid methyl ester
FAs	fatty amides
FDCA	furandicarboxylic acid
FHAs	fatty hydroxamic acids
FNCs	fatty nitrogen compounds
FTS	Fischer-Tropsch synthesis
GHG	greenhouse gas
GPC	gel permeation chromatography
GPTPS	glycerol-plasticized thermoplastic dried starch
GWP	global warming potential
H_2O_2	hydrogen peroxide
HBDBr	hexaethyl-butane-1,4-diammonium dibromide
HCs	hydrocarbons
HDI	hexmethylene diisocyanate
HDPE	high-density polyethylene
HER	hydrogen evolution reaction
HFCs	hydro fluorocarbons
HPDBr	hexaethyl-propane-1,3-diammonium dibromide
IB	isobutylene
ILED	ionic liquid-based extractive distillation
ILs	ionic liquids
IPCC	Intergovernmental Panel on Climate Change
IR	infrared

IV	intrinsic viscosity
LA	lactide
LA	levulinic acid
LC-MS	liquid chromatography-mass spectroscopy
LODs	limits of detection
MAA	methyl acetoacetate
MCL	medium chain length
MCRs	multi-component reactions
MDC	methylene dichloride
MDG	millennium development goals
MDR	multidrug-resistant
ME	matrix effects
MFC	microfibrillated cellulose
MILs	magnetic ionic liquids
MINLP	mixed-integer nonlinear programming
MTB	*M. tuberculosis*
MW	microwave
N	nitrogen
N_2O	nitrous oxide
NAC	*N*-acetylcysteine
Nbupy	*N*-butyl pyridinium
NHCs	*N*-heterocyclic carbenes
NMP	*N*-methyl-2-pyrrolidone
NOAA	National Oceanic and Atmospheric Administration
NPs	nanoparticles
NSCSs	nitrogen-sulfur-doped carbon spheres
O_3	ozone
OMMT	organically modified montmorillonite
P(3HB-co-3HV)	poly(3-hydroxybutyrate-co-3-hydroxyvalerate)
P3HB	poly(3-hydroxybutyrate)
PAHS	polycyclic aromatic hydrocarbons
PBAT	poly(butylene adipate-co-terephthalate)
PBSu	poly(butylene succinate)
PBSuPSus	poly(butylene succinate-co-propylene succinate)s
PCL	poly-ε-caprolactone
PCL/PPSu	poly(ε-caprolactone)/poly(propylene succinate)
PCR	polymerase chain reaction
PCS	photon correlation spectroscopy
PDI	polydispersibility index

PE	polymer electrolytes
PEF	polymer polyethylene furandicarboxylate
PEG	poly(ethylene glycol)
PEI	polyethyleneimine
PEO	polyethylene oxide
PEP	phosphoenol pyruvate
PESu	poly(ethylene succinate)
PET	poly(ethylene terephthalate)
PET-PESu	PET-poly(ethylene succinate)
PEU	poly(ester urethane)
PGA	polyglycolic acid
PHA	polyhydroxyalkanoate
PHB	polyhydroxybutyrate
PILs	protic ionic liquids
PLA	polylactic acid
PMHS	polymethylhydrosiloxane
PPC	poly(propylene carbonate)
PPG	poly(propylene glycol)
PPI	poly(propylene isophthalate)
PPSu	poly(propylene succinate)
PPTA	poly-p-phenyleneterephthalamide
PSHTMA	poly(*p*-methylstyrene)-3-(5,5-dimethylhydantoin) -*co*-trimethyl ammonium chloride
PSu	propylene succinate
p-TSA	p-tolunesulfonic acid
PUFA	polyunsaturated fatty acids
QD	quantum dot
RBC	red blood corpuscles
RCM	ring-closing metathesis
RE	recoveries
ROP	ring-opening polymerization
RT	room temperature
RTILs	room temperature ionic liquids
RWGS	reverse gas shift reaction
SA	succinic acid
SAILs	surface-active ionic liquids
SAR	structure-activity relationship
SCL/MCL	short-chain-length/medium-chain-length
SEM	scanning electron microscopy

SK-MEL-2	skin melanoma cells
SPE	solid polymer electrolytes
STM	scanning tunneling microscope
T	temperature
T	trolox
TAC	total annual cost
TBA^+	tetra-n-butylammonium
Tc	crystallization temperature
Td	decomposition temperature
TD	tapped density
TEC	triethyl citrate
TG	triglycerides
TGA	thermogravimetric analysis
THF	tetrahydrofuran
TMA	trimethylamine
TMB	trimethoxybenzaldehyde
TOC	total organic carbon
TPS	thermoplastic starch
TS	tensile strength
UAE	ultrasonic-assisted extraction
UNFCCC	United Nations Framework Convention on Climate Change
UV	ultraviolet
UZ	Üzülmez
VOCs	volatile organic compounds
WAXD	wide-angle x-ray diffraction
WF	wood flour
WPC	wood polymer composites
WWTPs	wastewater treatment plants
ZnO NFs	zinc oxide nanoflowers
ZnPHM	zinc-proline hybrid material
$ZrCl_4$	zirconium tetrachloride

Preface

This volume brings together innovative research, new concepts, and novel developments in the application of new tools for chemistry postgraduate students and chemicals. It is an immensely research-oriented, comprehensive, and practical book. Postgraduate chemistry students would benefit from reading this book as it provides a valuable insight into sustainable technology and innovations. It should appeal most to chemists and engineers in the chemical industry and in research, who should benefit from the technological, scientific, and economic interrelationships and their potential developments. It contains significant research, reporting new methodologies and important applications in the fields of green chemistry as well as the latest coverage of chemical databases and the development of new methods and efficient approaches for chemists.

In the first chapter of this book, ionic liquid as green solvents is discussed in detail.

The synthesis of cross-linked polysiloxane polymer electrolytes (PE) with pendant 3-(2-(2-methoxyethoxy)ethoxy)propyl groups as internally plasticizing chains and investigation of their electric-physical properties is conducted in Chapter 2.

In Chapter 3, we have elaborated on the efforts made toward the potentially active novel series of hybrid 3-(4, 6-dichloro-1,3,5-triazin-2-yl)-2-phenylthiazolidin-4-one derivatives as potent anti-inflammatory and antimicrobial agents via efficient synthetic methodology. This study suggests the role of halogenated electron-withdrawing groups in generating highly potent anti-inflammatory agents from the title hybrid skeleton. Thus the presence of lipophilic -Cl and -F at 5th position tolerates the procytokine activity. Also the presence of nonhalogenated electron-donating group (EDG) or electron-donating group (EDG) type of (4-OMe and 4-NO_2) group's present at 4-positions of terminal benzene ring found to be effective potent antimicrobial agents. Our optimization studies on the hybrid derivatives of 3-(4,6-dichloro-1,3,5-triazin-2-yl)-2-phenylthiazolidin-4-one underway and will be reported consequently in the future.

Chapter 4 focuses on the treatment of cancer MCF-7 cell line with a high dose of ascorbate, followed by cupric ions themselves and by a mixture of Cu(II) ions and ascorbate. For comparison, we selected normal 3T3 cells.

The oxidative system composed of Cu(II) ions and ascorbate applied in both 3T3 and cancer MCF-7 cells did not result in significant differences in (low) viability of these two cell lines. We examined the effect of N-acetylcysteine on oxidatively damaged MCF-7 and 3T3 cells.

Further, we investigated the effect of the addition of glutathione and glutathione disulfide on 3T3 and VH10 fibroblast cell lines exposed to oxidative damage by Cu(II) ions and ascorbate. The results showed that ascorbate only at the highest examined millimolar concentrations decreased the viability of MCF-7 cells more markedly than did 3T3 cells. The viability of 3T3 cell line was not affected significantly in the presence of N-acetylcysteine and cancer MCF-7 cells were a bit more vital after addition of N-acetylcysteine. Glutathione or glutathione disulfide did not prevent oxidative damage in both 3T3 and VH10 cell lines.

In Chapter 5, we express the synthesis and biological evolution of α-aminophosphonate derivatives, which is useful for the researcher to further study on α-aminophosphonates. Recently, scientist tried to develop the synthesis of biologically active compounds such as novel α-aminophosphonates by using ionic liquids and other environmental friendly as catalysts and solvents. Apart from being relatively nontoxic and environmentally friendly, the catalyst offers other advantages such as greater substrate compatibility, high reaction yields, short reaction times, recyclable, reusable, and the ability to tolerate functional groups, making it an important in addition to the reported methods. Aromatic amines and aldehydes are lipophilic in nature; utilized in therapeutic applications. These methods have been implemented for the synthesis of α-aminophosphonates because they are ecofriendly approach; maintain atom economy; provide an easy work-up procedure and provide an excellent yield of products.

An overview on polyelectrolyte based nanoplex is presented in detail in Chapter 6.

Chapter 7 widens the scientific inquiry and deep profundity in the field of nanotechnology, agricultural sciences, food technology, and biological sciences. Green revolution and biotechnology are today changing the face of scientific progress in developing and disadvantaged nations around the world. This chapter pinpoints the success, the scientific divination, and the vast scientific needs of research in nanotechnology and biological sciences in developing as well as highly industrialized nations around the world. The vision of nanobiotechnology and green sustainability are depicted profoundly in this chapter.

A detailed review on ecofriendly polymers is presented in Chapter 8.

An environmentally benign synthesis of 2,4,5-Triaryl-1H-Imidazoles via multi-component reactions and its medicinal importance is investigated in Chapter 9.

In Chapter 10, the author mentioned observations and other literature evidences, and an attempt has been made to summarize the significance of microwave techniques and their use in various chemical transformations.

Green synthesis, characterization, and biological studies of 1,3,4-thiadiazole derived Schiff base complexes are discussed in Chapter 11.

A study on the effect of global warming and greenhouse gases on environmental system is presented in Chapter 12.

This volume should also be useful to every chemist or chemical engineer involved directly or indirectly with sustainable chemistry. With clear explanations, and real-world examples, this volume emphasizes the concepts essential to the practice of chemical science, engineering, and technology while introducing the newest innovations in the field.

KEY FEATURES:

- The book serves a spectrum of individuals, from those who are directly involved in the chemical industry to others in related industries and activities. It provides not only the underlying science and technology for important industry sectors but also broad coverage of critical supporting topics. Industrial processes and products can be much enhanced through observing the tenets and applying the methodologies covered in individual chapters.
- This authoritative reference source provides the latest scholarly research on the use of applied concepts to enhance the current trends and productivity in sustainable technology. Highlighting theoretical foundations, real-world cases, and future directions, this book is ideally designed for researchers, practitioners, professionals, and students of materials chemistry and chemical engineering.
- The volume explains and discusses new theories and presents case studies concerning green material and sustainable technology.
- This book is an ideal reference source for academicians, researchers, advanced-level students, and technology developers seeking innovative research in chemistry and chemical engineering.
- This volume is on the application of the physical, chemical, and biological principles in the pharmaceutical sciences and it relates to

problems in biology and medicine. It also helps students, teachers, researchers, and industrial pharmaceutical scientists use elements of biology, physics, and chemistry in their work and study.

- This new book provides a systematic coverage in a single-source volume on the application of materials science techniques to the pharmaceutical field offering a comprehensive program for the physical characterization of raw materials, drug substances, and formulated products.
- This volume presents the principles and applications of physical chemistry as they are used to solve problems in biology and medicine. The volume is structured to provide overviews with historical perspectives on the evolution of ideas and on the future of physical biology and biological complexity as well.
- This new title lies at the heart of the behavior of those macromolecules and molecular assemblies that have vital roles in all living organisms and determines the stability of proteins and nucleic acids, the rate at which biochemical reactions proceed, and the transport of molecules across biological molecules. They allow us to describe structure and reactivity in complex biological systems, and make sense of how these systems operate.
- The book provides a balanced presentation of the concepts of physical chemistry, and their extensive applications to biology and biochemistry. It is written to show students how the tools of physical chemistry can elucidate and illuminate biological questions.

CHAPTER 1

Ionic Liquid as Green Solvents

AVINASH KUMAR RAI,[1] SEEMA KOTHARI,[1] RAKSHIT AMETA,[2] and SURESH C. AMETA[1]

[1]PAHER University, Udaipur–313003, Rajasthan, India, E-mail: ameta_sc@yahoo.com (S. C. Ameta)

[2]Department of Chemistry, J. R. N. Rajasthan Vidyapeeth (Deemed to be University), Udaipur–313001, Rajasthan, India

ABSTRACT

To date, most of the chemical reactions are carried out in molecular solvents, which are many time toxic to the environment. Recently, a new class of solvent has emerged as ionic liquids (ILs). The term ILs cover one of the broadest classes of chemical compounds of salt type with low melting temperature and appreciable wide liquid range. In recent years, ILs has become mainstream solvents in different fields of chemistry. They have attracted quite justifiably, enormous attention as media for green synthesis. This has been possible owing to environmentally advantageous properties and possibility of adaptation of their structure to a specific task. In many chemical reaction processes, ILs are suggested as solvents, catalysts, reagents, or combinations of these. Due to ecofriendly, behavior, and easy design, ILs will replace traditional organic solvents. A wide variety of anions and cations are available and as such, solvent properties can be fine-tuned. A wide range of reactions have already demonstrated in this medium. ILs are being investigated for both; the dissolution of a variety of biomaterials and for their processing into higher-value products. The ability of ILs to dissolve and stabilize enzymes, proteins, DNA, and RNA is also extremely valuable in biotechnological applications. A small introduction of ILs, properties, and their applications in synthesis, extraction, and biocatalysis has been presented here.

1.1 INTRODUCTION

Solvents are a necessary part of chemical reactions. There are many industries like agrochemical, pharmaceutical, dye, pulp, paper, chemical, food, automotive, etc., which are regularly using volatile organic solvents. The main disadvantage of using an organic solvent is its vapor pressure at room temperature (RT), which helps it to be released to the environment. Therefore, they contribute strongly to the pollution (air, water) as a major source in the chemical industries. Therefore, there is a difficulty in recycling such solvents and they can directly have adverse health effects. Thus, there is an urgent need to search for alternative solvents, which are green chemical in nature and ecofriendly.

Efforts are made to develop new cleaner and greener technologies with production of less amount of waste, which are relatively less harmful to the environment also. In recent years, ionic liquid (IL) has been considered as a promising candidate, which can be definitely used as alternatives to toxic VOCs. ILs have advantages because of their unique and favorable physical properties. These are summarized in Figure 1.1. Therefore, typical ILs cations can be applied in synthesis, biocatalysis, extraction, energy storage, etc. These solvents can easily recycle a number of times.

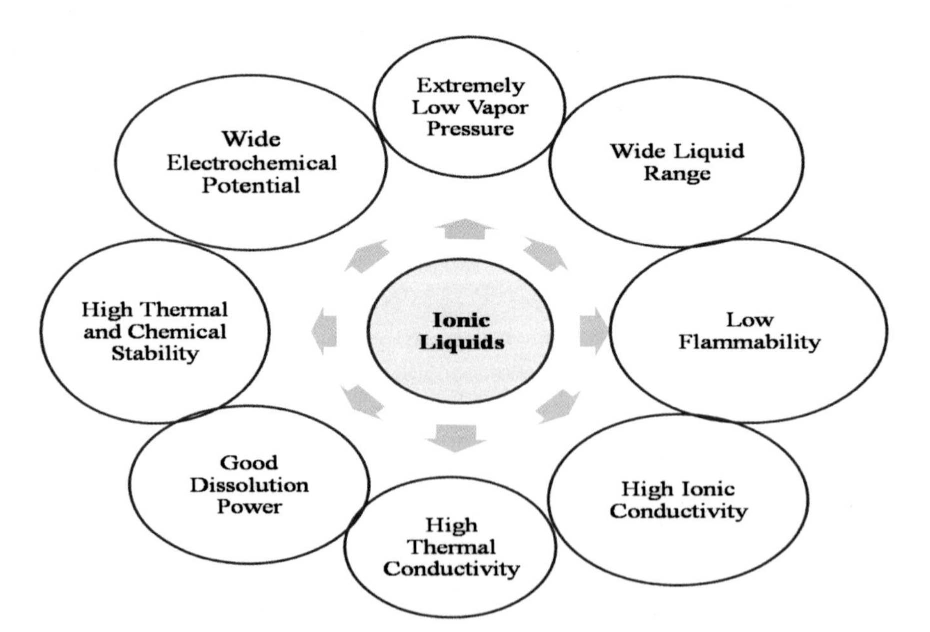

FIGURE 1.1 Physical properties of ionic liquids.

The first IL was reported in the mid-19th century, when a separate 'red oil' phase was formed during Friedel-Crafts reactions [1]. This separated red oil was thought to be salt, but it was identified as a stable intermediate complex formed through Friedel-Crafts reactions as evident from NMR spectroscopy [2]. Synthesis of the first IL comes to light shortly, just at the start of the 20th century, when ethyl ammonium nitrates ($EtNH_3$); (NO_3) was synthesized with a melting point of only 12°C [3]. The extraordinarily remarkable property of ILs is that such salts are liquids at or close to RT.

Most of the ILs are composed by two parts:

- First is a large massive unsymmetrical cation; and
- Second is smaller equally shaped anion.

ILs are not treated as common salts. Melt down ILs have lesser lattice energies due to incapacity to form efficient ion-ion packing and as a result, electrostatic interactions are reduced. They have reduced columbic interactions due to the diffuse charge present in the unsymmetrical cation attached with the weakly coordinating anion. ILs have low melting points, because of mainly two factors. These are:

- Weak electrostatic forces; and
- Delocalized diffuse charge.

Due to this unique property, interest in ILs is developing regularly.

ILs are mainly used as solvents and their solubility in various reagents can be tuned by selecting a particular anion or cation. ILs are now commonly used as alternative eco-friendly solvents due to large potential of combinations of various cations and anions. Their solubility also varies in chemical reactions. These are harmless and more environmentally friendly solvents. Most of synthesized ILs are miscible in water but some of them are hydrophobic also. They are excellent and alternate solvents for replacing more harmful volatile organic compounds (VOCs). ILs have four important properties. Such as:

- Rate of reaction is improved;
- Yield is increased;
- Isolation of product is improved; and
- Reduced formation of by products.

Use of ILs are based on the twelve basic principles of green chemistry, which were proposed by Clark and Macquarie in 2002 [4].

1.2 NOMENCLATURE OF IONIC LIQUIDS (ILS)

Because many ILs are composed of complex organic molecules, short names have been developed to name ILs (Figure 1.2).

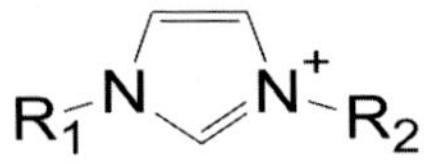

FIGURE 1.2 Imidazolium cation.

Here, 1 and 2 refer to the first letter name of the R_1 and R_2 carbon chain. e.g., butyl, methyl imidazolium is abbreviated to bmim. For quaternary ammonium systems, a system of the type N 1234^+ is used, where the subscripts indicate the chain length of each of the four alkyl chains of the quaternary ammonium cation (Figure 1.3).

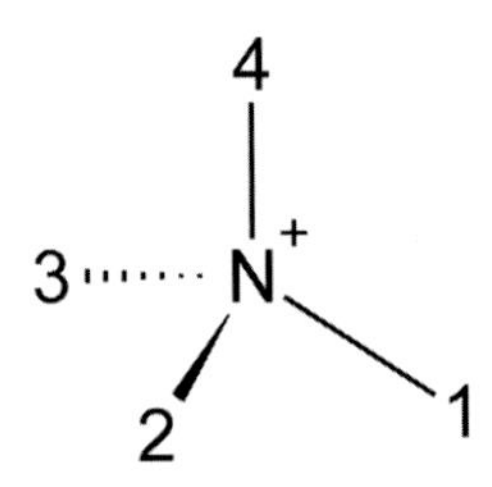

FIGURE 1.3 Quaternary ammonium cation (N_{1234}^+ = methylethylpropylbutylammonium).

ILs are composed of large unsymmetric imidazolium or quaternary ammonium systems. Comparatively, small, and symmetric ions do not form RTILs. These cations are coupled with large anions, which can either be symmetric, such as hexafluorophosphate [PF_6^-] or tetrafluoroborate (BF_4^-), or less symmetric, such as bis(trifluoromethanesulfonyl) imide [$TFSI^-$] or amides.

$TFSI^-$ anion forms ILs with many cations, excluding the small symmetric quaternary ammonium [R_4N^+] systems. However, the asymmetric amide, $TSAC^-$, [$CF_3CONSO_2CF_3^-$] can form RTILs with even small symmetric quaternary ammonium ions such as tetraethyl ammonium cation, [N $2222]^+$ [5]. These so-called second-generation RT ILs considerably widened the range of applications, where these ILs could be applied. This is because of the fact that these not only share the advantages of their precursors, but also remain stable, when exposed to water and air [6–8].

These systems include salts based upon poorly nucleophilic anions such as $[BF_4^-]$, $[PF_6^-]$, $[CF_3CO_2^-]$, $[CF_3SO_3^-]$, etc. Low melting point salts, established on other cations such as complex polycationic amines, [9] and heterocyclic containing drugs, have also been prepared [10, 11]. Most commonly used ILs are based around the 1-alkyl-3-methylimidazolium cation, (Figure 1.4) especially the 1-ethyl-3-methylimidazolium, $[emim]^+$ and N-butyl pyridinium $(Nbupy)^+$ cations.

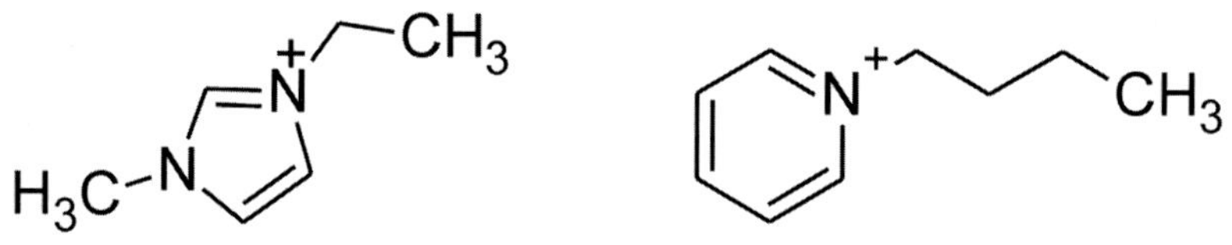

FIGURE 1.4 Structure of 1-alkyl-3-methylimidazolium $[Cnmim]^+$ and N-butyl pyridinium $(N\ bupy)^+$ cation.

Imidazolium and pyridinium based ILs are the most versatile used ILs due to their low volatility, good thermal stability, negligible vapor pressure, wide liquid range, electrolytic conductivity, low viscosity, etc [12–14]. They are used as both; solvents and catalysts [15–17].

1.3 SYNTHESIS OF IONIC LIQUIDS (ILS)

The process of synthesizing ILs can be divided in two sections. These are:

- Formation of the preferred cation; and
- Anion exchange.

These two methods are commonly used to obtain desired ILs by changing cation and anion. Many cations are available on commercially basis; thus, requiring only the anion exchange. There are two simple methods for the synthesis of ILs and these are quaternization and anion exchange reactions [18].

1. **Quaternization Reactions:** It is also called alkylation process. Here, an amine or a phosphine derivative is stirred and heated with the haloalkane. This process has the benefits like commercially available low cost haloalkanes are utilized and it does not require extra utility. In this process, minimum by products are formed. Apart

from this, usually, minimum temperature and time are required. The reaction kinetics is greatly dependent on the haloalkane used. The reactivity of the haloalkane decreases on increasing alkyl chain length. Usually, this reaction is carried out without using other solvents, as the reagents are normally liquids and miscible. The final desired halide salts are generally not soluble in the starting materials and these are mostly dense than solvents; so it can be easily isolated by layer separation and purification.

The isolated halide salts are commonly solids at RT and they are very hygroscopic and soluble in water. Their melting points mainly depend on alkyl chain substituents. 1-Alkyl-3-methylimidazolium halide salts can obtain as very hard solids [18].

2. **Anion-Exchange Reactions:** These can be divided into two categories:
   - Formation of ILs by anion metathesis; and
   - Direct treatment of halide salts with Lewis acids.

   These reactions are usually carried out in aqueous solution with the appropriate anion, ammonium salt, or an alkali metal salt. Then the product was extracted with an organic solvent and the organic phase is washed with water [18].

   Apart from these two methods, solvent-free microwave (MW) initiation is also for synthesizing ILs.
3. **Solvent-Free Microwave Initiation:** Generally, anion metathesis reactions are carried out with a large amount of methylene dichloride (MDC) or acetone as solvent and take a number of hours or even some days [19]. MW activation has been developed as a powerful and unique technique for different types of chemical reactions and it has become a valuable technology in organic chemistry [20]. Solvent-free conditions in combination with MW irradiation resulted in improving the reaction rate, reduced reaction time, improved conversions, as well as selectivity along with certain environmental advantages [21]. This combination is also successfully applied to the synthesis of several imidazolium-based ILs [22].

1.4 CHEMICAL AND PHYSICAL PROPERTIES OF IONIC LIQUIDS (ILS)

Different physicochemical properties of ILs vary with a change in its cation or anion. One of the main important characteristics of ILs is their negligible vapor pressure. Because of this unique physicochemical property of ILs, they are useful in controlling ever-increasing air pollution or adverse health effects as they generally do not evaporate in reaction vessels [23].

1.4.1 MELTING POINTS

ILs have been defined to have melting points around 100°C or below. Most of the ILs are liquid at RT. Both cations and anions affect the melting points of ILs. Melting points of the compounds are controlled by three factors;

- Intermolecular forces;
- Molecular symmetry; and
- Conformational degrees of freedom of molecules [24].

The increase in anion size will lead to a decrease in the melting point [25]. It has been reported that an increase in the degree of splitting within an alkyl chain of the imidazolium ring increases its melting point.

1.4.2 THERMAL STABILITIES

Since most of the ILs are non-volatile, the limit for higher range of degradation can be determined by the thermal decomposition temperature of the IL. The thermal stability of the ILs depends upon the nucleophilicity of the anion. One of the key characteristics of ILs is that they can be used at elevated temperatures also. Thus, it is important to understand the rate of degradation of ILs at constant elevated temperatures [26].

1.4.3 CHEMICAL STABILITY

The chemical stability is based on the fact that reagent is chemically stable, and it does not react or decay under reaction conditions. Stability can be increased towards bases. Phosphonium-based ILs are more stable even under strong basic condition [27]. Further, most of the ILs are considered as inert solvents, but some ILs may act as catalyst or nucleophile and react with some reagents.

1.4.4 VISCOSITY

ILs are widely used because of their viscous behavior. Both the anion and the cation play a significant role in determining the viscosity of an IL. Although, these ILs are more viscous than water, this difference is much reduced at higher temperatures [28]. It was revealed that dicationic ILs (DILs) have higher viscosities than monocationic ILs [29].

1.4.5 DENSITY

Density is one of the most often measured properties of ILs, probably because nearly every application requires knowledge of the material density. In general, ILs are denser than water with values ranging from 1 to 1.6 g mL^{-1}. The densities of ILs decrease with an increase in the length of the alkyl chain of the cation. However, there are some ILs possessing densities lower than 1 g mL^{-1}, such as those ILs based on dicyanoamide anion [30]. The densities of ILs are also affected by the anions. Anions with higher molecular weight increase the density of an IL for a specific cation.

1.4.6 SOLUBILITY IN WATER

ILs can generally be divided into two groups according to their solubility in water. These are:

- Hydrophobic nature; and
- Hydrophilic nature.

Hydrophobicity can often be modified by a suitable choice of anion. The hydrophilic and hydrophobic behavior is key for the solubility properties of ILs as it is important to dissolve the product and again isolate it by extraction or filtration.

ILs are gaining attention of scientists as potential solvents for many reasons:

- They have relatively high viscosities (20–1000 cP) at RT;
- These are non-flammable below their decay temperature;
- They have excellent solvation properties for a variety of chemical reactions;
- They have very good electric conductivities;
- They have excellent thermal stabilities and sometimes stable up to more than 250°C;

- They have chemical stability;
- They have good ionic conductivity and therefore, can be combined with some electrochemical processes;
- They possess adjustable coordination properties, as ILs have potential to be weakly coordinating toward transition metal complexes. They may improve reaction rates containing cationic electrophilic intermediates;
- They have high heat conductivity, which simplifies the management of large reactors, and allowed a very fast removal of heat of the reaction;
- They may be possibly non-protic or protic;
- They can exhibit Lewis acidity as they can act as both; co-catalyst and solvent;
- They have adjustable miscibility; which will permit them to act as both; hydrophobic or hydrophilic;
- They have high attraction for ionic intermediates. Ionic metal-catalysts can be immobilized without modification;
- They have no measurable vapor pressures.

1.5 APPLICATIONS

ILs have been considered for a variety of industrial applications due to their unique and tunable properties. They have a great potential for replacing volatile organic solvents (VOCs) in synthetic chemical processes, where the latter are considered as a source of environmental pollution problems. This allows ILs to be excellent candidates for industrial applications compared to volatile organic solvents.

Organic solvents have been used for several centuries, and obviously occupy most of the solvent market in industry. However, by comparing the properties of ILs and organic solvents, it could be anticipated that industry may be a suitable environment for the use of ILs (Figure 1.5).

Although, there are several possible applications of ILs. Here, focus is on ILs as extraction solvents, as reagents or medium for synthesis and biocatalysis.

1.5.1 SYNTHESIS

Room-temperature ILs based on 1-butylimidazolium salts with different anions were synthesized by Palimkar et al. [31]. These were used for the preparation of biologically active substituted quinolines and fused polycyclic quinolines using the Friedlander heteroannulation reaction.

1-Butylimidazolium tetrafluoroborate [Hbim]BF_4 was found to be the best IL for such heteroannulation reaction. These reactions proceed smoothly under relatively mild conditions and that too, without any addition of catalyst. IL acts as a promoter for this regiospecific synthesis and it can be recycled. Various quinolines have been prepared using this green approach with excellent yields and purity.

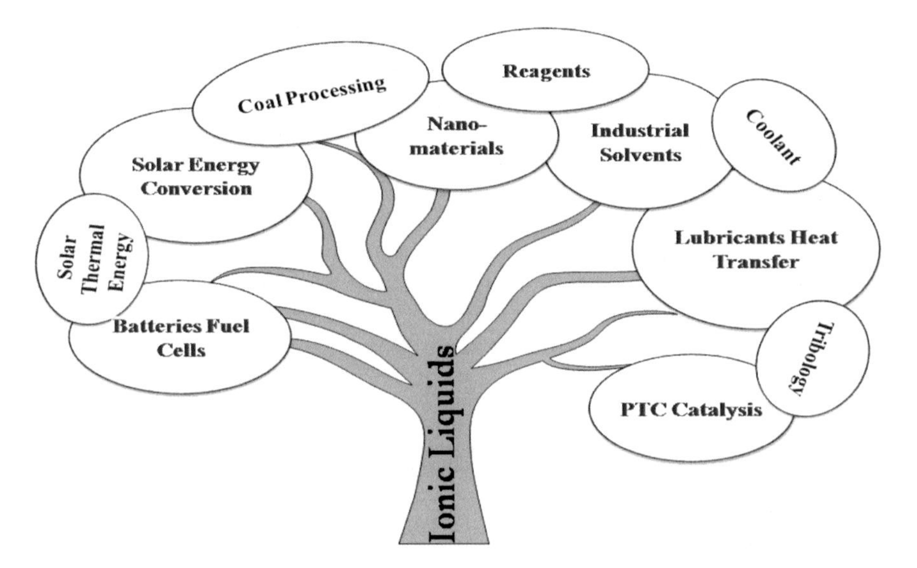

FIGURE 1.5 Applications of ionic liquids.

$$\text{2-aminoaryl ketone } (R_1,\ NH_3) + R_3CH_2COCH_2R_2 \xrightarrow{[\text{Hbim}][BF_4]} \text{quinoline } (R_1, R_2, CH_2R_3)$$

98% yield

Forbes et al. [32] conducted a study involving the scope of substrate in the Knoevenagel reaction in an IL medium. The formation of the condensation product was favored using electron deficient aryl aldehydes. When electron rich aldehydes were used lower levels of conversion was there while no measurable amounts of condensation products was obtained using ketones. A reusability study indicated that the IL medium could be used number of times affording the desired condensation product in excess of

90% conversion. Post-run analyses of the IL were also conducted, which revealed that IL medium remains unchanged upon reuse.

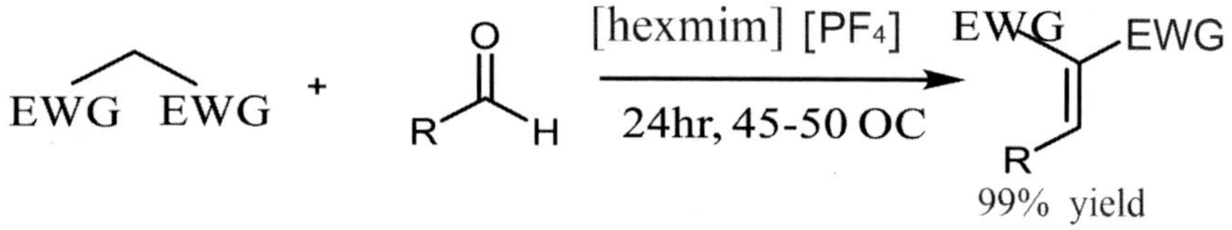

EWG EWG + R(C=O)H —[hexmim] [PF4], 24hr, 45-50 OC→ EWG EWG / R
99% yield

Konnerth and Prechtl [33] tested N-heterocyclic compounds in the selective hydrogenation catalyzed by small Ru nanoparticles (NPs) of 1–3 nm-sized; embedded in various imidazolium-based ILs. It was reported that a diol-functionalized IL showed the best performance in the hydrogenation of quinoline to 1,2,3,4-tetrahydroquinoline (1THQ) having 99% selectivity.

Heck coupling of aryl halides or benzoic anhydride with alkenes was carried out by Carmichael et al. [34]. Excellent yields were obtained in room-temperature ILs. ILs provided a medium for dissolution of palladium catalyst and it allowed easy separation of product and by-products. As a result, the catalyst and IL can be recycled and reused again.

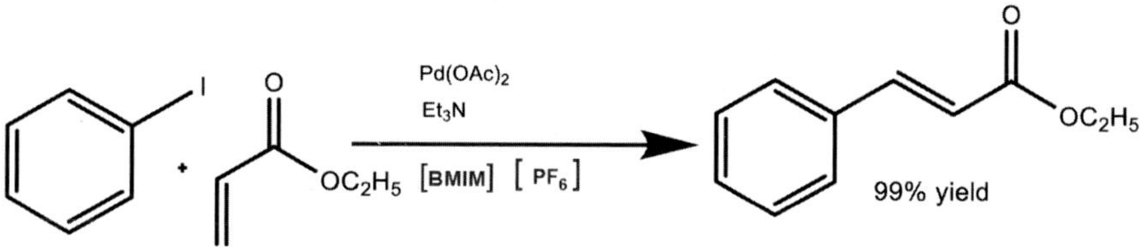

I + O / OC_2H_5 —$Pd(OAc)_2$, Et_3N, [BMIM] [PF_6]→ O / OC_2H_5
99% yield

Palladium imidazole complexes have been used by Mathews et al. [35] as catalyst precursors for carrying out Suzuki cross-coupling reaction in 1-butyl-3-methylimidazolium-based RT ILs. Such a system provided a stable and recyclable method for iodo- and bromoarenes. They obtained better reaction conditions and also evaluated the effect of change in IL components. Palladium imidazole complexes were used as catalyst precursors for the Suzuki cross-coupling reaction in 1-butyl-3-methylimidazolium-based ILs.

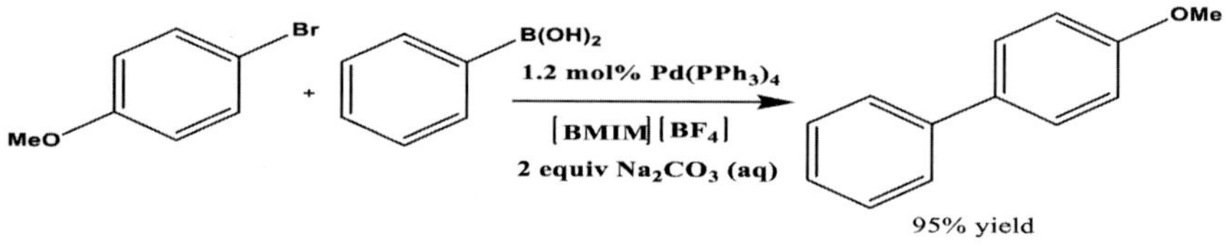

MeO / Br + $B(OH)_2$ —1.2 mol% $Pd(PPh_3)_4$, [BMIM] [BF_4], 2 equiv Na_2CO_3 (aq)→ OMe
95% yield

IL 1-butyl-3-methylimidazoliumhexafluorophosphate ([bmim]PF_6), was used by Buijsman et al. [36] as an effective medium for ring-closing metathesis (RCM) using Grubbs catalysts. RCM showed high conversions and a broad substrate tolerance on using [bmim]PF_6 as solvent. [bmim]PF_6 and the ruthenium catalyst can be recycled at least three times after extracting the product in the organic phase.

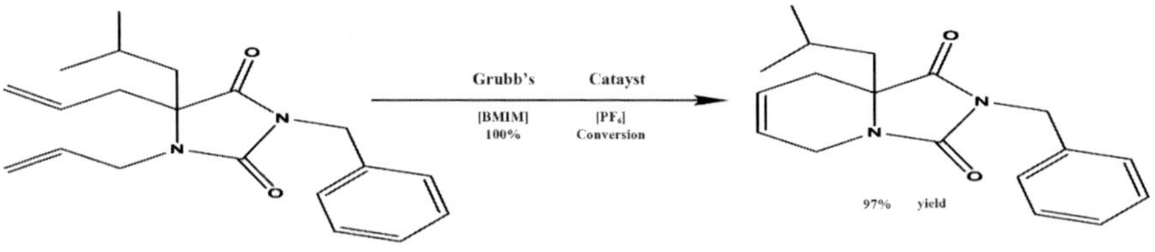

Song et al. [37] observed that ILs act as powerful media (or additives) in scandium triflate catalyzed Diels-Alder reactions, which not only facilitated catalyst recovery but it also accelerated the reaction rate and improved selectivity.

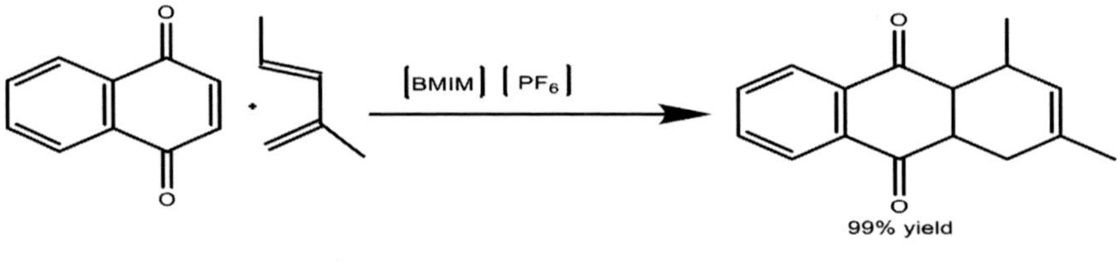

1-Butyl-3-methylimidazolium chloroaluminate ILs were used as media and as Lewis acid catalyst by Nara et al. [38] for Friedel–Crafts sulfonylation reaction of benzene and substituted benzenes with 4-methyl benzenesulfonyl chloride. These substrates exhibited increased reactivity with almost quantitative yields of diaryl sulfones at ambient conditions. The effect of Lewis acidity of the IL on initial extent of conversion of this reaction was also studied. The predominance of $[Al_2Cl_7]^-$ species in [Bmim]Cl-$AlCl_3$, acidic ionic liquid (AIL), in the presence of 4-methyl benzenesulfonyl chloride was confirmed by 27Al NMR spectral studies. It was reported that $[AlCl_4]^-$ species predominated in the reaction with the aromatic hydrocarbon. Such changes in speciation of aluminum may be attributed to the interaction of the Lewis acidic species $[Al_2Cl_7]^-$ of the IL with the formed HCl during the sulfonylation reaction as evident from the control experiment.

SO$_2$Cl

[BMIM] [Cl-AlCl$_3$]

89% yield

The relative nucleophilicity of chloride, bromide, and iodide anions in [bmim][BF_4] IL has been evaluated by Lancaster et al. [39] by their reactions with methyl p-nitrobenzenesulfonate. They found that iodide is the most nucleophilic halide, while chloride and bromide have almost equal nucleophilicities (However; Cl^- is slightly more nucleophilic compared to Br^-) in [bmim][BF_4]. Activation energies were also calculated for the reaction of chloride and bromide with methyl p-nitrobenzenesulfonate.

Proline-catalyzed asymmetric direct aldol reaction of different aromatic aldehydes with acetone and several other ketones are carried out by Kotrusz et al. [40] in RT IL 1-n-butyl-3-methylimidazolium hexafluorophosphate. They achieved good yields of aldolization products with reasonable enantioselectivities, while using only 1–5% of proline as the catalyst. The immobilization of the catalyst in an IL phase has other advantages like simple product isolation and reuse of the catalyst further in subsequent reactions.

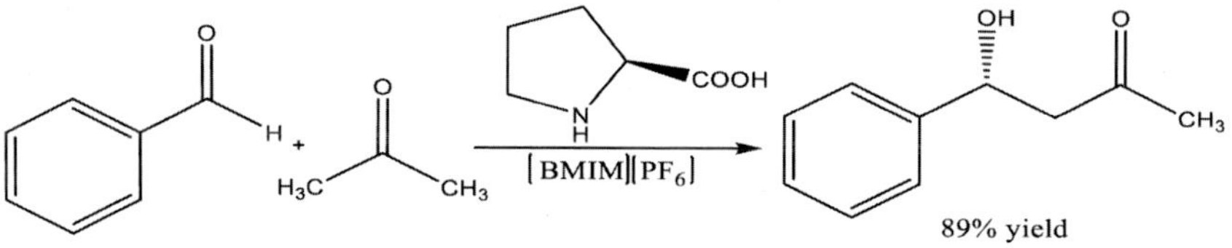

COOH

[BMIM][PF_6]

89% yield

Nazarov cyclization of two types of pyrrole derivatives was effectively catalyzed by Fujiwara et al. [41] using 5 mol% alumina-supported iron(III) perchlorate [$Fe(ClO_4)_3 \cdot Al_2O_3$]. Desired cyclization with high trans selectivity was achieved in good to excellent yields. The cyclized product was then treated with a vinyl ketone in the presence of the same iron salt to give the corresponding Michael product. A sequential type Nazarov/Michael reaction of pyrrole derivatives was carried out and 4,5-dihydrocyclopenta[b] pyrrol-6(1H)-one derivative or 5,6-dihydrocyclopenta[b]pyrrol4(1H)-one derivative were obtained.

Ngo et al. [42] synthesized polar bisphosphonic acid-derived Ru(BINAP)(DPEN)Cl_2 precatalysts. It was immobilized in room temperature ionic liquids (RTILs) for asymmetric hydrogenation of aromatic ketones with enantiomeric excess (ee) values of about 98.7%. It was revealed that performance of the Ru catalysts is highly dependent on the nature of imidazolium ILs. ILs (imidazolium ILs without acidic protons) and Ru catalysts may be recycled by simple extraction and reused. Such a simple method of immobilizing catalyst prevented the leaching of Ru (and Ru catalysts) into the chiral secondary alcohol products, which is desirable for the production of pharmaceutical intermediates free from metal contaminants.

The enantioselective hydrogenation of methyl acetoacetate (MAA) was investigated by Ochsner et al. [43] using a ruthenium-monodentate binaphthophosphepine complex in a homogeneous solution, which is formed by different ILs and methanol. IL additives not only opened a smooth path of recycling catalyst in repetitive batch-mode but it also led to significantly enhanced catalytic activity as compared to pure methanol. About 95% enantioselectivities have been achieved in mixed IL/methanol systems, which is almost at par to the values obtained in pure methanol. A group of ILs increasing the activity of ruthenium-phenyl-4,5-dihydro-3H-dinaphtho[2,1-c;1′2′-e]phosphepine complex were synthesized and used as a co-solvent in the asymmetric hydrogenation of MAA in methanol. Here also, enantioselectivities were found to be in the same range (90–96%) as in pure methanol.

Chiral Cu (II) complexes were generated in situ by Khan et al. [44] from C2-symmetric chiral secondary bis-amines 1′–4′ based on 1,2-diaminocyclohexane structure with H, t-Bu, and Cl substituents at 3, 3′, 4, 4′ and 5, 5′ with copper acetate. They used these catalysts for a green protocol for highly enantioselective nitroaldol reaction of various aldehydes. Nitromethane was used in the presence of different ILs at 0°C. Excellent yields (up to 90% with respect to aldehyde) of β-nitroalcohols were obtained with high enantioselectivity (ee, up to 94%) on using [emim]BF_4 as IL. The present IL mediated nitroaldol protocol is recyclable (up to five cycles) with no significant loss in its performance.

A new method has been developed by Gruttadauria et al. [45] for chiral catalysts recycling. It involves the treatment of a monolayer of covalently attached IL on the surface of silica gel with add additional IL. These layers serve as the reaction phase, where the homogeneous chiral catalyst was dissolved. L-Proline-catalyzed aldol reaction has been carried out. Good yields and enantiomeric excess (ee) values were obtained, comparable with homogeneous conditions. This material also shows high regenerability.

Li et al. [46] synthesized novel chiral dimeric salen Mn(III) complexes bridged with N,N-dialkylimidazolium ILs at the C-5 position of the salicylaldehyde moieties. These ILs-bridged complexes exhibited excellent enantioselectivity and catalytic activity in the oxidative kinetic resolution of α-methyl benzyl alcohol when diacetoxyiodobenzene was used as an oxidant. The complexes combined the two factors: (i) High catalytic efficiency of the active centers; and (ii) Special solubility of the ILs. They showed good efficacy of oxidative kinetic resolution as well as easy recovery from the reaction system by choosing a proper solvent. Such complexes proved to be efficient catalysts for the oxidative kinetic resolution of racemic secondary alcohols. These could be recycled easily by choice of solvent.

Kude et al. [47] reported that iron-catalyzed homocoupling reaction of aryl Grignard reagent was completed rapidly in 5 min when the reaction was carried out in a phosphonium salt IL solvent system at 0°C.

MgBr
R
$FeCl_3$ (1 mol %)
$ClCH_2CH_2Cl$(DCE) (1.2 eq.)
IL - THF (>5 : <1), 0 °C or 25 °C
R
R

where, IL is [P_{4441}][NTf_2], [P_{444ME}][NTf_2] or [bdmim][NTf_2].

2,5-Furandicarboxylic acid (FDCA) is an important biobased monomer for the degradable polymer polyethylene furandicarboxylate (PEF). In order to reduce the high costs of starting material and catalysts, a novel approach for the direct conversion of fructose into FDCA was developed by Yan et al. [48] using [Bmim]Cl as a solvent with non-noble metal (Fe-Zr-O) as a catalyst. It was revealed that the oxidation of intermediate FFCA to FDCA has highest activation energy, which indicates that this step is affected by reaction temperature. Other biomass sources like glucose, galactose, mannose, starch, and cellulose can also be directly converted, in the IL-promoted reaction system, but of course, with lower FDCA yield as compared to that of fructose, may be due to the ineffective isomerization of aldohexoses into fructose.

Four new free-halide-ILs derived from ammonium were synthesized by Olivarres-Xometl et al. [49]. These compounds were evaluated as corrosion inhibitors (CIs) of API 5L X60 steel in 1 M sulfuric acid. It was observed that the inhibition efficiency displayed by these four ILs is a function of their concentration and molecular structure. Range 51–89% inhibition efficiency (η) was obtained, which confirmed that this new class of ILs is good for corrosion inhibition purposes. These compounds inhibit the corrosion of steel and classified as mixed-type CIs. The adsorption process of these new CIs on the API 5L X60 steel surface, obeyed the Langmuir isotherm.

A simple and effective route for the production of levulinic acid (LA) from cellulose has been developed by Shen et al. [50] in SO_3H-functionalized ILs. The influences of structure of IL, reaction conditions, and combination of metal chlorides with ILs on the yield of LA were evaluated. The highest yield of 39.4% was achieved for 120 min in the presence of 1-(4-sulfonic acid) butyl-3-methylimidazolium hydrogen sulfate ([BSMim]HSO_4) with addition of H_2O. The catalytic activity of ILs was found to depend on the anions and it decreased in the order:

$$CF_3SO_3 > HSO_4 > OAc$$

It is in good agreement with their acidity order. ILs play a dual role here as solvent and acid for the cellulose conversion into LA. It was revealed that catalytic activity is retained even after four repeated runs.

Gas sensing properties of nanostructures can be reduced on reducing the particle size. Van Dao et al. [51] synthesized small CeO_2 NPs (3–5 nm) with the support of [EMIM][Tf_2N] IL. As-obtained ceria NPs showed relatively large surface area (68.29 m^3 g^{-1}) and high content of cerous ions (13.5%). It increases the chemisorption of reagents and oxygen leading to an enhancement in its gas sensing performance. The gas response of synthesized CeO_2 sensor to 100 ppm of ethanol at the optimal operating temperature of 400°C was 2.3, which was about 1.83 times higher than that of commercial available method (1.26). The type of superior gas sensing mechanism of synthesized ceria was proposed through electrical resistance change.

Conversion of carbon dioxide to hydrocarbons (HCs) can prove to be a key solutions to both; the reduction of the greenhouse effect as well as and the sustainable production of fuels and lubricants. Bimetallic Ru/Ni NPs (2–3 nm), Ru-rich shell, and Ni-rich core in a hydrophobic IL promote the direct hydrogenation of CO_2 to light HCs under very mild reaction conditions [52]. It was reported that reaction of CO_2 with hydrogen (1:4, 8.5 bar) at 150°C with Ru/Ni NPs (3:2) in bis((trifluoromethyl)sulfonyl)amide (BMI. NTf2) hydrophobic IL afforded C^{2+} HCs (79% alkanes and 16% olefins) with 5% CH_4 at 30% conversion. On the contrary, reaction performed in the hydrophilic IL 1-n-butyl-3-methyl-1H-imidazol-3-ium tetrafluoroborate afforded mainly CO as the major product. The catalytic hydrogenation of CO_2 proceeds in two-steps:

- Initial conversion of CO_2 into CO by reverse gas shift reaction (RWGS); and
- Fischer-Tropsch Synthesis (FTS).

It was interesting to observe that bimetallic NPs have higher catalytic efficiencies than their monometallic counterparts, which may be due to strong synergy between the metals. It was observed that the presence of Ni in the bimetallic NPs yielded a more active RWGS catalyst, while the Ru increased the FTS towards heavier HCs.

Kimura et al. [53] synthesized polysaccharide chemical gels in the absence of any cross-linking reagents, but using ionizing radiation in carboxylate-based RT ILs. Although, the maximum gel fraction of the cellulose gel produced using these techniques was relatively poor (14%) and also its mechanical properties were insufficient so as to use them as soft gel electrodes. The present methodology can overcome both these problems. A radiation-cross linked cellulose/chitosan hybrid gel was obtained, in the process of RTIL with an increase in the gel fraction having absorbed dose of up to 80%. The mechanical properties, biodegradability, and thermal stability of these hydrogels were also measured, which were found to be sufficient for their practical use as soft gel electrodes. The electronic voltage response of such hybrid gel was evaluated in terms of electronic conductivity, curvature, and displacement. It was predicted that this gel may find application in bio-devices and soft actuators in future.

A series of substituted piperazinium, piperidinium, and pyrrolidinium acetates were synthesized by Kasprzak et al. [54]. Highly effective cellulose solvents were designed by mixing an aprotic polar solvent; dimethylsulfoxide (DMSO), with these synthesized ILs. They prepared different weight ratios of corresponding IL/DMSO mixed solvents to know the effect of co-solvent on solubility of cellulose. It was measured at 25, 50, and, 80°C. The effect of different factors such as solvent composition, process temperature, type of IL's cation, and alkyl chain length of IL's cation on cellulose solubility was evaluated. N,N'-Dimethyl-N-ethylpiperazinium acetate ([DMEPpz][Ac]/DMSO) mixed solvent exhibited the best performance for dissolving cellulose, which reached up to 13.5 wt% at 80°C. This dissolved cellulose was regenerated, when water was used as an antisolvent.

Catalytic activities of surface active ionic liquids (SAILs) were investigated by Vieira et al. [55] for cyclic carbonate synthesis composed of cations and anions:

- 1-Butyl-3-methylimidazolium ([$bmim^+$]);
- Tetra-n-butylammonium ([TBA^+]);
- Lauryl sulfate ([$C_{12}SO_4^-$]);
- Lauryl ether sulfate ([$C_{12}ESO_4^-$]);
- Lauryl benzene sulfonate ([$C_{12}BSO_3^-$]);
- Lauryl sarcosinate ([$C_{12}SAR^-$]).

It was found that [TBA^+] was more active as a catalyst due to its higher molecular volume increasing the cation-anion distance and weakening the electrostatic interaction resulting in a more electrophilic cation. [TBA] [$C_{12}BSO_3$] showed better catalytic activity for styrene carbonate synthesis, which reached 81.4% of conversion and 87.0% of selectivity. High recycling capacity for the syntheses of different cyclic carbonates was also reported: Glycidyl isopropyl ether carbonate and epichlorohydrin carbonate.

A cationic polymerization of isobutylene (IB) was investigated in a 1-butyl-3-methylimidazolium hexafluorophosphate ([Bmim][PF_6]) IL at −10°C. Different initiating systems tried for IB polymerization such as titanium tetrachloride, boron trichloride, and ethylaluminum sesquichloride, in [Bmim][PF_6]. A highly reactive polyisobutylene (HR PIB) was synthesized by Li et al. [56] with a high exo-olefin end group content (more than 80%) using a $H_2O/TiCl_4$ initiating system in the [Bmim][PF_6] IL. It was revealed that polymerization proceeded at the interface of IL particles in a mild exothermic manner and $[PF_6]^-$ anions were found to promote the ionization of the initiating system and stabilizing the carbocation active center.

Heteroatom-doped carbon materials have a potential as metal-free catalysts for the hydrogen evolution reaction (HER) at low voltage with high durability. Many active sites introduced by heteroatom doping are hardly accessible within the bulk carbon structure and as a result, contribute little to the catalytic activity. Zhao et al. [57] reported a facile IL-assisted method for the synthesis of mesoporous nitrogen-doped carbons, enabling enrichment of nitrogen atoms at the pore surface. As-synthesized metal-free nitrogen-doped mesoporous carbons exhibited a remarkable electrocatalytic activity in HER. Accessibility and efficient utilization of nitrogen atoms are considered responsible for the better HER catalytic activity.

Aluminum hydride is reported as one of the most promising hydrogen storage materials with a high theoretical hydrogen storage capacity (10.08 wt.%) and relatively low dehydriding temperature (100–200°C). A cost-effective route to synthesize α-AlH_3 nano-composite was developed by Duan et al. [58] using low-cost metal hydrides and aluminum chloride as starting materials. The $LiH/AlCl_3$ and $MgH_2/AlCl_3$ reaction systems were systemically studied. It was observed that the α-AlH_3 nanocomposite can be successfully synthesized by reactive milling of commercial $AlCl_3$ and LiH in a neutral IL ([2-Eim] OAc). An average grain size of 56 nm could be obtained by mechanochemical process as evident from XRD and TEM. By setting the isothermal condition; the As-synthesized α-AlH_3 nanocomposite exhibited an advantage in hydrogen desorption capacity as well as has fast dehydriding in kinetics. The hydrogen desorption content of 9.93 wt.% was achieved at 160°C, indicating the possible utilization of this nanocomposite in hydrogen storage applications.

1.5.2 EXTRACTANT

Extraction of DNA is a bottleneck in analysis of nucleic acid. Clark et al. [59] synthesized hydrophobic magnetic ionic liquids (MILs) and used as solvents for the efficient and fast extraction of DNA from aqueous solutions. DNA enriched micro-droplets were manipulated using a magnetic field. Three MILs exhibited high DNA extraction capabilities on application to a variety of DNA samples and matrices. These are:

- Benzyltrioctylammonium, bromotrichloroferrate(III) for smaller single-stranded and double-stranded DNA;
- Dicationic 1,12-di (3-hexadecylbenzimidazolium) and dodecane bis [(trifluoromethyl)sulfonyl]imide bromotrichloroferrate(III) MIL for higher extraction efficiencies for larger DNA molecules.

This MIL-based method was also utilized for the extraction of DNA from a complex matrix containing albumin. It was revealed that a competitive extraction behavior was there for the trihexyl(tetradecyl)phosphonium tetrachloroferrate(III); $[P_{6,6,6,14}{}^{+}][FeCl_4{}^{-}]$ MIL in contrast to the $[(C_8)_3BnN^{+}]$ $[FeCl_3Br^{-}]$ MIL with significantly less co-extraction of albumin. The MIL-DNA method was also used for the extraction of plasmid DNA from bacterial cell lysate. DNA of sufficient quality and quantity for polymerase chain reaction (PCR) amplification was recovered from the MIL extraction phase, which demonstrates that MIL-based DNA sample preparation is also feasible prior to downstream analysis.

Poly-p-phenyleneterephthalamide (PPTA) is an aramid polymer, which has high tensile strength (TS) and is presently synthesized on industrial scale in a solvent mixture of N-methylpyrrolidone (NMP) and $CaCl_2$. As NMP is toxic and it needs a salt to increase the solubility, ILs have been considered suitable as alternative solvents. A series of ILs were studied by Dewilde et al. [60] for their solubilization strength towards PPTA. They synthesized small PPTA oligomers and used as model compounds for their solubility with ILs to get an insight in the types of cations and anions needed for optimal dissolving behavior. It was observed that ILs with coordinating anions are a requirement to solubilize PPTA because they disrupt the intermolecular hydrogen bond network, as in the case of cellulose dissolution. Interaction of the anions with the hydrogen atoms of the secondary amides of the aramid chains was ascertained with infrared (IR) and NMR-spectroscopic studies. It was reported that cations with hydrogen atoms capable of forming hydrogen bond are poor solvents for PPTA like imidazolium cations. These cations hamper the anions to use their full potential for coordination with the oligomers. Ammonium and phosphonium ILs containing only sp^3-bonded hydrogen atoms on the cation do not have a tendency to form hydrogen bonds and hence dissolve PPTA oligomers much better as compared to their imidazolium analogs. It was confirmed by substituting methyl groups in place of hydrogen atoms on imidazolium and pyridinium cations with significant improvement of solvent power of the ILs. Several types of ILs were formed able to dissolve larger amounts of the PPTA oligomers on a molar basis than industrial solvent NMP/$CaCl_2$ used earlier.

Deferm et al. [61] developed a sustainable solvent extraction process for purification of indium from a chloride aqueous feed solution using the ILs like Cyphos® IL 101 and Aliquat® 336. It was observed that indium(III) has a high affinity for the IL phase and an extraction percentages above 95% could be achieved keeping HCl concentration ranging between 0.5 to 12 M. Based on the relationship between the viscosity of the IL phase and the loading with indium(III) ions, an extraction mechanism was proposed. It was revealed that equilibrium was reached within 10 min even for loadings as high as 100 g L^{-1}. They reported that stripping of indium(III) from the IL phase was quite difficult with aqueous or acid solutions because of very high distribution ratio for indium(III). Indium could be easily recovered as $In(OH)_3$ by precipitation stripping using NaOH solution. Precipitation stripping has an added advantage that there is no loss of IL components to the aqueous phase and apart from it, IL can be regenerated for direct reuse. The extraction of some other metal ions, which that are commonly found as impurities in industrial indium process solutions like copper(II), cadmium(II), nickel(II), zinc(II), iron(III), manganese(II), and tin(IV) has also been investigated. It was reported that indium(III) can be purified efficiently by a combination of extraction, scrubbing, and stripping stages.

An extractive distillation process was investigated, by Zhu et al. [62] comparing two ILs; imidazolium-based ILs [EMIM][BF_4] and [BMIM][BF_4] as solvents for ethanol dehydration. They used an Aspen Plus process simulator to simulate the feasibility of the ionic liquid-based extractive distillation (ILED) process. The main operating conditions of anhydrous ethanol production were also determined. Derived IL components with required thermodynamic and physical property parameters were created in Aspen. [EMIM][BF_4] was used as the solvent to evaluate the potential of extractive distillation for ethanol dehydration. Results indicated that the distillate purity of ethanol was > 99.9 mol%. It was revealed that there is an advantage of decreased energy requirements as compared to the extractive distillation process using other conventional solvents. This process design obtained from the sensitivity analysis for minimum energy consumption was found to have a lower total annual cost (TAC).

ILs are widely used as solvents for extractive desulfurization (EDS) of fuel oils. A mixed-integer nonlinear programming (MINLP) problem was formulated for the purpose of computer-aided ionic liquid design (CAILD) by Song et al. [63]. This MINLP problem was solved to optimize the liquid-liquid extraction performance of ILs in a multi-component model EDS system. It is based on consideration of constraints like IL structure, thermodynamic, and physical properties. Top five ILs were preidentified from CAILD and then these were evaluated by means of process simulation using ASPEN Plus. Finally, [C_5MPy][$C(CN)_3$] was identified as the most suitable solvent for EDS.

Yu et al. [64] used three hydrophobic MILs containing the tetrachloromanganate(II) ($MnCl_4^{2-}$) anion. These are:

- Aliquat tetrachloromanganate(II) ([$Aliquat^+]_2[MnCl_4^{2-}$]);
- Methyltrioctylammonium [$MnCl_4^{2-}$] ([$N_{1,8,8,8}^+]_2[MnCl_4^{2-}$]); and
- Trihexyltetradecylphosphonium tetrachloromanganate(II) ([$P_{6,6,6,14}^+]_2$ [$MnCl_4^{2-}$]).
- These were used as extraction solvents. MILs were developed with the three main features:
- Magnetic susceptibility to permit rapid retrieval of the extraction solvent;
- Hydrophobicity to allow for phase separation from water; and
- Mobile phase compatibility with reversed phase HPLC.

These were customized so as to minimize hydrolysis of the anionic component in aqueous media and also in reducing absorbance on subjecting to HPLC. These MILs were applied for the extraction of some pharmaceutical drugs, phenolics, insecticides, and polycyclic aromatic hydrocarbons (PAHS). Optimal extraction efficiency for each MIL was achieved by variation of disperser solvent type, volume, mass of MIL, extraction time, the pH of the sample solution, and salt concentration. Out of these [$P_{6,6,6,14}^+]_2[MnCl_4^{2-}$] MIL exhibited the extraction efficiencies for most of the target analytes as compared to other MILs. The limits of detection (LODs) of all analytes also ranged between 0.25–1.00 g L^{-1}. The relative recovery was studied in lake and river water. It was reported that relative recovery in lake water was 53.8–114.7% at a spiked concentration of 20 g L^{-1} (5 g L^{-1} for phenanthrene) and between 52.1–106.7% at 150 g L^{-1} (37.5 g L^{-1} for phenanthrene) while it varied between 44.6–110.7% at a spiked concentration of 20 g L^{-1} (5 g L^{-1} for phenanthrene) and 42.9–83.6% at 150 g L^{-1} (37.5 g L^{-1} for phenanthrene) in river water.

Omega-3 poly unsaturated fatty acids (PUFA) particularly eicosapentaenoic acid (EPA), and docosahexaenoic acid (DHA) are having many health benefits such as reducing the risk of cancer and cardiovascular disease. ILs were used by Matlagh et al. [65] in lipid extraction from microalgae, which provided a potential to overcome most of the common drawbacks offering some other benefits. Fast screening of ILs was carried out with the help of a conductor like screening model for real solvents (COSMO-RS). ILs having higher capacity used in extraction of EPA was compared. 352 screened cation/anion were used, but the highest capacity for EPA extraction was found with combinations of [TMAm][SO_4]. It was observed that ILs with small anions was observed having higher capacities, and they possess higher charge density as compared to larger ones. Hence, they are more preferred for extraction. Shorter alkyl chain cations were preferred, when using imidazolium-based ILs were used.

De Boeck et al. [66] reported that ILs can be used for the quantification of a large group of antidepressants in whole blood samples. The sample was prepared by adding 1.0 mL aqueous buffer pH 3.0 and 60 μL of IL (1-butyl-3-methylimidazolium hexafluorophosphate) to 1.0 mL whole blood. Then, a 5-min rotary mixing step was performed, which was followed by centrifugation. The lower IL phase was collected, diluted in 1:10 methanol and 10 μL of it was injected into the liquid chromatography-mass spectroscopy (LC-MS)/MS. The following eighteen (18) analytes were used: (Agomelatine, Amitriptyline, Bupropion, Clomipramine, Dosulepin, Doxepin, Duloxetine, Escitalopram, Fluoxetine, Imipramine, Maprotiline, Mianserin, Mirtazapine, Nortriptyline, Paroxetine, Reboxetine, Trazodone, and Venlafaxine). Selectivity was checked for ten different whole blood matrices. Interferences of deuterated standards or other antidepressants were also evaluated and, no interferences were found. They also evaluated accuracy and precision over eight days at three concentration levels (n = 2). LOD values were found within a range of 1–2 ng mL^{-1} for most analytes. Recoveries (RE) and matrix effects (ME) were also evaluated for five types of donor whole blood, at two concentration levels.

RE values were observed within a range of 53.11–132.98% while, ME values were within a range of 61.92–123.24%. The applicability of ILs as extraction solvents for a large group of antidepressants was proved in complex whole blood matrices.

Polyunsaturated fatty acids (PUFAs) play a significant role in the modulation and prevention of different diseases. Thraustochytrids are marine heterokonts, which exhibit robust growth rates, high PUFA content, and more specifically, a large percentage of omega-3 fatty acids like DHA. Zhang et al. [67] evaluated two ILs, the imidazolium 1-ethyl-3-methylimidazolium ethylsulfate $[C_2mim][EtSO_4]$ IL and the phosphonium (tetrabutylphosphonium propanoate $[P_{4444}][Prop]$) IL for their ability for extraction of PUFA-containing lipids from a *Thraustochytrium* sp. (T18) through efficient cell wall disruption. They characterized the oil extracted after IL pretreatment with respect to fatty acid methyl ester (FAME) composition, while the effects of various process parameters, like ratio of IL to co-solvent, the mass ratio of microalgae to the mixture of IL, and type of co-solvent were also investigated for both ILs. It was indicated that these ILs can disrupt the cells of *Thraustochytrium* sp. on mixing with a co-solvent (methanol), while, facilitated the recovery of oils over a large degree of dewatered *Thraustochytrium* biomass (0–77.2 wt% water) in 60 min at ambient temperature. Thus, it is a water compatible, low-energy, and lipid recovery method. It was observed that lipid recovery was not affected by repeated usage of recycled ILs up to five times.

Banda et al. [68] separated yttrium/europium oxide, into its individual elements which was obtained by the processing of fluorescent lamp waste powder using solvent extraction. They used two undiluted ILs, trihexyl(tetradecyl) phosphonium thiocyanate, [C101][SCN], and tricaprylmethylammonium thiocyanate, [A336][SCN] for this purpose. The best extraction performance was obtained with [C101][SCN] on using an organic-to-aqueous volume ratio of 1/10 and four counter-current extraction stages. A loaded organic phase was scrubbed with a solution of 3 mol L^{-1} $CaCl_2$ + 0.8 mol L^{-1} NH_4SCN for removing the co-extracted europium. Yttrium was then quantitatively stripped from the scrubbed organic phase using deionized water. Yttrium and europium were finally recovered as their corresponding hydroxides by precipitation with ammonia, followed by calcination to the respective oxides. They also tested yttrium/europium separation from synthetic chloride solutions on a leachate obtained from the dissolution of a real mixed oxide. The purity of Y_2O_3 with respect to the rare-earth content was found to be 98.2% while the purity of Eu_2O_3 with respect to calcium was 98.7%.

Three DILs N,N,N,N′,N′,N′-hexaethyl-ethane-1,2-diammonium dibromide (HEDBr), N,N,N,N′,N′,N′-hexaethyl-propane-1,3-diammonium dibromide (HPDBr) and N,N,N,N′,N′,N′-hexaethyl-butane-1,4-diammonium dibromide (HBDBr) were designed by Ji et al. [69] and used to extract phenolic compounds from model oil and coal tar oil. The effects of various parameters were evaluated on the extraction of phenol such as stirring time, temperature, DIL:phenol mole ratio, and initial phenol concentration. It was reported that the DIL:phenol mole ratio was only around 0.3, when the highest extraction efficiency of phenol was obtained. This extraction process was completed within 5 min at RT, and it does not depend on temperature. It was interesting to note that ultimate phenol concentration remained virtually constant, even if initial phenol concentrations is different. It was observed that HPDBr showed the lowest ultimate phenol concentration of 3.9 g dm^{-3}, and the extraction efficiency of phenol can reach up to 97.0%. These DILs can be regenerated by anti-solvent method and reused several times without significant loss in extraction efficiency. HPDBr was also demonstrated to extract phenolic compounds from real coal tar oil, with its extraction efficiency about 92.7%.

Sonmez et al. [70] extracted two Turkish coals with N-methyl-2-pyrrolidone (NMP) and NMP containing a small amount of ILs under reflux conditions at ambient pressure. The effects of coal type on extraction yield were investigated, such as IL type, IL-to-coal ratio, and ultrasonic interactions. The ILs used were:

- 1-butyl-3-methylimidazolium chloride ([Bmim]Cl);
- 1-butyl-4-methylpyridinium chloride ([Bmp]Cl);
- 1-ethyl-3-methylimidazolium chloride ([Emim]Cl);
- 1-butyl-2,3-dimethylimidazolium chloride ([Bdmim]Cl).

It was reported that the extraction yield of coals using NMP/ILs depends on type of coal, IL, and amount of IL. These ILs were found effective on the extraction of Afsin-Elbistan (AE) lignite, but not on the extraction of Üzülmez (UZ) coal. A significant increase in extraction yield was observed for AE lignite on addition of a small amount of IL to NMP. It was indicated that [Bmim]Cl was the most efficient IL used for the extraction of lignite as compared to the others. The extraction efficiency was increased on increasing the amount of IL into NMP.

The olive oil industry annually generates approximately.75–1.5 million tons of *Olea europaea* leaves on global level and it is used as waste typically burned for production of energy. This agricultural by-product is a rich source of oleanolic acid, which is a high value triterpenic acid with excellent pharmaceutical and nutraceutical activities. Claudio et al. [71] observed the extraction of oleanolic acid from dried *O. europaea* leaves with aqueous solutions of surface-active ILs as green solvents. A number of imidazolium-based ILs were synthesized with variable chain length, different anions, and optional side-chain functionalization and used in the extraction. It was reported that ILs with long alkyl chains remarkably enhanced the solubility of oleanolic acid in water and therefore, able to compete with the solubilities of other organic solvents. They were considered suitable alternatives for the solid-liquid extraction of triterpenic acids from natural matrices. They were able to improve extraction yields up to 2.5 wt.% oleanolic acid from olive tree leaves.

A new method has been proposed by Rodrigues et al. [72] combining ultrasonic assisted extraction (UAE) with ILs to extract phycobiliproteins from the microalgae *Spirulina* (Arthrospira) platensis. Extraction of these pigments was carried out by them in an ultrasonic bath at 25°C and at a frequency of 25 kHz. The effects of pH and solvent: biomass ratio was evaluated and the best extraction conditions were achieved. Solvents used include:

- Protic ionic liquids (PILs);
- 2-Hydroxy ethylammonium acetate (2-HEAA);
- 2-Hydroxy ethylammonium formate (2-HEAF);
- Equimolar mixture (2-HEAA +2-HEAF) (1:1 v/v);
- Commercial ionic liquid, [Bmim]Cl.

Sodium phosphate buffer (0.1 M) was used as a control. It was revealed that the PILs were able to extract the phycobiliproteins from microalgae. pH and solvent: biomass ratio were found to be significant in the extraction process. The highest concentrations of phycobiliproteins extracted were achieved on using 2-HEAA +2-HEAF as solvent, at pH 6.50 and solvent: biomass ratio 7.93 mL g^{-1} within 30 min. Pigment were extracted in the following order:

$$\text{Allophycocyanin } (6.34\ mg\cdot g^{-1}) > \text{Phycocyanin } (5.95\ mg\cdot g^{-1}) > \text{Phycoerythrin } (2.62\ mg\cdot g^{-1}).$$

It was revealed that the ultrasound affect the cellular structure of the microalgae as confirmed by scanning electron microscopy (SEM).

An alternative methodology using different PILs, mechanical agitation, and thermal heating was also proposed by them [73] to obtain phycobiliproteins from the *Spirulina* (Arthrospira) platensis microalgae. The PILs N-methyl-2-hydroxy-ethylammonium acetate (2-HEAA) PIL; N-methyl-2-hydroxyethylammonium formate (2-HEAF) PIL; the mixture of these PILs (2-HEAA + 2-HEAF) (1:1, v/v) were synthesized and their performance as solvent was compared to a control (sodium phosphate buffer). The results showed that all PILs were efficient in the extraction of phycobiliproteins. The most suitable conditions were: Temperature 35°C; pH 6.50 and ratio solvent: biomass 6.59 $mL \cdot g^{-1}$, using 2-HEAA + 2-HEAF as solvent after 150 min of extraction. The maximum concentrations of phycocyanin, allophycocyanin, and phycoerythrin were 1.65, 1.70 and 0.64 $g \cdot L^{-1}$, respectively. The obtained phycobiliproteins presented a purification index around 0.50 and can be applied in cosmetics and foods. The phycobiliproteins obtained by extraction using 2-HEAA + 2-HEAF exhibited appreciable antioxidant activity, 61.1% of DPPH scavenging activity and 77.4% of H_2O_2 scavenging activity, at a concentration of 20 mg mL^{-1}.

Yang et al. [74] prepared five novel ILs, 1,3-dibutylimidazolium bromide [BBMIm][Br], 1-pentyl-3-butylimidazolium bromide [BPMIm][Br], 1-hexyl -3-butylimidazolium bromide [BHMIm][Br], 1,1'-(butane-1,4-diyl)bis(3-butylimidazolium) bromide [C_4(BMIm)$_2$][Br_2], and 1,1'-(butane-1,4-diyl) bis(3-methylimidazolium) bromide [C_4(MIm)$_2$][Br_2]. They used this ILs *in situ* to react with bis(trifluoromethane)sulfonamide lithium salt to extract following fungicides from water samples such as myclobutanil, tebuconazole, cyproconazole, and prothioconazole. It was reported that mono-cationic ILs were much better than dicationic ILs from recovery point of view. Mono-imidazolium IL bearing butyl groups at N-1 and N-3 sites were reported to achieve the best recovery. It was observed that if the length of the alkyl substituent group was more than four carbons at N-3 site, the recovery decreased with increase in length of alkyl chain 1-butylimidazolium IL. The extraction efficiency order of triazoles found in following decreasing order:

[BBMIm][Br] > [BPMIm][Br] > [BHMIm][Br] > [BMIm][Br] (1-butyl-3-methylimidazolium bromide) > [C_4(BMIm)$_2$]Br_2 > [C_4(MIm)$_2$]Br_2.

They also established *in situ* IL dispersive liquid-liquid microextraction combined with ultrasmall superparamagnetic Fe_3O_4 as a pretreatment method for enrichment of triazole fungicides in water samples using the synthetic [BBMIm][Br] as the cationic IL. Then analytes were detected by high-performance liquid chromatography. Present proposed method showed a good linearity within a range of 5–250 μg L^{-1} under the optimized conditions. High mean enrichment factors were achieved in the range from 187 to 323, and the RE of the target analytes from real water samples at spiking levels of 10.0, 20.0, and 50.0 μg L^{-1} were obtained between 70.1–115.0%. The LODs for the analytes were reported to be

0.74–1.44 μg L^{-1} with an intra-day relative standard deviations from 5.23–8.65%.

1.5.3 BIOCATALYSIS

An interesting feature of ILs is phase behavior with molecular solvents, e.g., organic solvents and water as these can be fully miscible, partially miscible, or non-miscible with them depending on ions. ILs are normally composed of poorly coordinating ions, making them highly polar but non-coordinating solvents. As a result, they are non-miscible with most of the hydrophobic organic solvents; thus, providing a non-aqueous and polar media to develop two-phase systems. They may be fully or partially miscible also with polar organic solvents depending on the alkyl chain length of the cation. Similarly, most ILs are non-miscible with water, and hence, they can be used to develop biphasic systems with polar characteristics.

Papadopoulou et al. [75] used neoteric and biodegradable ILs based on various hydroxyl ammonium cations and formic acid anion as media for biocatalytic oxidation-reduction catalyzed by different metalloproteins. The effect of these ILs on the biocatalytic behavior and structure of solubilized enzymes was studied using cytochrome c (cyt c) as a model protein. The tolerance of cyt c against the denaturing effect of H_2O_2 and its catalytic efficiency increases almost twenty folds as compared to that in buffer on using ILs-based media. It was revealed that the use of hydroxyl ammonium-based ILs increased the decolorization activity of cyt c by four times. All ILs used can be recycled and reused successfully at least three times, which indicates these ILs have a potential as environmentally friendly media for some biocatalytic processes of industrial interest.

Biomass has been widely used in nature as a source of energy as well as chemical building blocks, but recalcitrance towards traditional chemical processes and solvents is a significant barrier for wide acceptance on global. Brogan et al. [76] discovered solvent-induced substrate promiscuity of glucosidase by optimizing enzyme solubility in ILs, which demonstrates a beautiful example of homogeneous enzyme bioprocessing of cellulose. It was reported that chemical modification of glucosidase for solubilization in ILs can increase its thermal stability to up to 137°C, which allowed 30 times greater enzymatic activity as compared to aqueous media. It was established that biocatalytic capability of enzymes can be increased via a synergistic combination of chemical biology (enzyme modification) and reaction engineering (solvent choice).

Low solubility of L-methionine and low activity of enzyme are the major drawbacks in production of L-methionine by the enzymatic conversion approach. Various ILs were used by Mai and Koo [77] as additives for the enzyme-catalyzed production of L-methionine from O-acetyl L-homoserine and methyl mercaptan. It was found that tetraalkylammonium hydroxide ILs increased the solubility of L-methionine as well as the activity of the enzyme. They reported that methionine solubility decreased with increasing alkyl chain length, but it increased with increasing concentration of IL. L-methionine (232 g l^{-1}) could be dissolved up to 10% in tetramethylammonium hydroxide solution. The enzyme O-acetylhomoserine aminocarboxypropyltransferase showed its maximum activity, when concentration of IL was kept 2.5% (3 times higher than without using ILs), but it significantly decreased with increasing IL concentration. It was observed that stability of the enzyme also decreased rapidly after 2 h of incubation whether ILs is present or not. It was also revealed that 74 g L^{-1} of L-methionine could be produced in a 2.5% tetraethylammonium hydroxide as compared to 35 g L^{-1} of L-methionine without ILs.

Zarski et al. [78] esterified potato starch with oleic acid, using [Bmim] Cl as a reaction medium. They used immobilized lipase from *Thermomyces lanuginosus* as a catalyst. The degree of substitution (DS) of the products was determined by the volumetric method. The effect of different reaction parameters on the DS was also observed, like time and temperature. Highest DS (0.22) was found in product, when the reaction was carried out at 60°C for 4 h. Esterification of the potato starch was confirmed on the bases of FTIR and NMR. It was also revealed that the crystallinity and morphology of the native potato starch was slightly affected during its partial gelatinization in the IL, but it was completely destroyed, when esters are formed. The thermal stability of the starch oleate was found to decrease in comparison to the unmodified starch, as evident from a thermal gravimetric analysis (TGA).

Binary mixtures of ILs have been used by Lozano et al. [79] for developing one-pot systems suitable for the direct extraction and biocatalytic transformation of algal oil to biodiesel. These mixtures were based on the combination of sponge-like ILs such as 1-hexadecyl-3-methylimidazolium bis(trifluoromethylsulfonyl)imide, [C16mim][NTf2]) with the IL [Bmim]Cl. Here, first IL has excellent suitability to carry out the biocatalytic synthesis of biodiesel, while latter has the ability for dissolving cellulosic biomass. The extraction of oils was carried out by incubating two dry microalgae (*Chlorella vulgaris* or *Chlorella protothecoides*) in IL binary mixture at 110°C. It was then being cooled at 60°C, which allowed transformation to biodiesel by an immobilized lipase. This resulted in a fast and efficient biodiesel synthesis, up to almost 100% yield in 2 h at 60°C. The subsequent cooling until RT and centrifugation of the resulting semi-solid systems at lower temperature (20 and 18°C) led to the separation of the liquid algae biodiesel from the solid IL mixture. It can be recovered and reused again.

Lozano et al. [80] reported that hydrophobic ILs based on cations with long alkyl side-chains, e.g., N,N,N,N-hexadecyltrimethylammonium bis (trifluoromethylsulfonyl)imide ([C16tma][NTf_2]), are temperature switchable IL/solid phases and they behave as sponge-like systems (sponge-like ionic liquid, SLILs). These SLILs have been used to develop straightforward and clean approaches for producing nearly pure synthetic compounds with added value such as geranyl acetate, anisyl acetate, methyl oleate, etc. in two steps:

- An enzymatic synthetic step as liquid phase; and
- Product separation step involving simple centrifugation as a solid phase.

Whole cells of *Escherichia coli* over expressing a glucosyl transferase from *Vitis vinifera* were used for the glucosylation of geraniol to geranyl glucoside. A high cell density cultivation process for the production of whole-cell biocatalysts was developed by Schmideder et al. [81] with a process time of 48 h gaining a dry cell mass concentration of up to 67.6 ± 1.2 g L^{-1} and a glucosyl transferase concentration of up to 2.7 ± 0.1 g protein L^{-1}. Highest conversion of geraniol was achieved at pH 7.0 although the pH optimum of the purified glucosyl transferase was at 8.5. It was reported that performance was improved significantly on addition of a water-immiscible ILs (N-hexylpyridinium bis(trifluoromethylsulfonyl)imid). Highest final geranyl glucoside concentration (291 ± 9 mg L^{-1}) and conversion (71 ± 2%) was reported for whole-cell biotransformations of geraniol, when 5% (v/v) of the IL was added.

Ether-functionalized ILs usually have low viscosities, and these can be designed to be compatible with enzymes. Zhao et al. [82] evaluated some new ether-functionalized ILs having different cation cores (pairing with Tf2N− anions) in two Novozym 435-catalyzed reactions:

- Transesterification of ethyl sorbate with 1-propanol at 50°C; and
- Ring-opening polymerization (ROP) of ε-caprolactone at 70°C.

The lipase played two different roles:

- In the first reaction, [$CH_3OCH_2CH_2$-Et_3N][Tf_2N] and [$CH_3OCH_2CH_2$-Py][Tf_2N] gave the highest reaction rates; and

- In the second reaction, $[CH_3OCH_2CH_2\text{-}PBu_3][Tf_2N]$ produced the highest molecular mass (Mw up to 25,400 Da).

The thermal stability of lipase in $[CH_3OCH_2CH_2\text{-}Et_3N][Tf_2N]$ was also found much higher than in t-butanol. It was concluded that lipase activity is not only dependent on the structure of IL but also on the substrate and other reaction conditions.

ILs are now recognized as solvents for use in lipase-catalyzed reactions; however, there still remains a serious drawback in that the rate of reaction in an IL is slow. Attempts have been made to overcome the problem of slower rate in IL than that in a conventional organic solvent. Abe et al. [83] developed phosphonium ILs for lipase-catalyzed reactions and evaluated their capability for use as solvent for the lipase-catalyzed reactions. It increases the lipase PS-catalyzed transesterification of secondary alcohols on using 2-methoxyethyl(tri-n-butyl)phosphonium bis(trifluoromethanesulfonyl)imide ($[MEBu_3P][NTf_2]$) as solvent. This affords an example of superior reaction rate than that in diisopropyl ether.

To establish an efficient glycerolysis approach for enzymatic production of diglycerides (DG), Guo et al. [84] reported a novel concept to improve the yield of DG by applying a binary IL system, where one IL system has a better DG production selectivity while another IL is able to achieve higher conversion of triglycerides (TG). A combination of $TOMA.Tf_2N$/Ammoeng 102 was used for optimization by Response Surface Methodology. They could obtain 80–85% mol% of oil conversion and up to 90% (mol%) of total DG yield (73%, wt%), which is significantly higher than reported earlier. It was proposed that coupling the technical advantage (high TG conversion and high DG production) of individual ILs into a binary system can prove to be beneficial.

Candida rugosa lipase was immobilized by Jiang et al. [85] on magnetic NPs supported ILs with different cation chain lengths (C1, C4, and C8) and anions (Cl^-, BF_4^- and PF_6^-). They obtained magnetic NPs supported ILs by covalent bonding of ILs-silane on magnetic silica NPs. The particles were found to be superparamagnetic with an average diameter of about 55 nm. Large amount of lipase (63.89 mg 100 mg^{-1} carrier) was loaded on the support through ionic adsorption. Activity of this immobilized lipase was determined by the catalysis of esterification between oleic acid and butanol. The activity of bound lipase was higher (about 118.3%) as compared to that of the lipase. Immobilized lipase was able to maintain 60% of its initial activity, even if the temperature was up to 80°C. It was revealed that immobilized lipase retained 60% of its initial activity after completing eight batches reaction, but no activity was detected even after 6 cycles for the free lipase.

1.6 CONCLUSION

VOCs are considered the main culprits in creating air pollution as these are commonly used reaction media for many chemical processes. VOCs cannot be easily separated from the desired reaction products and they are difficult to recycle. The range of reactions in ILs gives a flavor of what can be achieved in these neoteric solvents. Because the properties and behavior of the IL can be adjusted to suit an individual reaction type, they can truly be described as designer solvents also. ILs have been implemented as solvent systems in chemical reactions, separations, extractions, electroanalytical, and chemical sensing, among many other applications. Also, they have a high ionic character that enhances the reaction rates to a great extent in many reactions. Although ILs has been already proved to present advantageous industrial applications, they are also attractive for analytical chemists. Liquid/liquid and aqueous two-phase systems based on ILs appear to be very useful for that purpose. Often the IL can be recycled, and this leads to a reduction in the costs of the processes. It must be emphasized that reactions in ILs are not difficult to perform and usually require no special apparatus or methodologies. The reactions are often quicker and easier to carry out than in conventional organic solvents.

KEYWORDS

- **biocatalysis**
- **eco-friendly solvents**
- **green solvents**
- **ionic liquids**
- **melting points**
- **viscosity**

REFERENCES

1. Wier, Jr. T., & Hurley, F., (1948). *Ionic Liquids in Synthesis* (Vol. 4, No. 446, p. 349). US Patent.
2. Wilkes, J. S., (2002). A short history of ionic liquids-from molten salts to neoteric solvents. *Green Chem.*, *4*, 73–80.
3. Walden, P., (1914). Molecular weights and electrical conductivity of several fused salts. *Bull. Acad. Imper. Sci.*, 404–422.
4. Clarke, J., & Macquarie, D., (2002). *Handbook of Green Chemistry and Technology*. Oxford.
5. Plechkovo, N. V., & Seddon, K. R., (2008). Applications of ionic liquids in the chemical industry. *Chem. Soc. Rev.*, *37*, 123–150.
6. Earle, M. J., & Seddon, K. R., (2000). Expanding the polarity range of ionic liquids. *Pure Appl. Chem.*, *72*, 1391–1404.
7. Wilkes, J. S., (2002). Ionic liquids in perspective: The past with an eye toward the industrial Future, *American Chem. Soc.*, *818*, 214–229.
8. Seddon, K. R., (1997). Ionic liquids for clean technology. *J. Chem. Tech. Biotechnol.*, *68*, 351–356.
9. Lall, S. I., Mancheno, D., Castro, S., Behaj, V., Cohen, J. L. I., & Enge, R., (2000). A new category of room temperature ionic liquid based on polyammonium salts. *Chem. Commun.*, 2413–2414.
10. Davis, J. H., Forrester, K. J., & Merrigan, T., (1998). Novel organic ionic liquids (OILs) incorporating cations derived from the antifungal drug miconazole. *Tetrahedron Lett.*, *39*, 8955–8958.
11. Abbott, A. P., Capper, G., Davies, D. L., Rasheed, R., & Tambyrajah, V., (2003). Ionic liquid analogues formed from hydrated metal salts. *Chem. Commun.*, *70*, 136–144.
12. Shepard, E. R., & Shonle, H. A., (1947). Imidazolium and imidazolinium salts as topical antiseptics. *J. Am. Chem. Soc.*, *69*, 2269–2277.
13. Chan, B. K. M., Chang, N. H., & Grimmett, M. R., (2005). The synthesis and thermolysis of imidazole quaternary salts. *Aust. J. Chem.*, *30*, 1977–2013.
14. Ricciardi, F., Romanchick, W. A., & Joullié, M. M., (1983). 1,3-dialkylimidazolium salts as latent catalysts in the curing of epoxy resins. *J. Polym. Sci. Polym. Lett.*, *21*(8), 633–638.
15. Welton, T., (2004). Ionic liquids in catalysis. *Coord. Chem. Rev.*, *248*, 2459–2477.
16. Deetlefs, M., Raubenheimer, H. G., & Esterhuysen, M. W., (2002). Stoichiometric and catalytic reactions of gold utilizing ionic liquids. *Catal. Today*, *72*, 29–41.
17. Carmichael, A. J., Earle, M. J., Holbrey, J. D., Mc-Cormac, P. B., & Seddon, K. R., (1999). The Heck reaction in ionic liquids: A multiphasic catalyst system. *Org. Lett.*, *1*, 997–1000.
18. Wasserscheid, P., & Welton, T., (2008). *Ionic Liquids in Synthesis* (p. 1). WILEY-VCH.
19. Suarez, P. A. Z., Dullius, J. E. L., Einloft, S., Souza, R. F., & Dupont, J., (1996). The use of new ionic liquids in two-phase catalytic hydrogenation reaction by rhodium complexes. *Polyhedron*, *15*, 1217–1219.
20. Alterman, M., & Hallberg, A., (2000). Fast microwave-assisted preparation of aryl and vinyl nitriles and the corresponding tetrazoles from organo-halides. *J. Org. Chem.*, *65*, 7984–7989.
21. Polshettiwar, V., & Varma, R. S., (2008). Microwave-assisted organic synthesis and transformations using benign reaction media. *Chem. Res.*, *41*, 629–639.

22. Varma, R. S., & Namboodiri, V. V., (2002). An improved preparation of 1,3-dialkylimidazolium tetrafluoroborate ionic liquids using microwaves. *Tetrahedron Lett.*, *43*, 5381–5383.
23. Freemantle, M., (2010). Properties of ionic liquids. *Royal Soc. Chem.*, *4*, 44–50.
24. Zhou, Z., Matsumoto, H., & Tatsumi, K., (2005). Low melting, low viscous, hydrophobic ionic liquids: Aliphatic quaternary ammonium salts with perfluoroalkyltrifluoroborates. *Eur. J. Chem.*, *11*, 752–760.
25. Endres, F., & El Abedin, S. Z., (2006). Air and water stable ionic liquids in physical chemistry. *Phy. Chem. Chem. Phy.*, *8*, 2101–2128.
26. Chan, B. K. M., Chang, N. H., & Grimmett, M. R., (2005). The synthesis and thermolysis of imidazole quaternary salts. *Aust. J. Chem.*, *30*, 1977–1984.
27. Ramnial, T., Ino, D. D., & Clyburne, J. A. C., (2005). Phosphonium ionic liquids as reaction media for strong bases. *Chem. Comm.*, 325–330.
28. Chiappe, C., & Pieraccini, D., (2005). Ionic liquids: Solvent properties and organic reactivity. *J. Phys. Org. Chem.*, *18*, 275–279.
29. Anderson, J. L., Ding, R., Ellern, A., & Armstrong, D. W., (2005). Structure and properties of high stability geminal dicationic ionic liquids. *J. Am. Chem. Soc.*, *127*, 593–600.
30. Galinski, M., Lewandowski, A., & Stepniak, I., (2006). Ionic liquids as electrolytes. *Electrochimica. Acta.*, *51*, 5567–5575.
31. Palimkar, S. S., Siddiqui, S. A., Daniel, T., Lahoti, R. J., & Srinivasan, K. V., (2003). Ionic liquid-promoted regiospecific Friedlander annulation: Novel synthesis of quinolines and fused polycyclic quinolines. *J. Org. Chem.*, *68*(24), 9371–9378.
32. Forbes, D. C., Law, A. M., & Morrison, D. W., (2006). The Knoevenagel reaction: Analysis and recycling of the ionic liquid medium. *Tetrahedron Letters*, *47*(11), 1699–1703.
33. Konnerth, H., & Prechtl, M. H. G., (2017). Selective hydrogenation of N-heterocyclic compounds using Ru nanocatalysts in ionic liquids. *Green Chem.*, *19*(12), 2762–2767.
34. Carmichael, A. J., Earle, M. J., Holbrey, J. D., McCormac, P. B., & Seddon, K. R., (1999). The heck reaction in ionic liquids: A multiphasic catalyst system. *Org. Lett.*, *1*(7), 997–1000.
35. Mathews, C., Smith, P. J., & Welton, T., (2004). N-donor complexes of palladium as catalysts for Suzuki cross-coupling reactions in ionic liquids. *J. Mol. Catal. A: Chem.*, *214*(1), 27–32.
36. Buijsman, R. C., Van, V. E., & Sterrenburg, J. G., (2001). Ruthenium-catalyzed olefin metathesis in ionic liquids. *Org. Lett.*, *3*(23), 3785–3787.
37. Song, C. E., Roh, E. J., Lee, S., Shim, W. H., & Choi, J. H., (2001). Ionic liquids as powerful media in scandium triflate catalysed Diels-Alder reactions: Significant rate acceleration, selectivity improvement and easy recycling of catalyst. *Chem. Commun.*, *12*, 1122–1123.
38. Nara, S. J., Harjani, J. R., & Salunkhe, M. M., (2001). Friedel-crafts sulfonylation in 1-butyl-3-methylimidazolium chloroaluminate ionic liquids. *J. Org. Chem.*, *66*(25), 8616–8620.
39. Lancaster, N. L., Welton, T., & Young, G. B., (2001). A study of halide nucleophilicity in ionic liquids. *J. Chem. Soc., Perkin Trans.*, *2*(12), 2267–2270.
40. Kotrusz, P., Kmentová, I., Gotov, B., Toma, Š., & Solčániová, E., (2002). Proline-catalysed asymmetric aldol reaction in the room temperature ionic liquid [bmim]PF_6. *Chem. Commun.*, *21*, 2510–2511.

41. Fujiwara, M., Kawatsura, M., Hayase, S., Nanjo, M., & Itoh, T., (2008). Iron(III) salt-catalyzed nazarov cyclization/Michael addition of pyrrole derivatives. *Adv. Synth. Catal.*, *351*(1/2), 123–128.
42. Ngo, H. L., Hu, A., & Lin, W., (2005). Catalytic asymmetric hydrogenation of aromatic ketones in room temperature ionic liquids. *Tetrahedron Lett.*, *46*(4), 595–597.
43. Öchsner, E., Schneiders, K., Junge, K., Beller, M., & Wasserscheid, P., (2009). Highly enantioselective Ru-catalyzed asymmetric hydrogenation of β-keto ester in ionic liquid/methanol mixtures. *Appl. Catal. A: Gen.*, *364*(1/2), 8–14.
44. Khan, N. H., Prasetyanto, E. A., Kim, Y. K., Ansari, M. B., & Park, S. E., (2010). Chiral Cu(II) complexes as recyclable catalysts for asymmetric nitroaldol (henry) reaction in ionic liquids as greener reaction media. *Catal. Lett.*, *140*(3/4), 189–196.
45. Gruttadauria, M., Riela, S., Lo Meo, P., D'Anna, F., & Noto, R., (2004). Supported ionic liquid asymmetric catalysis. A new method for chiral catalysts recycling. The case of proline-catalyzed aldol reaction. *Tetrahedron Lett.*, *45*(32), 6113–6116.
46. Li, C., Zhao, J., Tan, R., Peng, Z., Luo, R., Peng, M., & Yin, D., (2011). Recyclable ionic liquid-bridged chiral dimeric salen Mn(III) complexes for oxidative kinetic resolution of racemic secondary alcohols. *Catal. Commun.*, *15*(1), 27–31.
47. Kude, K., Hayase, S., Kawatsura, M., & Itoh, T., (2011). Iron-catalyzed quick homo-coupling reaction of aryl or alkynyl Grignard reagents using a phosphonium ionic liquid solvent system. *Heteroatom Chem.*, *22*(3/4), 397–404.
48. Yan, D., Wang, G., Gao, K., Lu, X., Xin, J., & Zhang, S., (2018). One-pot synthesis of 2, 5-furandicarboxylic acid from fructose in ionic liquids. *Ind. Eng. Chem. Res.*, *57*(6), 1851–1858.
49. Olivares-Xometl, O., Álvarez-Álvarez, E., Likhanova, N. V., Lijanova, I. V., Hernández-Ramírez, R. E., Arellanes-Lozada, P., & Varela-Caselis, J. L., (2017). Synthesis and corrosion inhibition mechanism of ammonium-based ionic liquids on API 5L X60 steel in sulfuric acid solution. *J. Adhes. Sci. Technol.*, *32*(10), 1092–1113.
50. Shen, Y., Sun, J. K., Yi, Y. X., Wang, B., Xu, F., & Sun, R. C., (2015). One-pot synthesis of levulinic acid from cellulose in ionic liquids. *Bioresourc. Technol.*, *192*, 812–816.
51. Van Dao, D., Nguyen, T. T. D., Majhi, S. M., Adilbish, G., Lee, H. J., Yu, Y. T., & Lee, I. H., (2019). Ionic liquid-supported synthesis of CeO_2 nanoparticles and its enhanced ethanol gas sensing properties. *Mater. Chem. Phys.* doi: 10.1016/j.matchemphys.2019.03.025.
52. Qadir, M. I., Bernardi, F., Scholten, J. D., Baptista, D. L., & Dupont, J., (2019). Synergistic CO_2 hydrogenation over bimetallic Ru/Ni nanoparticles in ionic liquids. *Appl. Catal. B: Environ.* doi: 10.1016/j.apcatb.2019.04.005.
53. Kimura, A., Nagasawa, N., & Taguchi, M., (2019). Synthesis of polysaccharide hybrid gel in ionic liquids via radiation-induced crosslinking. *Polym. Degrad. Stab.*, *159*, 133–138.
54. Kasprzak, D., Krystkowiak, E., Stępniak, I., & Galiński, M., (2019). Dissolution of cellulose in novel carboxylate-based ionic liquids and dimethyl sulfoxide mixed solvents. *Europ. Poly. J.* doi: 10.1016/j.eurpolymj.2019.01.053.
55. Vieira, M. O., Monteiro, W. F., Neto, B. S., Chaban, V. V., Ligabue, R., & Einloft, S., (2019). Chemical fixation of CO_2: The influence of linear amphiphilic anions on surface active ionic liquids (SAILs) as catalysts for synthesis of cyclic carbonates under solvent-free conditions. *React. Kinet. Mech. Catal.* doi: 10.1007/s11144-019-01544-6.
56. Li, X., Wu, Y., Zhang, J., Li, S., Zhang, M., Yang, D., & Yan, P., (2019). Synthesis of highly reactive polyisobutylenes via cationic polymerization in ionic liquid: Characteristics and mechanism. *Polym. Chem.*, *10*, 201–208. doi: 10.1039/c8py01141a.

57. Zhao, X., Li, S., Cheng, H., Schmidt, J., & Thomas, A., (2018). Ionic liquid-assisted synthesis of mesoporous carbons with surface-enriched nitrogen for the hydrogen evolution reaction. *ACS Appl. Mater. Interfaces*, *10*(4), 3912–3920.
58. Duan, C. W., Hu, L. X., & Ma, J. L., (2018). Ionic liquids as an efficient medium for the mechanochemical synthesis of α-AlH_3 nano-composites. *J. Mater. Chem. A*, *6*(15), 6309–6318.
59. Clark, K. D., Nacham, O., Yu, H., Li, T., Yamsek, M. M., Ronning, D. R., & Anderson, J. L., (2015). Extraction of DNA by magnetic ionic liquids: Tunable solvents for rapid and selective DNA analysis. *Anal. Chem.*, *87*(3), 1552–1559.
60. Dewilde, S., Dehaen, W., & Binnemans, K., (2016). Ionic liquids as solvents for PPTA oligomers. *Green Chem.*, *18*(6), 1639–1652.
61. Deferm, C., Van De Voorde, M., Luyten, J., Oosterhof, H., Fransaer, J., & Binnemans, K., (2016). Purification of indium by solvent extraction with undiluted ionic liquids. *Green Chem.*, *18*(14), 4116–4127.
62. Zhu, Z., Ri, Y., Li, M., Jia, H., Wang, Y., & Wang, Y., (2016). Extractive distillation for ethanol dehydration using imidazolium-based ionic liquids as solvents. *Chem. Eng. Process*, *109*, 190–198.
63. Song, Z., Zhang, C., Qi, Z., Zhou, T., & Sundmacher, K., (2017). Computer-aided design of ionic liquids as solvents for extractive desulfurization. *AIChE Journal*, *64*(3), 1013–1025.
64. Yu, H., Merib, J., & Anderson, J. L., (2016). Faster dispersive liquid-liquid microextraction methods using magnetic ionic liquids as solvents. *J. Chromatogr. A*, *1463*, 11–19.
65. Motlagh, S. R., Harun, R., Biak, D. A., Hussain, S., Wan, A. K. G. W., Khezri, R., Wilfred, C. D., & Elgharbawy, A., (2019). Screening of suitable ionic liquids as green solvents for extraction of eicosapentaenoic acid (EPA) from microalgae biomass using COSMO-RS model. *Molecules*, *24*(4), 713.
66. De Boeck, M., Dubrulle, L., Dehaen, W., Tytgat, J., & Cuypers, E., (2018). Fast and easy extraction of antidepressants from whole blood using ionic liquids as extraction solvent. *Talanta*, *180*, 292–299.
67. Zhang, Y., Ward, V., Dennis, D., Plechkova, N., Armenta, R., & Rehmann, L., (2018). Efficient extraction of a docosahexaenoic acid (DHA)-rich lipid fraction from thraustochytrium sp. using ionic liquids. *Materials*, *11*(10), 1986.
68. Banda, R., Forte, F., Onghena, B., & Binnemans, K., (2019). Yttrium and europium separation by solvent extraction with undiluted thiocyanate ionic liquids. *RSC Advances*, *9*(9), 4876–4883.
69. Ji, Y., Hou, Y., Ren, S., Yao, C., & Wu, W., (2018). Highly efficient extraction of phenolic compounds from oil mixtures by trimethylamine-based dicationic ionic liquids via forming deep eutectic solvents. *Fuel Processing Technology*, *171*, 183–191.
70. Sönmez, Ö., Yıldız, Ö., Çakır, M. Ö., Gözmen, B., & Giray, E. S., (2018). Influence of the addition of various ionic liquids on coal extraction with NMP. *Fuel*, *212*, 12–18.
71. Cláudio, A. F. M., Cognigni, A., De Faria, E. L. P., Silvestre, A. J. D., Zirbs, R., Freire, M. G., & Bica, K., (2018). Valorization of olive tree leaves: Extraction of oleanolic acid using aqueous solutions of surface-active ionic liquids. *Sep. Purif. Technol.*, *204*, 30–37.
72. Rodrigues, R. D. P., De Castro, F. C., Santiago-Aguiar, R. S. D., & Rocha, M. V. P., (2018). Ultrasound-assisted extraction of phycobiliproteins from Spirulina (Arthrospira) platensis using protic ionic liquids as solvent. *Algal Res.*, *31*, 454–462.

73. Rodrigues, R. D. P., De Lima, P. F., Santiago-Aguiar, R. S. D., & Rocha, M. V. P., (2019). Evaluation of protic ionic liquids as potential solvents for the heating extraction of phycobiliproteins from Spirulina (Arthrospira) platensis. *Algal Res., 38*, 101391. doi: 10.1016/j.algal.2018.101391.
74. Yang, J., Fan, C., Kong, D., Tang, G., Zhang, W., Dong, H., Liang, Y., Wang, D., & Cao, Y., (2018). Synthesis and application of imidazolium-based ionic liquids as extraction solvent for pretreatment of triazole fungicides in water samples. *Anal. Bioanal. Chem., 410*(6), 1647–1656.
75. Papadopoulou, A. A., Tzani, A., Alivertis, D., Katsoura, M. H., Polydera, A. C., Detsi, A., & Stamatis, H., (2016). Hydroxyl ammonium ionic liquids as media for biocatalytic oxidations. *Green Chem., 18*(4), 1147–1158.
76. Brogan, A. P. S., Bui-Le, L., & Hallett, J. P., (2018). Non-aqueous homogeneous biocatalytic conversion of polysaccharides in ionic liquids using chemically modified glucosidase. *Nature Chem., 10*(8), 859–865.
77. Mai, N. L., & Koo, Y. M., (2018). Enhanced enzyme-catalyzed synthesis of L-methionine with ionic liquid additives. *Proc. Biochem.* doi: 10.1016/j.procbio.2018.11.020.
78. Zarski, A., Ptak, S., Siemion, P., & Kapusniak, J., (2016). Esterification of potato starch by a biocatalysed reaction in an ionic liquid. *Carbohydr. Polym., 137*, 657–663.
79. Lozano, P., Bernal, J. M., Gómez, C., Álvarez, E., Markiv, B., García-Verdugo, E., & Luis, S. V., (2019). Green biocatalytic synthesis of biodiesel from microalgae in one-pot systems based on sponge-like ionic liquids. *Catal. Today*. doi: 10.1016/j.cattod.2019.01.073.
80. Lozano, P., Bernal, J. M., Gómez, C., García-Verdugo, E., Isabel, B. M., Sánchez, G., & Luis, S. V., (2015). Green bioprocesses in sponge-like ionic liquids. *Catal. Today, 255*, 54–59.
81. Schmideder, A., Priebe, X., Rubenbauer, M., Hoffmann, T., Huang, F. C., Schwab, W., & Weuster-Botz, D., (2016). Non-water miscible ionic liquid improves biocatalytic production of geranyl glucoside with *Escherichia coli* over expressing a glucosyl transferase. *Bioproc. Biosyst. Eng., 39*(9), 1409–1414.
82. Zhao, H., Kanpadee, N., & Jindarat, C., (2019). Ether-functionalized ionic liquids for nonaqueous biocatalysis: Effect of different cation cores. *Proc. Biochem.* doi: 10.1016/j.procbio.2019.03.018.
83. Abe, Y., Kude, K., Hayase, S., Kawatsura, M., Tsunashima, K., & Itoh, T., (2008). Design of phosphonium ionic liquids for lipase-catalyzed transesterification. *J. Mol. Catal. B: Enzym., 51*, 81–85.
84. Guo, Z., Kahveci, D., Ozcelik, B., & Xu, X. B., (2009). Improving enzymatic production of diglycerides by engineering binary ionic liquid medium system. *New Biotechnol., 26*, 37–43.
85. Jiang, Y., Guo, C., Xia, H., Mahmood, I., Liu, C., & Liu, H., (2009). Magnetic nanoparticles supported ionic liquids for lipase immobilization: Enzyme activity in catalyzing esterification. *J. Mol. Catal. B: Enzym., 58*, 103–109.

CHAPTER 2

Interpenetrating Network on the Basis of Methylsiloxane Matrix

O. MUKBANIANI,[1] W. BROSTOW,[2] J. ANELI,[3] T. TATRISHVILI,[1,3] E. MARKARASHVILI,[1,3] and N. JALAGONIA[3]

[1]*Department of Chemistry, Iv. Javakhishvili Tbilisi State University, I. Chavchavadze Ave., 1, Tbilisi 0179, Georgia, E-mail: omar.mukbaniani@tsu.ge (O. Mukbaniani)*

[2]*Laboratory of Advanced Polymers and Optimized Materials (LAPOM) Department of Materials Science and Engineering and Department of Physics, University of North Texas, 3940 North Elm Street, Denton, TX 76207, USA, E-mail: wkbrostow@gmail.com*

[3]*Institute of Macromolecular Chemistry and Polymeric Materials, IV. Javakhishvili Tbilisi State University, I. Chavchavadze Ave., 13, Tbilisi 0179, Georgia*

ABSTRACT

The hydrosilylation reactions of α,ω−bis(trimethylsiloxy)methylhydrosiloxane with 3-(2-(2-methoxyethoxy)ethoxy)prop-1-ene and vinyltrieth-oxysilane catalyzed by Karstedt's ($Pt_2[(VinSiMe_2)_2O]_3$) catalyst, platinum hydrochloric acid (0.1 M solution in THF) and platinum on carbon have been studied. In the presence of Karstedt's catalyst the reaction order, activation energies and rate constants have been determined. The synthesized oligomers were analyzed with FTIR, ^1H, ^{13}C, and ^{29}Si NMR spectroscopy. Synthesized polysiloxanes were investigated by wide-angle X-ray scattering, gel-permeation chromatography, and DSC methods. Solid polymer electrolyte membranes have been obtained via sol-gel processes of oligomer systems doped with lithium trifluoromethylsulfonate (triflate) or lithium bis(trifluoromethylsulfonyl) imide. The dependence of ionic conductivity as a function of temperature and salt concentration has been investigated. It has been found that the electric conductivity of the polymer electrolyte membranes at room temperature (RT) changes in the range $4 \times 10^{-5} - 6 \times 10^{-7}$ S/cm.

2.1 INTRODUCTION

Currently, solid polymer electrolytes (SPE) on the basis of polysiloxanes present an interest because of their wide functional possibilities. The reason lies in properties of polysiloxanes such as strong heat resistance, elastomeric behavior, biocompatibility, thermal-, UV-and oxidative stabilities, low surface energy, good weather ability, low melting points and glass transition temperatures, convenient rheological properties, and outstanding electrical properties. The polysiloxanes, with very low glass transition temperatures, extremely high free volumes, are expected to be good hosts for Li^+ transport, when electron donor units are introduced into the polymer backbone. Double-comb-type polysiloxane compounds prepared via the condensation reaction of bis[oligo(ethylene glycol) ether propyl] dichlorosilane show conductivity of 4.5×10^{-4} $S \cdot cm^{-1}$ [1]. It must be noted that the level of conductivity of these polymers is very close to value necessary for practical applications ($10^{-3} \times S \cdot cm^{-1}$) [2–6]. Polyethylene oxide (PEO) substituted polysiloxanes as ionic conductive polymer hosts have been previously investigated [7, 8]. It was discovered ionic conductivity in a PEO/Na^+ complex, the research and development effort has become quite active on SPE, in particular, on the improvement of the ionic conductivity.

In the presented work the synthesis of cross-linked polysiloxane polymer electrolytes (PE) with pendant 3-(2-(2-methoxyethoxy)ethoxy)propyl groups as internally plasticizing chains and investigation of their electric-physical properties are conducted. This polysiloxane conductive material doped with lithium trifluoromethylsulfonate ($LiSO_3CF_3$) or lithium bis(trifluoromethylsulfonyl) imide ($LiN(SO_2CF_3)_2$) shows high room temperature (RT) conductivity, arising from the vigorous segmental motion of the pendant propyl butyrate chains as well as the high flexibility of the siloxane backbone.

2.2 EXPERIMENTAL PART

2.2.1 MATERIALS

Polymethylhydrosiloxane (PMHS) (with degree of polymerization (DP), n=35), platinum hydrochloric acid, Karstedt's catalyst ($Pt_2[(VinSiMe_2)_2O]_3$), Pt/C (10%) and vinyltriethoxysilane) were used from Aldrich. 3-(2-(2-methoxyethoxy)ethoxy)prop-1-ene have been prepared by known method [3]. Lithium trifluoromethylsulfonate (triflate) and Lithium bis(trifluoromethyl-sulfonyl)imide were purchased from Aldrich and used after drying in vacuum. Toluene was dried and distilled from sodium under atmosphere of dry nitrogen. Tetrahydrofuran (THF) was dried over and distilled from K-Na alloy under an atmosphere of dry nitrogen. 0.1 M solution of platinum hydrochloric acid in THF was prepared and kept under nitrogen at low temperature.

2.2.2 HYDROSILYLATION REACTION OF PMHS WITH 3-(2-(2-METHOXYETHOXY)ETHOXY)PROP-1-ENE AND VINYLTRIETHOXYSILANE IN THE PRESENCE OF KARSTEDT'S CATALYST

PMHS 1.6000 g (0.7073 mmole) were transferred into a 100 mL flask under nitrogen using standard Schlenk techniques. High vacuum was applied to the flask for half an hour before the addition of 3-(2-(2-methoxyethoxy) ethoxy)prop-1-ene (2.8519 g, 0.0198 mole) and 0.9424 g (0.00495 mole) vinyltriethoxysilane. The mixture was then dissolved in 3 mL of toluene, and 3 μL Karstedt's catalyst was syringed into the flask. The homogeneous mixture was degassed and placed into an oil bath, which was previously set to 60°C and reaction continued at 60°C. The reaction was controlled by the decrease of intensity of active ≡Si-H groups.

Subsequently, 0.1% activated carbon was added and refluxed for 12 h for deactivation of catalysts. All volatiles were removed by rotary evaporation and the polymer was precipitated at least three times into pentane to remove side products. Finally, all volatiles were removed under vacuum and further evacuated under high vacuum for 24 h to isolate the colorless viscous polymer, 4.75 g (88%).

2.2.3 PREPARATION OF CROSS-LINKED POLYMER ELECTROLYTES (PE)

The 0.75 g of the base compound I was dissolved in 4 mL of dry THF and thoroughly mixed for half an hour before the addition of the catalytic amount of acid (one drop of 0.1 N HCl solution in ethyl alcohol) to initiate the cross-linking process. After stirring for another 3 h the required amount of lithium triflate from the previously prepared stock solution in THF was added to the mixture and further stirred for 1 h. The mixture was then poured onto a Teflon mold with a diameter of 4 cm and the solvent was allowed to evaporate slowly overnight. Finally, the membrane was dried in an oven at 70°C for 3 days and at 100°C for 1 h. Homogeneous and transparent films with an average thickness of 200 μm was obtained in this way. These films were insoluble in all solvents, only swollen in THF.

2.2.4 METHODS OF INVESTIGATION

The structure of the obtained materials was investigated with use of FTIR method (spectra was detected on a Nicolet Nexus 470 spectrometer), ^{1}H, ^{13}C and ^{29}Si NMR spectra were recorded on a Varian Mercury 300VX NMR spectrometer, using dimethylsulphoxide and CCl_4 as the solvent and an internal standard. Differential scanning calorimetric (DSC) measurements were performed on a Netzsch DSC 200 F3 Maia apparatus. The heating and cooling scanning rates were 10 K/min.

Gel-permeation chromatographic studies were carried out with the use of a Waters Model 6000A chromatograph with an R 401 differential refractometer detector. The column set comprised of 10^3 and 10^4 Å Ultrastyragel columns. Sample concentration was approximately 3% by weight in toluene and a typical injection volume for the siloxane was 5 μL flow rate –1.0 ml/min. Standardization of the GPC was accomplished using styrene or polydimethylsiloxane standards with known molecular weight. Wide-angle X-ray analyses were performed on a DRON-2 (Burevestnik, Saint Petersburg, Russia) instrument. CuK_α radiation was used with graphitic monochromator; the angular velocity of the motor was $\omega \approx 2°/min$. The content of active ≡Si-H groups in oligomers was calculated according to literature data [9].

The total ionic conductivity of samples was determined by locating an electrolyte disk between two 10 mm diameter brass electrodes. The electrode/electrolyte assembly was secured in a suitable constant volume support which allowed extremely reproducible measurements of conductivity to be obtained between repeated heating-cooling cycles. The cell support was located in an oven and the sample temperature was measured using a thermocouple positioned close to the electrolyte disk. The bulk conductivities of electrolytes were obtained during a heating cycle using the impedance technique (Impedance meter BM 507-TESLA for frequencies 50 Hz-500 kHz) over a temperature range between 20 and 100°C.

2.3 RESULTS AND DISCUSSIONS

Methylsiloxane polymers with 3-(2-(2-methoxyethoxy)ethoxy)propyl and ethoxy side groups have been synthesized via hydrosilylation reaction of α,ω-bis(trimethylsiloxy)methylhydrosiloxane with 3-(2-(2-methoxyethoxy)ethoxy)prop-1-ene and vinyltriethoxysilane in the presence of platinum catalysts: (platinum hydrochloric acid, Karstedt's catalyst and platinum on carbon) in concentrated solution of dry toluene.

Preliminary heating of initial compounds in the temperature range of 50–80°C in the presence of catalysts showed that in these conditions polymerization of 3-(2-(2-methoxyethoxy)ethoxy)prop-1-ene and destruction of siloxane backbone do not take place. No changes in the NMR and FTIR spectra of initial compounds were found.

The above-mentioned reactions were carried out without solvent at 50°C temperatures. It was found that the reactions proceeded vigorously (especially in case of platinum hydrochloric acid and Karstedt's catalyst) and at initial stages of conversion of ≡Si-H bonds (~25%) gelation takes place via intermolecular dehydrocoupling reaction. To prevent gelation and to investigate the kinetic parameters of the reaction we have carried out hydrosilylation in dry toluene (C 0.06301 mole/l) solution in the presence of Karstedt's catalyst.

In general, hydrosilylation of PMHS with 3-(2-(2-methoxyethoxy)ethoxy)prop-1-ene proceeds according to the following Scheme 2.1:

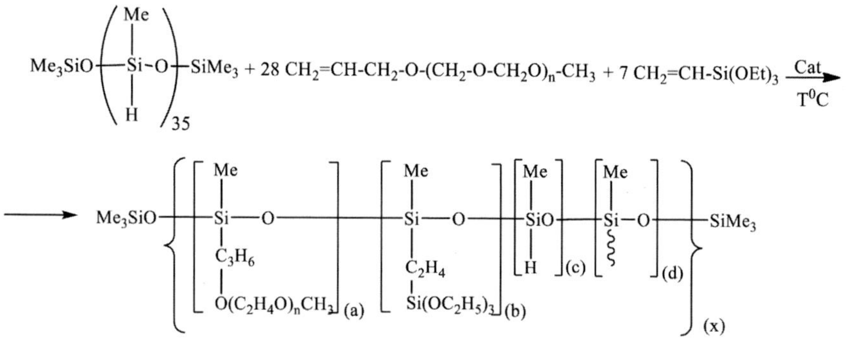

$$Me_3SiO\left(\begin{array}{c}Me\\|\\Si{-}O\\|\\H\end{array}\right)_{35}SiMe_3 + 28\ CH_2{=}CH{-}CH_2{-}O{-}(CH_2{-}O{-}CH_2O)_n{-}CH_3 + 7\ CH_2{=}CH{-}Si(OEt)_3 \xrightarrow[T^0C]{Cat}$$

$$\longrightarrow Me_3SiO\left\{\left[\begin{array}{c}Me\\|\\Si{-}O\\|\\C_3H_6\\|\\O(C_2H_4O)_nCH_3\end{array}\right]_{(a)}\left[\begin{array}{c}Me\\|\\Si{-}O\\|\\C_2H_4\\|\\Si(OC_2H_5)_3\end{array}\right]_{(b)}\left[\begin{array}{c}Me\\|\\SiO\\|\\H\end{array}\right]_{(c)}\left[\begin{array}{c}Me\\|\\Si{-}O\\ \wr\end{array}\right]_{(d)}\right\}_{(x)}SiMe_3$$

SCHEME 2.1 Hydrosilylation reaction of PMHS with 3-(2-(2-methoxyethoxy)ethoxy)prop-1-ene and vinyltriethoxysilane.

where, [(a)+(b)+(c)+(d)](x)=m≈35; n=2; Cat-Karstedt's $(Pt_2[(VinSiMe_2)_2O]_3$ catalyst –60°C (I^1), 70°C (I^2) and 80°C (I^3); H_2PtCl_6–80°C (I^4), Pt/C –80°C (I^5) (ratio =1:28:7).

The synthesized oligomers are vitreous liquid products, which are well soluble in organic solvents with specific viscosity $\eta_{sp} \approx 0.07–0.10$. Structures and compositions of the oligomers were determined by elemental and functional analyses, FTIR, ^{1}H, ^{13}C and ^{29}Si NMR spectral data. Synthesized polysiloxanes were characterized by wide-angle X-ray scattering analyses. Some characteristics of the hydrosilylation reaction and the synthesized oligomers are presented in Table 2.1.

In FTIR spectra of oligomers I^3 one can observe absorption bands characteristic for asymmetric valence oscillation of linear ≡Si-O-Si≡ bonds at 1025 cm^{-1}, absorption bands at 1265, 1103, 2152 and 2738–2969 cm^{-1} is characteristic for ≡Si-CH_3, C-O-C bond in repeated elementary ring -OCH_2CH_2-, unreacted ≡Si-H bonds and for C-H bonds accordingly.

In ^{29}Si NMR spectra of oligomer I^3 there are resonance signals with chemical shifts $\delta = -22.07$, 37.64, –45.95 and –58.05 ppm characteristics for D and T structure inside triethoxysilane group.

In ^{1}H NMR spectra of oligomers the resonance singlet signals with chemical shift δ=0.2 and 3.26 ppm correspond to ≡Si-Me and OMe group. Multiple signals with chemical shift δ=0.55 ppm correspond to protons of ≡Si-CH_2-group (anti-Markovnikov addition). Multiple signals with chemical shifts δ= 0.97–1.0, 1.60, 3.36, 3.48 and 3.75 ppm corresponds to the proton in CH_3-CH= (Markovnikov addition), ≡Si-CH_2-CH_2-CH_2-, ≡Si-CH_2-CH_2-CH_2-O, and -CH_2-CH_2-O-CH_2-, -O-CH_2-CH_3 fragments accordingly.

In ^{13}C NMR spectra of oligomers I^3 one can see resonance signals with chemical shifts $\delta = -1.10$, –1.31, 12.2, 17.79, 20.86, 57.36, 57.86, 69.39–69.70, 69.79 and 71.24 ppm characteristic for carbon atoms in ≡Si-CH_3 fragment, in ≡Si-CH_2 (in ethoxyl fragment), ≡Si-CH_2 (di(tri)ethyleneglycol fragment), ≡Si-CH_2-CH_2-CH_2-, OCH_3, O-CH_2-CH_3, CH_3-O-CH_2-CH_2-O and O-CH_2-CH_2-O fragments accordingly.

The average molecular weights of the synthesized oligomers exceed several times the theoretical values of the molecular weights calculated in case of full hydrosilylation. It indicates that during hydrosilylation reaction inter-molecular branching processes on active ≡Si-H groups also takes place, which is in agreement with literature data [10]. Hence, the obtained oligomers are various linked branched systems.

During the hydrosilylation reaction, a decrease of active ≡Si-H groups' concentration with time was observed. The hydrosilylation reaction was performed in dry toluene solution ($C \approx 6.301 \cdot 10^{-2}$ mole/l). As it is evident from Table 2.1 not all active, ≡Si-H groups participate in the hydrosilylation reaction. Figure 2.1 shows the variation of the concentration of active

TABLE 2.1 Some Characteristic Data of the PMHS Hydrosilylation Reaction with Allyl Butyrate and the Resulting Oligomers

Oligomer №	Yield, %	Reaction temperature, °C	Ratio of Initial Compounds	≡Si-H%, Conversion	$\bar{M}_n \times 10^{-3}$ $\bar{M}_\omega \times 10^{-3}$ (P)	η_{sp}*	d_1, Å	T_g, °C
I^1	88	60	1:28:7	62	-	0.08	7.39	-
I^2	90	70	1:28:7	71	-	0.08	-	-
I^3	93	80	1:28:7	80	22.7177 32.2673 (1.42)	0.09	-	–106
I^4	92	80	1:28:7	79	-	0.10	7.42	-
I^5	85	80	1:28:7	63	-	0.07	-	-

*In 1% toluene solution, at 25°C.

≡Si-H groups with time during the hydrosilylation reaction of PMHS with 3-(2-(2-methoxyethoxy)ethoxy)prop-1-ene and vinyltriethoxysilane in the presence of Karstedt's catalyst at different reaction temperatures. The ratio of initial compounds was of 1:28:7 (see Scheme 2.1). From Figure 2.1 it is evident that the depth of hydrosilylation reaction rises with the increase of temperature.

From the dependence of the inverse concentration of ≡Si-H groups with time during hydrosilylation reaction of PMHS with allyl butyrate it was found that the reaction is of second order; the reaction rate constants at various temperatures were determined, yielding: $k_{60^\circ C} \approx 0.5543$, $k_{70^\circ C} \approx 1.00$ and $k_{80^\circ C} \approx 2.05030$ *l/mole· min*. From the dependence of the hydrosilylation reaction rate constants' logarithm on the reciprocal temperature the activation energy of hydrosilylation reaction was calculated, which is equal to $E_{act} \approx 64.36$ *kJ/mole.*

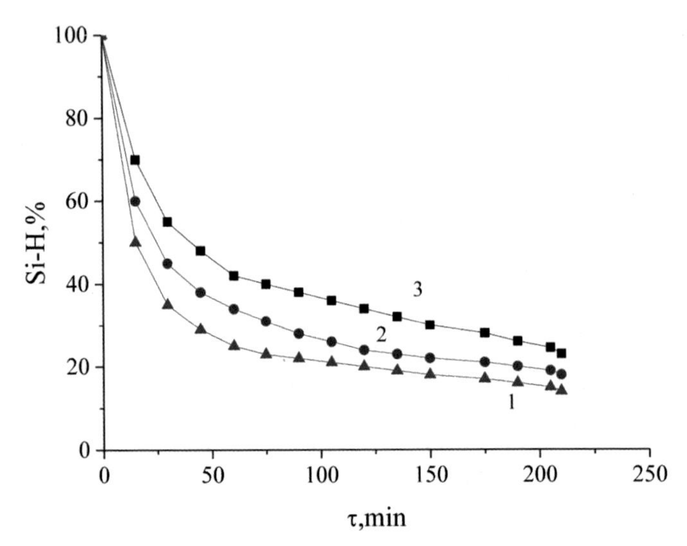

FIGURE 2.1 Dependence of changes of active ≡Si-H groups concentration on the time during hydrosilylation reactions of PMHS with 3-(2-(2-methoxyethoxy)ethoxy)prop-1-ene and vinyltriethoxysilane at 60°C (1), 70°C (2) and 80°C (3) (-Karstedt's catalyst).

The wide-angle X-ray scattering diffractograms of the investigated polymers showed two broad diffraction maximums at 2θ°≈11.75–12.00° and 2θ°≈ 20.50–20.75°. Results suggest that the polymers are represented as one phase amorphous systems. The first intensity maximum corresponds to interchain distances in the range d_1= 7.39–7.42 Å and the second one at

4.35–4.41 Å which characterized intra- and interchain atomic interactions respectively [10].

For oligomer I^3 DSC investigation has been carried out and glass transition temperature of oligomer I^3 is determined which is about $T_g = -106°C$.

Oligomers I^3 and I^4 were doped with two types of lithium salts: triflate (CF_3SO_3Li) - S_1 and (trifluoromethanesulfonyl)imide ($CF_3SO_2N(Li)SO_2CF_3$) – S_2 at two concentrations of these salts (10 and 20 wt.%).

AC impedance measurements were employed to investigate the variation of conductivity with temperature for all the obtained electrolytes. Figures 2.2 and 2.3 illustrate the Arrhenius plots of the dependence of electrical conductivity of PE based on I^3 and I^4 with salts S_1 and S_2 at different contents.

The curves in Figures 2.2 and 2.3 (see also data of Table 2.2) show that electrolytes on the basis of polymer I^3 containing salts S_1 and S_2 are characterized with relatively high conductivity in comparison to the electrolytes on the basis of polymer I^4. Probably the first polymer with triflate creates the friable structure significantly abetting ion transfer through the polymer matrix. On the other hand, ions on the basis of salt triflate apparently have higher mobility to some extent than second ones.

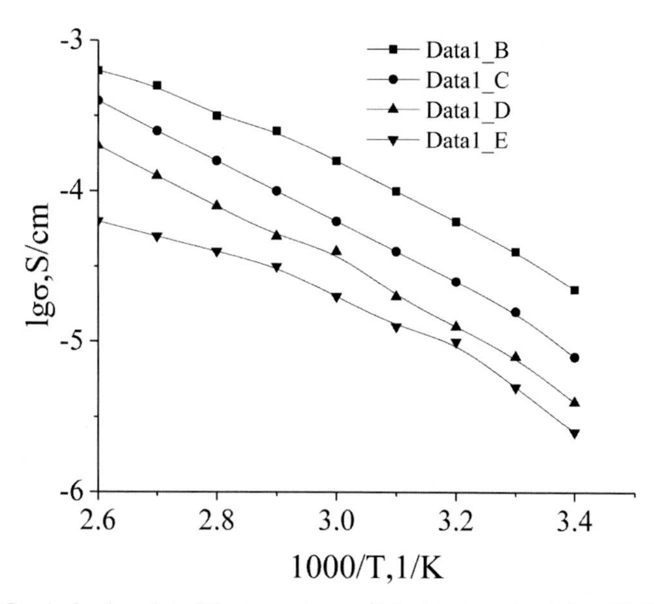

FIGURE 2.2 Arrhenius plot of the dependence of electrical conductivity of PE based on I^3 with salts S_1 at concentrations 20 wt% (B) and 10 wt% (C) and with S_2 at concentrations 10 wt% (D) and 20 wt% (E).

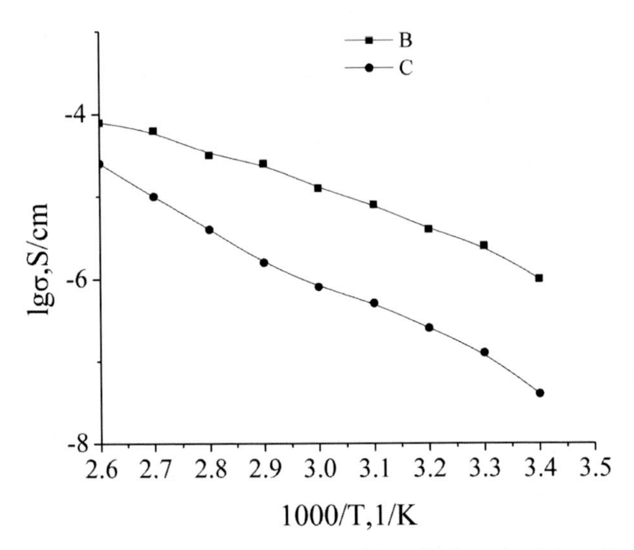

FIGURE 2.3 Arrhenius plot of the dependence of electrical conductivity of PE based on I^4 with salt S_2 at concentrations 10 wt% (B) and 20 wt% (C).

TABLE 2.2 Conductivity of Polyelectrolytes on the Basis of Polymers of Type I^3 and I^4-Containing the Salts S_1 and S_2

Polymer	Salt	Content of Salt, wt%	σ (30°C), S · cm^{-1}	σ (90°C), S · cm^{-1}
I^3	S_1	10	6.3×10^{-6}	4.2×10^{-4}
I^3	S_1	20	2.1×10^{-5}	6.1×10^{-4}
I^3	S_2	10	3.2×10^{-6}	1.6×10^{-4}
I^3	S_2	20	2.1×10^{-6}	6.0×10^{-5}
I^4	S_2	10	8.1×10^{-7}	8.0×10^{-5}
I^4	S_2	20	3.2×10^{-8}	5.1×10^{-5}

In general, the temperature dependencies of conductivity of the investigated membranes can be described by an equation close to the well-known VTF one [11, 12]. It must be noted that the conductivity of the electrolyte membrane I^3S_1, at salt concentration 20 wt.% is characterized with higher conductivity values than those for the one containing 10 wt.% of the same salt. However, the conductivity of the membrane IIS_2 containing 20 wt.% S_2 is lower than that of the one containing 10 wt.% of the same salt. This phenomenon may be explained by: (1) formation of ion pairs on the basis of molecules of salt S_2 which are characterized with relatively high interactions (usually this formation takes place at higher concentrations of the salt in the electrolytes [13], (2) increase of material density and consequently decrease of charge transfer in the polymer matrix due to the higher concentration of S_2 salt having bulkier molecules than S_1. Obviously, when considering the factors effectively influencing the charge transfer in PE it is necessary to take into account various aspects such as structure peculiarities, localization of the salt molecules in the polymer matrix, viscosity, and thermal properties. The viscosity, in particular, is a factor significantly affecting the conductivity of PE. Experimentally it is established that viscosity (η) and conductivity (σ) are in inversely proportional dependence and that their product is constant ($\eta \times \sigma$ = const). Besides, ion conductivity at a given temperature (T) is proportional to the sum of the products of the following parameters: number of free ions (n_i), charge carrier ions (q_i) and ion mobility (μ_i) [14]: ($\sigma(T) = \Sigma\, n_i\, q_i\, \mu_i$). Voltammograms of each electrolyte based on a different type and content of the salts are presented in Figure 2.4.

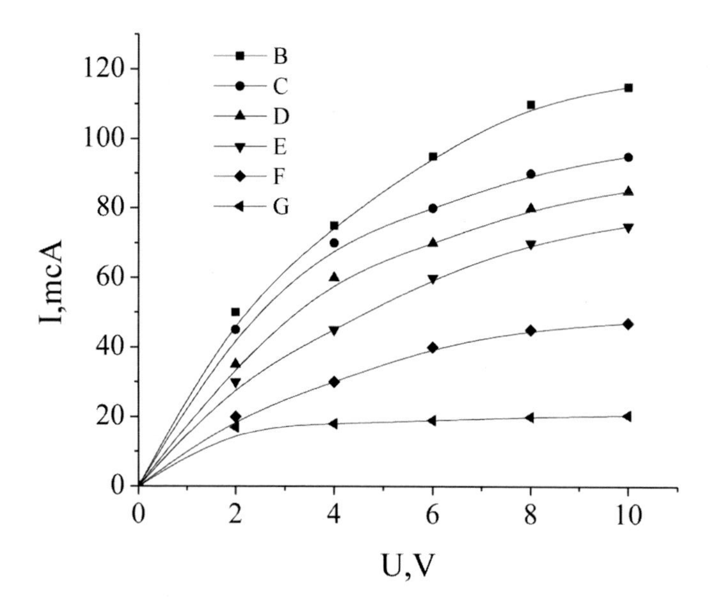

FIGURE 2.4 Voltammograms of the electrolytes based on I^3S_1 with content of S_1: 10 wt% (F) and 20 wt% (G); I^3S_2 with content of S_2: 10 wt% (E) and 20 wt% (D); I^4S_2 at concentrations of S_2 10 wt% (C) and 20 wt% (B).

In accordance with Figure 2.4, the saturation of the electrical current depends significantly on the composition of the electrolytes. The length of the initial linear part (up to bend) of the electrical current is the longer the higher is the conductivity. Deviations from the linear increase of intensity with voltage are partially related to charge-phonon interactions that are enhanced with increasing ion energy giving rise to a decrease of the electrolytes conductivity [15]. Phonon-charge carriers interactions would be promoted for higher densities of the polymer matrix density.

2.4 CONCLUSION

By hydrosilylation reaction of PMHS with 3-(2-(2-methoxyethoxy)ethoxy) prop-1-ene and vinyltriethoxysilane in the presence of Karstedt's catalysts branchiate comb-type methylsiloxane polymers with electro donor diethylene glycol and triethoxysilyl fragments in the side chain has been obtained.

Via sol-gel processes of polymers doped with lithium trifluoromethylsulfonate or lithium bis(trifluoromethylsulfonyl)imide solid polymer electrolyte membranes have been obtained.

The conductivity of the membranes in the range 30–90°C can be described with an equation close to VTF type. The membranes containing triflate type salt are characterized with higher conductivity in the range of temperatures investigated with respect to electrolytes containing (trifluoromethylsulfonyl) imide presumably due to the higher mobility of the former salt ions.

ACKNOWLEDGMENT

The financial support of the Georgian National Science Foundation Grant #STCU 5055 is gratefully acknowledged.

KEYWORDS

- **cross-linking**
- **dichlorosilane**
- **hydrosilylation**
- **polymer electrolyte**
- **polysiloxanes**
- **solid polymer electrolytes**

REFERENCES

1. Hooper, R., Lyons, L. J., Mapes, M. K., Schumacher, D., Moline, D. A., & West, R., (2001). *Macro-Molecules, 34*(4), 931.
2. Fish, D., Khan, I. M., Wu, E., & Smid, J., (1988). *Brit. Polym. J., 20*, 281.
3. Jong-Wook, K., Kwang-Sun, J., Jae-Pil, L., & Jong-Wan, P., (2003). *J. of Power Sources 119–121*, 415.
4. Spindler, R., & Shriver, D. F., (1988). *Macromolecules, 21*(3), 648.
5. Morales, E., & Acosta, J. L., (1999). *Electrochim. Acta, 45*(7), 1049.
6. Kanga, Y., Lee, W., Suh, D. H., & Ch, L., (2003). *J. Power Sources, 119–121,* 448–453.
7. Hyun-Soo, K., Jung-Han, S., Seong-In, M., & Mun-Soo, Y., (2003). *J. Power Sources, 119–121*, 482.
8. Iwahara, T., Kusakabe, M., Chiba, M., & Yonezawa, K., (1993). *J. Polym. Sci., A, 31*(10), 2617.
9. Tatrishvili, T., Jalagonia, N., Gelashvili, K., Khachidze, M., Markarashvili, E., Aneli, J., & Mukbaniani, O., (2015). *Oxid. Comm., 38*(1), 13.
10. Mukbaniani, O., Tatrishvili, T., Titvinidze, G., Mukbaniani, N., Brostow, W., & Pietkiewicz, D., (2007). *J. Appl. Polym. Sci., 104*(2), 1176–1183.
11. Zhang, Z. C., Sherlock, D., West, R., West, R., Amine, K., & Lyons, L. J., (2003). *Macromol., 36*(24), 9176.
12. Zhang, Z. C., Simon, A., Jin, J. J., Lyons, L. J., Amine, K., & West, R., (2004). *Polym. Mater. Sci. & Engin., 91*, 587.
13. Zhang, L., Zhang, Z. C. L., Harring, S., Straughan, M., Butorac, R., Chen, Z., Lyons, L., Amine, K., & West, R., (2008). *J. Mater. Chem., 18*, 3713.
14. Lin, C., Kao, H., Wu, R., & Kuo, P., (2002). *Macromolecules, 35*(9), 3083–3096.
15. Ziman, J. M., (1964). *Principles of the Theory of Solids* (p. 221). Cambridge University Press.

CHAPTER 3

Design and Synthesis of Thiaolidinone-Based Triazines Derivatives as a New Potential Anti-Inflammatory and Anti-Microbial Agent

RAVINDRA S. SHINDE[1] and POONAM KHULLAR[2]

[1]*Department of Chemistry and Industrial Chemistry, Dayanand Science College, Latur–413512, Maharashtra, India*

[2]*Department of Chemistry, B. B. K. D. A. V. College for Women, Amritsar–143005, Punjab, India*

ABSTRACT

A frequent approach to synthesize a series of 3-(4,6-dichloro-1,3,5-triazin-2-yl)-2-phenylthiazolidin-4-one was developed by applying an efficient multi-component reaction (MCR). Also, the synthesized compounds were tested for their antimicrobial activity, anti-inflammatory (TNF-and IL-6) activity. The most admirable results were observed with the substituted phenylthiazolidin-4-one triazine analogs and it could be a potential starting point to develop innovative lead compounds fight against bacteria and fungi. The presence of lipophilic -Cl and -F at 5th position tolerates the procytokine activity All synthesized compounds were characterized by IR, ^{1}H NMR, ^{13}C NMR, mass, and elemental analysis.

3.1 INTRODUCTION

One of the principal objectives of medicinal and pharmaceutical chemistry is to planed, synthesize, and manufacture molecules possessing worth as human therapeutic agents. The compounds containing heterocyclic nucleus are of huge significance getting special consideration as they belong to a class of medicinal chemistry. As an example, five-membered ring heterocycles containing three carbon atoms, one sulfur atom, and one nitrogen atom recognized as thiazoles are of large attention in various areas of pharmaceutical chemistry. The thiazolidinone acts as a multifunctional center exhibiting a variety of biological activities.

Thiazolidinone has been measured as a supernatural moiety which posses roughly all types of biological and pharmaceuticals activities. The 4-carbonyl derivatives of thiazolidine are known as 4-thiazolidinones.

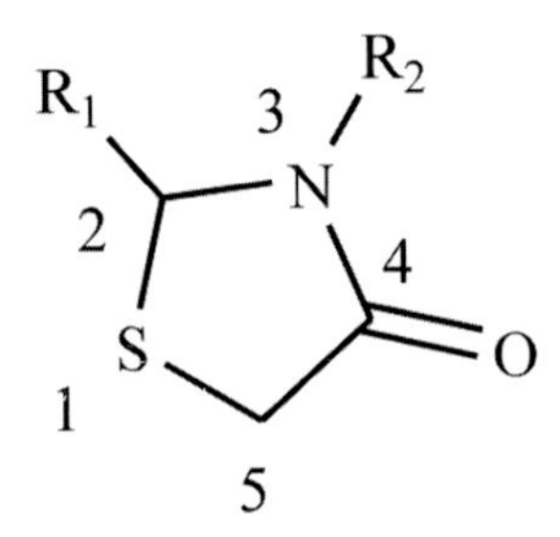

In order to reduce the rapid multidrug-resistance in pathogenic microbes, we require to find out and manufacture new drugs that act through a new mechanism of action [1]. Multidrug-resistant (MDR) Strain, a rapid growth of pathogens causing a severe resistance towards presently available standard drugs, poses the frightening threat by rising strict opportunistic microbial infections in the past decades [2]. In the visualization of the above trouble, the existing situation highlights the vital need to develop fresh agents with precise activity with improved potency to maintain a pool of novel bioactive entities. The synthesis of innovative compounds likely to be unchanged by existing resistance mechanisms is an area of enormous effect for medicinal chemists [3]. Recently, numerous heterocyclic compounds from the series of s-triazine and thiazolidinone have been synthesized and their pharmacological activity has therefore investigated. It has been accomplished that, s-triazine ring frame hold an extensive spectrum of biological and pharmaceutical activities [4–6]. The s-Triazine has played a fundamental role in unique drug innovation for modulating physical and biological properties of the molecule due to wide diversity of biological applications [7]. The s-triazine core has known and huge concentration linking chemists through victorious resource of pharmacological activities such as antibacterial, [8, 9] antimalarial, [10] antiprotozoal, [11] antifungal, [12] anticancer, [13] antimychobacterial, [14] and antiviral [15]. In addition to these s-triazine derivatives, thiazolidinone bearing s-triazine analogs are focused as potential bioactive molecules [16].

Moreover, thiazolidin-4-one derivatives are also reported to have important biological activities such as anti-inflammatory, [17] antituberculosis, [18] anticancer, [19] antitumor, [20] anti-HIV, [21] antibacterial, [22] antifungal, [23] antioxidant [24], antiviral [25], anticonvulsant [26], diuretics [27], nematicidal [28], antihistaminic activity [29] etc. This variety in the biological activity has paying attention of many researchers to discover this framework to its multiple potential against numerous activities. Here, we report the synthesis of 3-(4,6-dichloro-1,3,5-triazin-2-yl)-2-phenylthiazolidin-4-one as anti-inflammatory and antimicrobial agents. In the structure-activity relationship (SAR) studies, the biological activity of these molecules was compared with the standard reference.

3.2 REVIEW OF LITERATURE AND METHOD OF SYNTHESIS

The review is to present the chemistry, various biological activities, potential, pitfalls of thiazolidinediones and limitations.

Patel et al. [30] reported the synthesis of various new series of compounds containing 2-(2-substituted benzylidene hydrazinyl)-4-(4-(4-methoxyphenyl) piperazin-1-yl)-6-(4-tolyl oxy)-1,3,5-triazine derivatives. The produced compounds have been evaluated by physical methods and spectral data (IR and NMR). The resulting compounds were evaluated for antimicrobial activity.

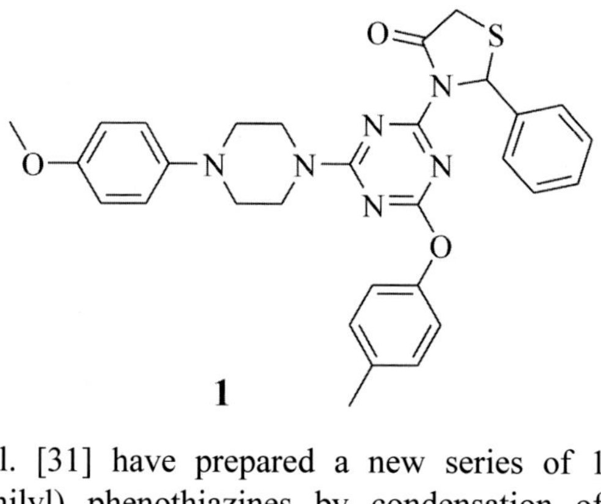

1

Hassan et al. [31] have prepared a new series of 10-(*p*-substituted arylidenesulphanilyl) phenothiazines by condensation of 10-sulphanilyl phenothiazine with substituted aromatic aldehydes. The interaction of mercaptoacetic acid with Schiff bases afforded 10-(p-thiazolidinonyl benzene sulphonyl) phenothiazines.

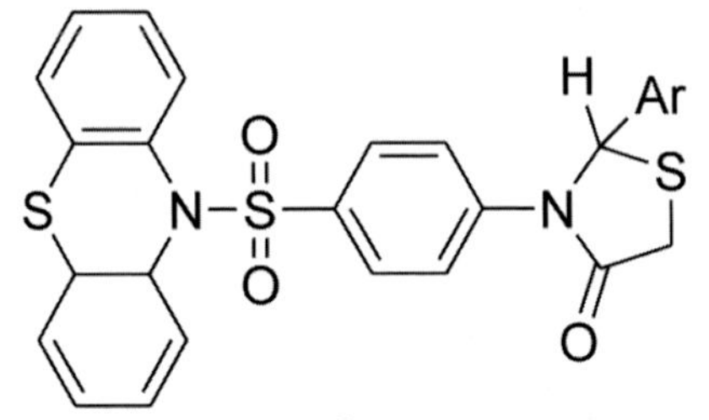

2

where, Ar = -Ph, furfuryl, p-OCH_3–C_6H_4, etc.

Pathak and co-workers et al. [32] have synthesized substituted thiazolidin-4-one derivatives and tested for their biological activity. The majority of the synthesized compounds possessed significant antimycobacterial activity (0.35e1 mg/mL) against M. tuberculosis H37Rv.

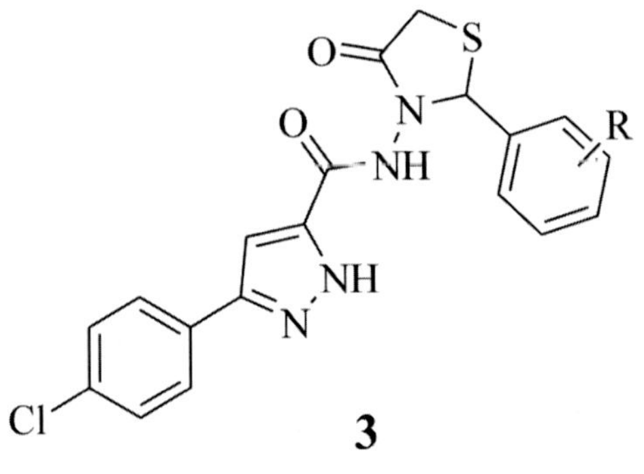

3

Baxi et al. [33] have previously studied the synthesis of novel series of 3-(3'-carboxyphenyl sulphonamido)-2-aryl-5*H*-4-thiazolidinone by condensation way and were tested for their antimicrobial activity.

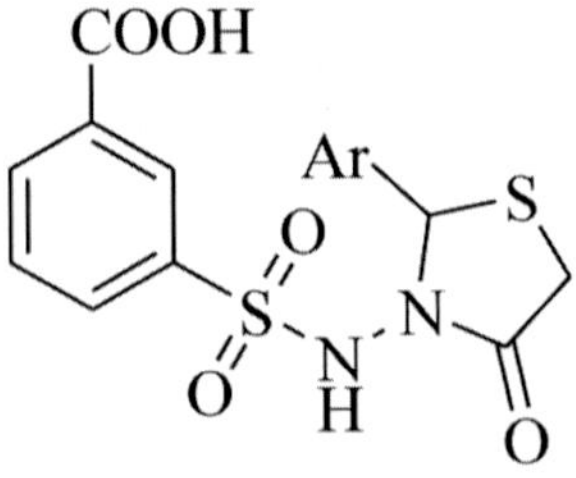

4

Jadhav et al. [34] synthesized a series of 2-(1-(8-ethoxy-2-oxo-2H-chromen-3-yl) ethylidene)hydrazine carbothioamide. The resultant compound reacted with chloroacetyl chloride, substituted aryl aldehydes were coupled in presence of alum catalyst to give 5-benzylidene-2-((E)-(1-(8-ethoxy-2-oxo-2Hchromen-3yl)ethylidene)hydrazono) thiazolidin-4-one by easy grinding technique at room temperature (RT) under solvent free condition. The prepared compounds were confirmed by spectral analysis.

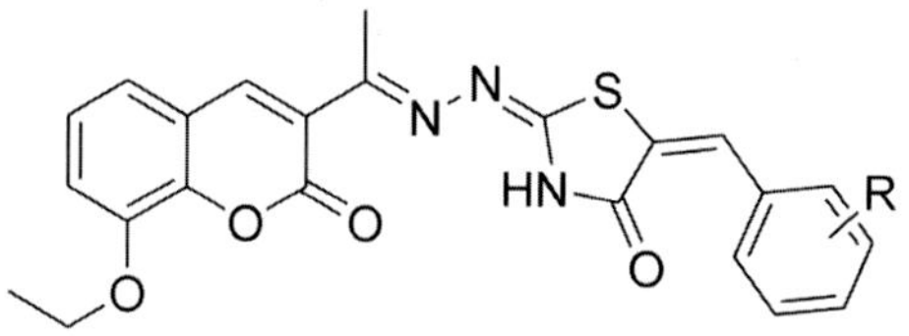

5

Sharma et al. [35] further developed synthesis of several thiazolidinones using montmorillonite K10 catalyst. The multi-component reaction (MCR) of an amine, an aldehyde, and thioglycolic acid in *N*,*N*-dimethyl formamide as solvent at reasonable (50°C–120°C) temperatures to form thiazolidinones. The reaction involves easy workup and purification. The campholenic aldehyde was used to synthesize potentially bioactive thiazolidinones. All products were characterized by spectral data. The anticancer screening tests can be taken and some compounds show anticancer activity.

Bhagwat et al. [36] have reported the synthesis of some thiazolidinone derivatives bearing 2,4-bithiazol moiety.

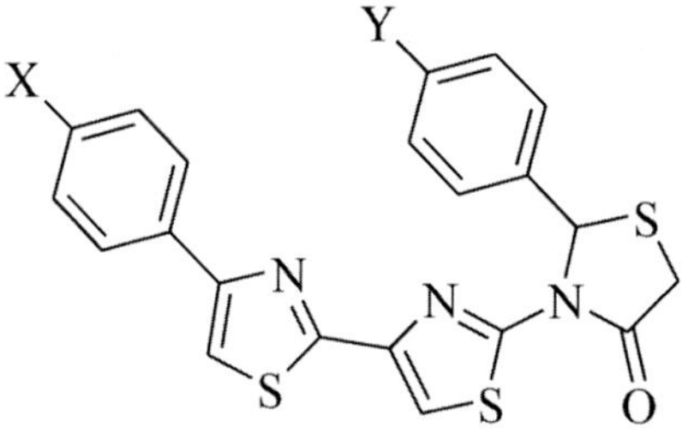

6

Zheng et al. [37] have prepared a new enzymatic multi-component synthesis of 4-thiazolidinones. The trypsin from porcine pancreas shows good catalytic activity to support the reaction of amine, mercaptoacetic acid, and aldehyde with excellent yields. The trypsin-catalyzed one pot translation technique provided a fresh approach to synthesize thiazolidinones and stretched the application of trypsin in organic heterocyclic production.

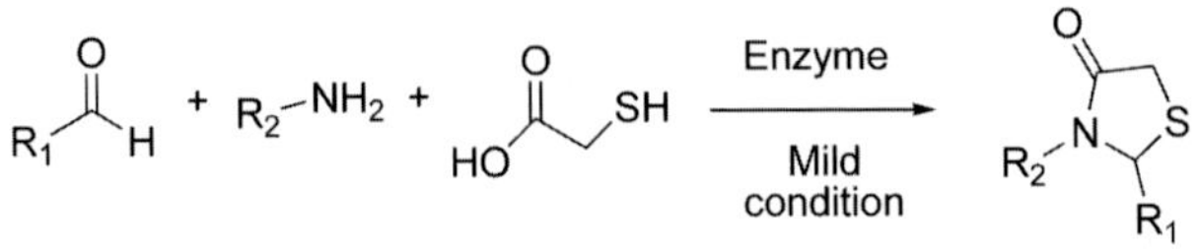

Enzyme
Mild condition

7 8 9 10

Srivastava et al. [38] accomplished designed and synthesis of new 4-thiazolidinones by DCC mediated MCR of aldehyde, amine, and mercaptoacetic acid. The final products were obtained in quantitative yields and agreeable to increase operations. The yields of the thiazolidinones were not depending on the nature of the reactants.

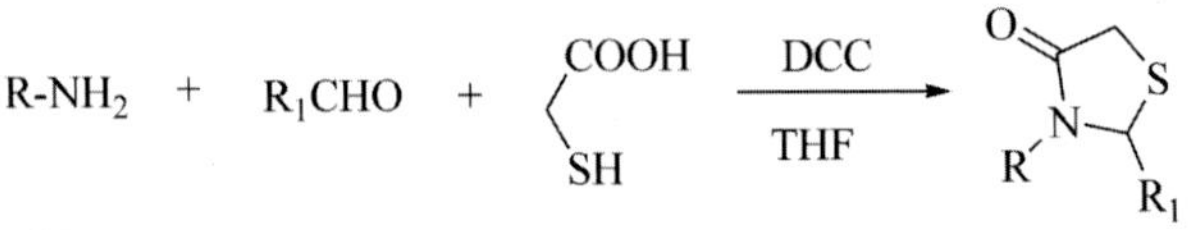

$R\text{-}NH_2$ + R_1CHO + COOH / SH —DCC / THF→

11 7 9 12

Joshi et al. [39] paying concentration on the synthesis of somc 4-thiazolidinones bearing benzo[b] thiophene core as potential antimicrobial and antitubercular agents.

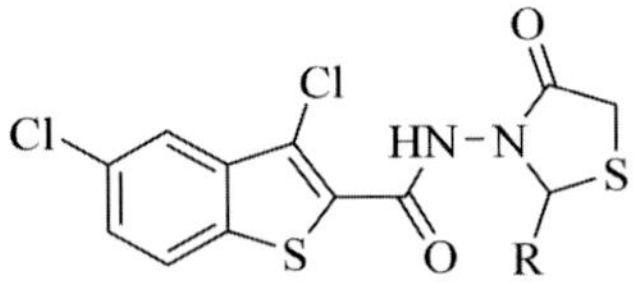

13

where, R = 3-Br-C_6H_4, 3-Cl-C_6H_4, etc.

Dubreuil et al. [40] identified a novel series library of 4-thiazolidinones by a MCR under microwave (MW) heating technique.

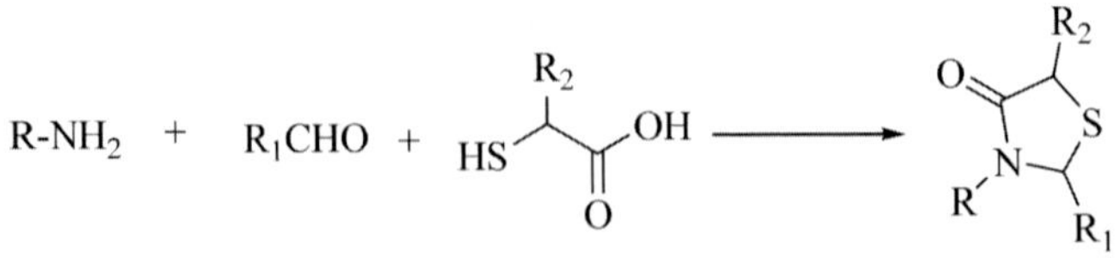

$R\text{-}NH_2$ + R_1CHO +

11 7 14 15

Bhagwat and co-workers et al. [41] have synthesized a variety of thiazolidinone based benzothiazolyl moiety tested against *B. Subtilis*, *E. Cartovara* and *E. Coli*.

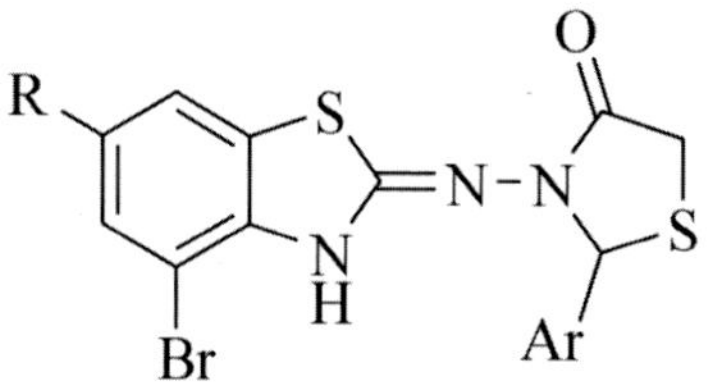

16

Srivastava et al. [42] have studied and reported a various 5-arylidene-2-aryl-3-(2-chloro phenothiazinoacetamidyl)-1,3*(H)*-thiazolidin-4-ones as an anticonvulsant and antifungal agents.

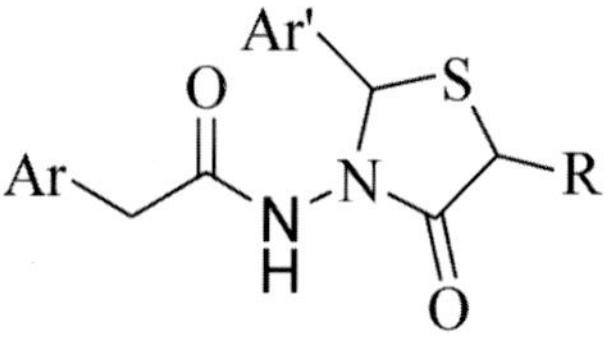

17

where, Ar = 2-chloro-10,10a-dihydro-4aH-phenothiazine.

Ar' = 2-Cl-C_6H_4, 3-Cl-C_6H_4, etc.

R = -H, -Cl, -Me, etc.

Pareek et al. [43] reported synthesis and bioevalution of 4-thiazolidinone derivatives inserting benzothiazole derivatives.

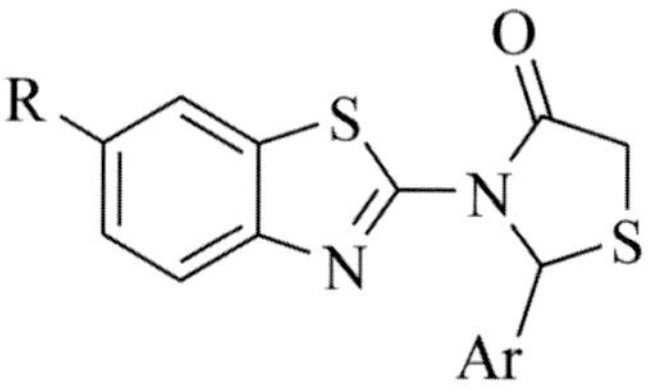

18

where, R = -OCH_3, -F, -COOH

Ar = -C_6H_5, o-C_6H_4-OH, p-C_6H_4-OCH_3, p-C_6H_4Cl

Khot et al. [43] reported the synthesis of thiazolidinone derivatives by reaction of different amines with carbon disulfide and chloroacetic acid in presence of $NaHCO_3$ with MW assisted MCRs (140 watts) in a scientific MW oven. The resulting compound promotes with various aldehydes and hydrazides provide the 1,3,6-Triaryl-hexahydro-pyrazolo[3,4-d]thiazole-5-thione derivatives. All compounds were subjected to PASS for the discovery of biological activity. The QSAR study of all compounds was done. Some of the compounds found to be most active as *insulin inhibitors and* Mcl-1 antagonists.

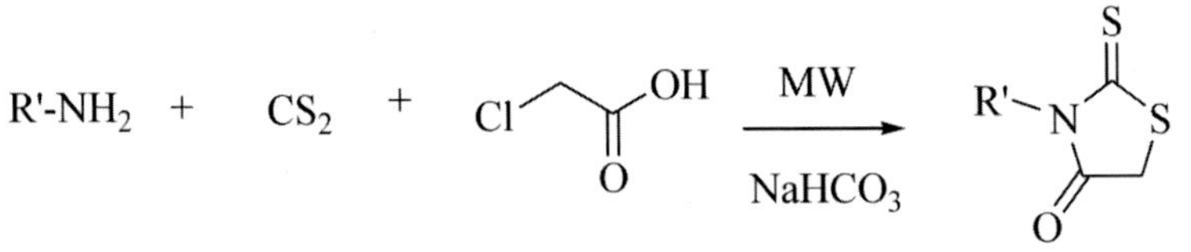

R'-NH_2 + CS_2 + Cl–CH_2–COOH $\xrightarrow[NaHCO_3]{MW}$ 22

19 **20** **21** **22**

R'=-$C_6H_5NH_2$, -C_6H_5-CH_2-NH_2, 4-Cl-$C_6H_5NH_2$, 2-Methyl-$C_6H_5NH_2$, C_6H_5 NH-NH_2, 2,4-Dinitrophenyl hydrazine, 2-amino-5-chlorobenzoacetophenone.

3.3 PRESENT WORK

Our synthetic strategy for novel compounds was outlined in Scheme 3.1. Thus as shown in Scheme 3.1, the synthesis of the designed target compounds were accomplished via one-pot MCR. The synthetic pathway used to obtain compounds (26a-l). The synthesis of thiazolidinone triazine derivatives was achieved by the reaction between substituted aromatic aldehyde, 4,6-dichloro-1,3,5-triazin-2-amine and ethyl 2-mercaptoacetate using dry ethanol (26a-l). The accuracy of the synthesis of compounds was confirmed on the basis of ^{1}H NMR, ^{13}C NMR, and Mass spectra (Figure 3.1).

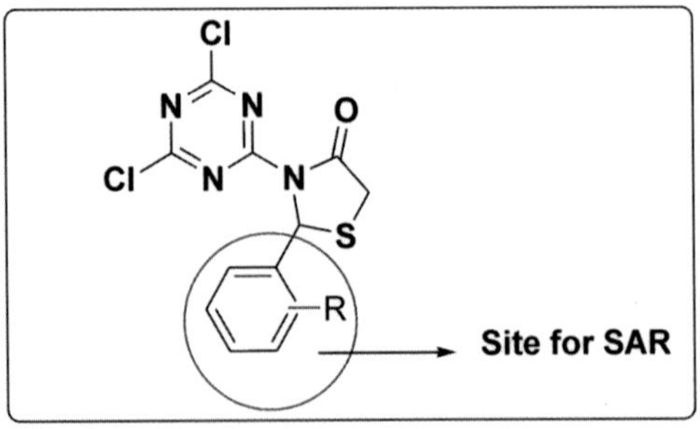

FIGURE 3.1 Design scaffold.

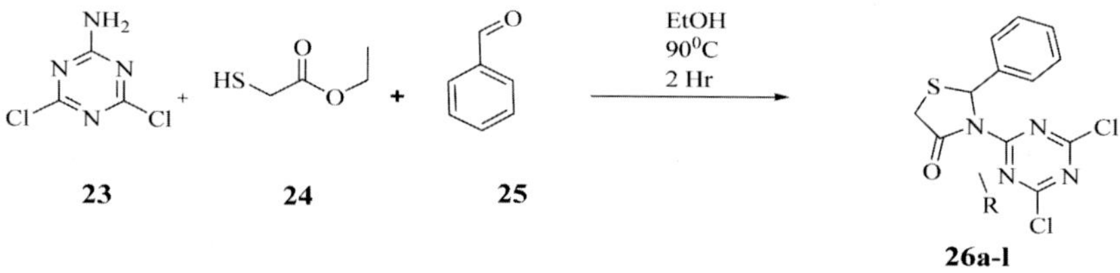

EtOH
90^0C
2 Hr

23 24 25 26a-l

R = 26a-l

26a = Phenyl	**26g = 3-methyl**
26b = 4-Methoxy	**26h = 4-Nitro**
26c = 2- fluoro	**26i = 2-Hydroxy**
26d = 2-fluoro-4-bromo	**26j = 3-Methoxy**
26e = 4-bromo	**26k = 2,5-difluoro**
26f = 2-fluoro-5-chloro	**26l = 3- CN**

SCHEME 3.1 Synthesis of *s*-triazine based thiazolidinones.

Mechanism of MCR can be represented as follows:

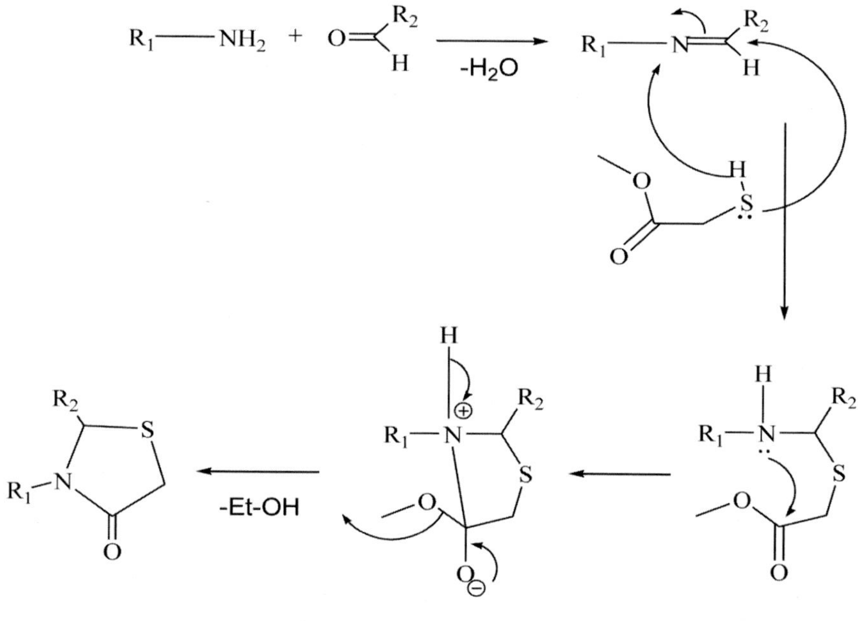

$-H_2O$

-Et-OH

R_1 = 4,6-dichloro-1,3,5-triazin-2-amine.
R_2 = substituted benzaldehyde.

3.4 EXPERIMENTAL

General method for the preparation of 3-(4,6-dichloro-1,3,5-triazin-2-yl)-2-phenyl thiazolidin-4-one (26a).

A mixture of 2-amino 4,6-dichloro-1,3,5-triazine (0.01 mol), aromatic aldehyde (0.01 mol) and ethyl-2-mercaptoacetate (0.01 mol) was refluxed for 2 hrs in dry ethanol (50 ml). The progress of reaction was monitored by TLC (Toluene: Acetone (4:6)). The excess alcohol was evaporated in vacuo. The resulting crude product was added to crushed ice. The solid obtained was washed with water, dried, and recrystallized from ethanol to afford the final respective compounds (26a-l).

Similarly, the remaining compound 26b-l was prepared.

Characterization data for compounds (26a-l) are given as follows:

Synthesis of 3-(4,6-dichloro-1,3,5-triazin-2-yl)-2-phenylthiazolidin-4-one (26a):

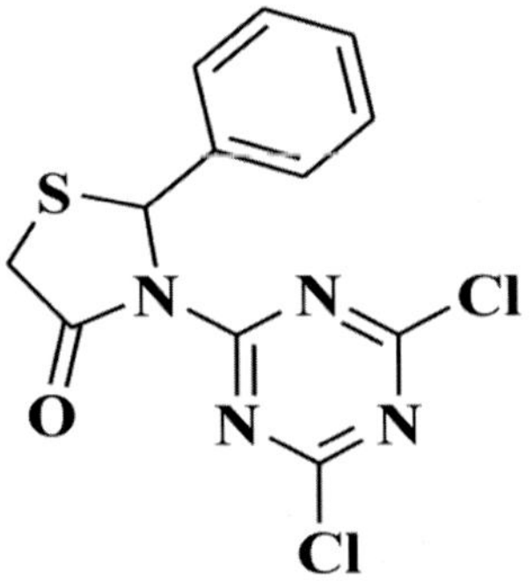

Yield: 80% M.P. 138–140°C

Mol. Formula: $C_{12}H_8Cl_2N_4OS$ Mol. Weight: 326

IR (KBr cm^{-1}): 2942 (Ar-C-H str.), 1606 (Ar-C=C- str.), 814(C-N- str.), 1688 (-C=O str.), 1300(-C=N-), 670 (Ar-C-H def.), 640 (Ar-C-Cl str.).

^{1}H NMR (400 MHz, CDCl3): δ 7.41–6.85 (m, 5H, Ar-H), 5.33 (s, 1H, -CH-), 3.40 (d, $J = 7.1$ Hz, 2H, $-CH_2-$).

MS: (m/z: RA%): 325.96 [M+., 80%], 328.1[M+2, 27%], 330.01 [M+4, 9%]$^+$.

Elemental analysis:	%C	%H	%N
Calculated:	44.05	2.46	17.12
Found:	43.98	2.35	17.00

Synthesis of 3-(4, 6-dichloro-1, 3, 5-triazin-2-yl)-2-(4-methoxyphenyl) thiazolidin-4-one (26b):

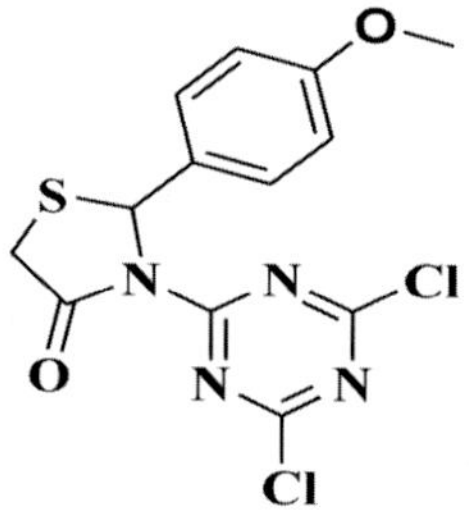

Yield: 82%		M. P. 140–142°C
Mol. Formula:	$C_{13}H_{10}Cl_2N_4O_2S$	Mol. Weight: 357.12

IR (KBr cm^{-1}): 2838 (Ar-C-H str.), 1615 (Ar-C=C- str.), 813(C-N- str.), 1692 (-C=O str.), 1345 (-C=N-), 710 (Ar-C-H def.), 650 (Ar-C-Cl str.).

^{1}H NMR (400 MHz, CDCl3): δ 7.41(m, 2H, Ar-H), 6.89(m, 2H, Ar-H), 5.34 (d, 1H, -CH-), 3.82 (s, 3H,-OCH$_3$), 3.43 (d, J = 7.2Hz, 2H, -CH$_2$-) (Spectrum 6.01).

MS: (m/z: RA%): 358.08 [M+1, 78%], 359.9 [M+2, 26%], 361.16 [M+4, 8%] (Spectrum 6.02).

Elemental analysis:	%C	%H	%N
Calculated:	43.71	2.82	15.68
Found:	43.30	2.45	15.48

Synthesis of 3-(4, 6-dichloro-1,3,5-triazin-2-yl)-2-(2-fluorophenyl) thiazolidin- 4-one (26c):

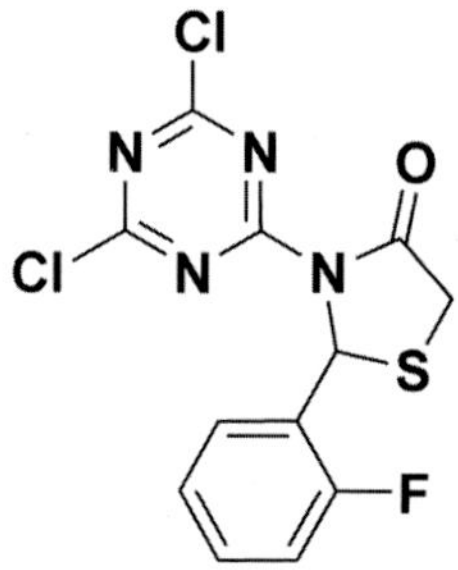

Yield: 78%		M.P.145–147°C
Mol. Formula:	$C_{12}H_7FCl_2N_4OS$	Mol. Weight: 344.12

IR (KBr cm^{-1}): 2836 (Ar-C-H str.), 1629 (Ar-C=C- str.), 809(C-N- str.), 1730 (-C=O str.), 1530 (-C=N-), 760 (Ar-C-H def.), 675 (Ar-C-Cl str.).

^{1}H NMR (400 MHz, CDCl3): δ 7.58 (td, *J* =7.5,1.9 Hz, 1H, Ar-H), 7.32 (td, *J* = 6.6, 5.8, 3.7 Hz, 1H, Ar-H), 7.18 (td, *J* = 7.6, 1.2 Hz, 1H, Ar-H), 7.06 (ddd, *J* = 9.5, 8.3.1.2 Hz, 1H, Ar-H), 5.94 (d, *J* = 7.4 Hz, 1H, -CH-), 4.25–4.10 (m, 2H, -CH_2-) (Spectrum 6.03).

MS:(m/z:RA%): 345.70 [M+1, 60%], 347.7 [M+2, 20%], 349.2 [M+4, 6%] (Spectrum 6.04).

Elemental analysis:	%C	%H	%N
Calculated:	41.75	2.04	16.23
Found:	41.55	2.00	16.10

Synthesis of 2-(4-bromo-2-fluorophenyl)-3-(4,6-dichloro-1,3,5-triazin-2-yl) thiazolidin-4-one (26d):

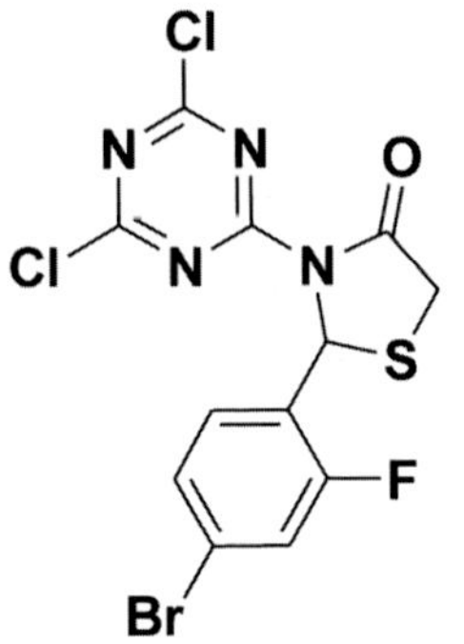

Yield: 76%		M. P. 145–147°C
Mol. Formula:	$C_{12}H_6$ F Br Cl_2N_4OS	Mol. Weight: 423.01

IR (KBr cm^{-1}): 3060 (Ar-C-H str.), 1640 (Ar-C=C- str.), 807(C-N- str.), 1720 (-C=O str.), 1345 (-C=N-), 739 (Ar-C-H def.), 670 (Ar-C-Cl str.).

^{1}H NMR (400 MHz, CDCl3): δ 7.48–7.42 (m, 1H, Ar-H), 7.33 (dd, *J* = 8.4, 2.0 Hz, 1H, Ar-H), 7.25 (dd, *J* = 9.4, 1.9 Hz, 1H, Ar-H), 5.86 (s, 1H, -CH-), 4.20–4.13 (m, 2H, -CH_2-) (Spectrum 6.05).

MS: (m/z: RA%): 424.12[M+1, 65%], 426.11 [M+2, 21%], 428.02 [M+4, 6%], 430.02 [M+6, 2%].

Elemental analysis:	%C	%H	%N
Calculated:	33.99	1.43	13.21
Found:	33.41	1.35	13.10

Synthesis of 2-(4-bromophenyl)-3-(4,6-dichloro-1,3,5-triazin-2-yl)thiazolidin-4-one (26e):

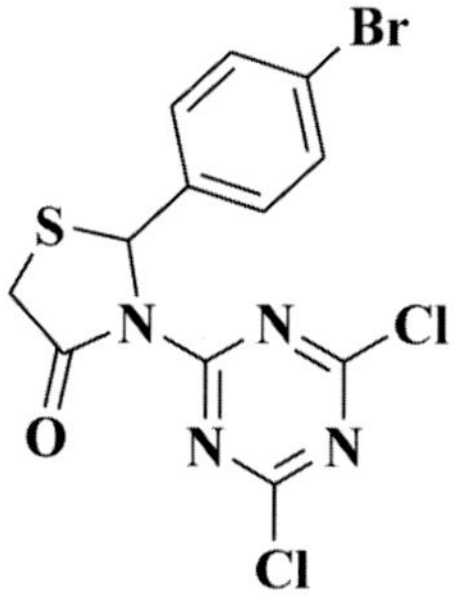

Yield: 80%		M.P. 160–162°C
Mol. Formula:	$C_{12}H_7BrCl_2N_4OS$	Mol. Weight: 403.12

IR (KBr cm^{-1}): 3060 (Ar-C-H str.), 1534 (Ar-C=C- str.), 801(C-N- str.), 1751 (-C=O str.), 1556 (Ar-C=C-), 780 (Ar-C-H def.), 677 (Ar-C-Cl str.).

^{1}H NMR (400 MHz, $CDCl_3$): δ 7.53–7.47(m, 2H, Ar-H), 7.41–7.34 (m, 2H, Ar-H), 5.34 (s, 1H, -CH-), 4.17 (q, J = 7.2Hz, 2H, $-CH_2-$).

MS: (m/z: RA%): 404 [M+1, 65%], 406.11 [M+2, 21%], 408.02 [M+4, 6%], 410.02 [M+6, 2%].

Elemental analysis:	%C	%H	%N
Calculated:	35.49	1.74	13.80
Found:	35.30	1.45	13.48

Synthesis of 3-(4,6-dichloro-1,3,5-triazin-2-yl)-2-(4-chloro-2-fluorophenyl) thiazolidin-4-one (26f):

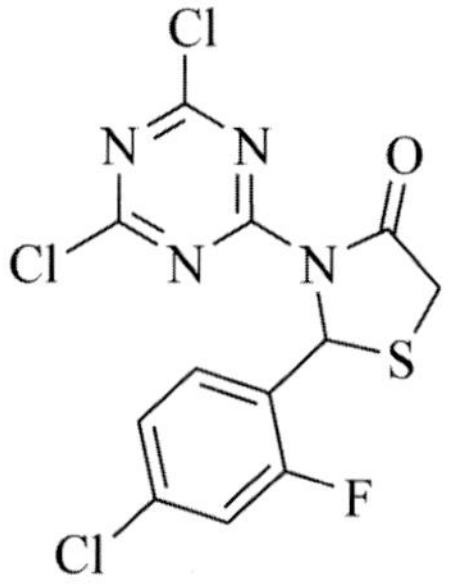

Yield: 81%		M. P. 155–157°C
Mol. Formula:	$C_{12}H_6FCl_3N_4OS$	Mol. Weight: 377.11

IR (KBr cm^{-1}): 2850–3010 (Ar-C-H str.), 1488 (Ar-C=C- str.), 812(C-N- str.), 1748 (-C=O str.), 872–779 (Ar-C-H def.), 1022 (Ar-C-F str.).

^{1}H NMR (400 MHz, $CDCl_3$): δ 7.69 (dd, J = 6.4, 2.6 Hz, 1H, Ar-H), 7.41 (ddd, J = 8.7, 4.6, 2.6 Hz, 1H, Ar-H), 6.95 (dd, J = 9.6, 8.7 Hz, 1H, Ar-H), 5.86 (s, 1H, -CH-), 4.17 (m, 2H, $-CH_2-$).

MS: (m/z: RA%): 378 [M+1, 79%]$^+$. 380.20 [M+2, 26%], 382.10 [M+4, 8%], 384.13 [M+6, 2%].

Elemental analysis:	%C	%H	%N
Calculated:	37.97	1.59	14.76\
Found:	37.41	1.35	14.30

Synthesis of 3-(4,6-dichloro-1,3,5-triazin-2-yl)-2-m-tolylthiazolidin-4-one (26g):

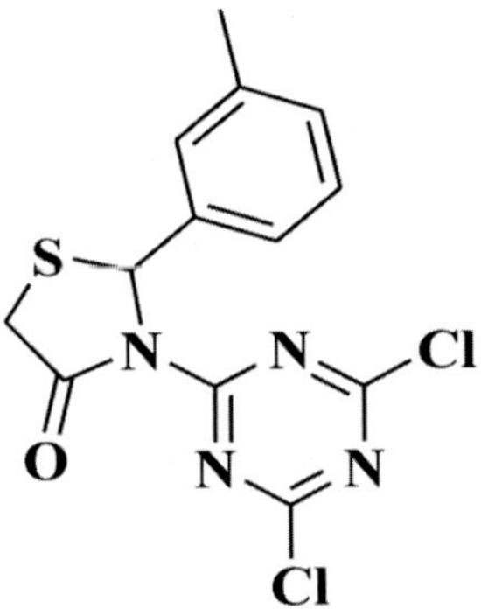

Yield: 86% M. P. 140–142°C

Mol. Formula: $C_{13}H_{10}Cl_2N_4OS$ Mol. Weight: 340

IR (KBr cm^{-1}): 2842 (Ar-C-H str.), 1534 (Ar-C=C- str.), 800(C-N- str.), 1760 (-C=O str.), 789 (Ar-C-H def.), 671–641(Ar-C-Cl str.), 814(C-N- str. s-triazine).

^{1}H NMR (400 MHz, $CDCl_3$): δ 7.37–7.33(m, 2H, Ar-H), 7.32–7.27 (m, 2H, Ar-H), 5.88 (s, 1H, -CH-), 2.21 (s, 3H, $Ar-CH_3$), 4.22–4.27(d, J= 7.2Hz, 2H, $-CH_2-$).

MS: (m/z: RA%): 341 [M+1, 67%], 342.67 [M+2, 22%], 344.90 [M+4, 7%].

Elemental analysis:	%C	%H	%N
Calculated:	45.76	2.95	16.42
Found:	45.56	2.65	16.10

Synthesis of 3-(4,6-dichloro-1,3,5-triazin-2-yl)-2-(4-nitrophenyl)thiazolidin-4-one (26h):

Yield: 78% M.P.170–172°C

Mol. Formula: $C_{12}H_7Cl_2N_5O_3S$ Mol. Weight: 370.10

IR (KBr cm^{-1}): 3060 (Ar-C-H str.), 1650–1577 (Ar-C=C- str.), 1758 (>C=O), 638–891 (Ar-C-H def.), 1320 (Ar-C-N str.), 1521 (Ar-C-NO_2 str.), 804(C-N-str. s-triazine), 621–771 (Ar-C-Cl str.).

^{1}H NMR (400 MHz, $CDCl_3$): δ 7.66 (m, 2H, Ar-H), 6.85 (m, 2H, Ar-H), 5.84 (s, 1H, -CH-), 4.19–4.12 (m, 2H, -CH_2-).

MS: (m/z: RA%): 370.9[M+1, 66%]$^{+}$., 372.67 [M+2, 22%], 374.20 [M+4, 7%].

Elemental analysis:	%C	%H	%N
Calculated:	38.72	1.90	18.82
Found:	38.41	1.65	18.40

Synthesis of 3-(4,6-dichloro-1,3,5-triazin-2-yl)-2-(2-hydroxyphenyl) thia-zolidin-4-one (26i):

Yield: 82% M. P. 160–162°C

Mol. Formula: $C_{12}H_8 Cl_2N_4O_2S$ Mol. Weight: 341.22

IR (KBr cm^{-1}): 3124–2859 (Ar-C-H str.), 1562 (Ar-C=C- str.), 1738 (>C=O), 640–886 (Ar-C-H def.), 625–778 (Ar-C-Cl str.), 810 (C-N- str. s-triazine).

^{1}H NMR (400 MHz, CDCl$_3$): δ 7.58 (d, *J* = 11 Hz, 1.8 Hz, 1H, Ar-H), 7.32 (d, *J* = 6.5, 5.7, 3.6 Hz, 1H, Ar-H), 7.18 (d, *J* = 7.5, 1.3 Hz, 1H, Ar-H), 7.06 (dd, *J* = 9.4, 8.2.1.2 Hz, 1H, Ar-H), 5.92 (d, *J* = 7.5 Hz, 1H, -CH-), 4.21–4.12 (m, 2H, -CH$_2$-).

MS: (m/z: RA%): 341.9 [M+1, 78%]$^+$., 344.67 [M+2, 26%], 346.20 [M+4, 8%].

Elemental analysis:	%C	%H	%N
Calculated:	42.00	2.35	16.33
Found:	41.90	2.10	16.10

Synthesis of 3-(4,6-dichloro-1,3,5-triazin-2-yl)-2-(3-methoxyphenyl) thia-zolidin-4-one (26j):

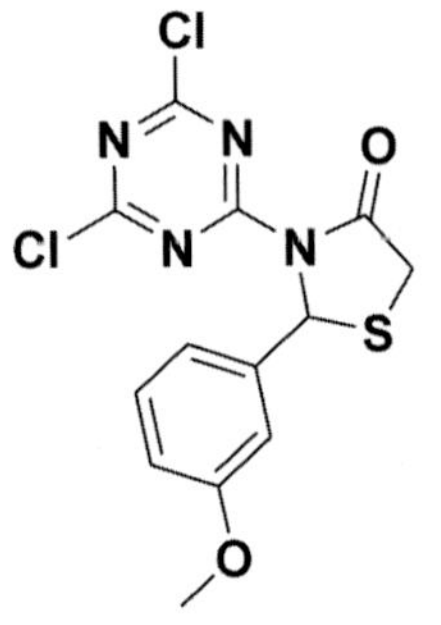

Yield: 80% M. P. 141–143°C

Mol. Formula: $C_{13}H_{10}Cl_2N_4O_2S$ Mol. Weight: 355.9

IR (KBr cm^{-1}): 3120–2880 (Ar-C-H str.), 1650–1500 (Ar-C=C- str.), 1751(>C=O), 680–830 (Ar-C-H def.), 620–754 (Ar-C-Cl str.), 803 (C-N- str. s-triazine).

^{1}H NMR (400 MHz, CDCl$_3$): δ 8.15 (s, 1H, Ar-H), 7.93 (m, 1H, Ar-H), 7.60 (m, 1H, Ar-H), 7.46 (m, 1H, Ar-H), 5.37 (d, *J* = 7.1 Hz, 1H, -CH-), 3.75(s, 3H, Ar-OCH$_3$), 3.38 (d, 2H, -CH$_2$-).

MS: (m/z: RA%): 356.7[M+1, 75%]$^+$., 358.67 [M+2, 25%], 360.20 [M+4, 10%].

Elemental analysis:	%C	%H	%N
Calculated:	43.71	2.82	15.68
Found:	43.50	2.75	15.42

Synthesis of 3-(4,6-dichloro-1,3,5-triazin-2-yl)-2-(2,5-difluorophenyl)thiazolidin-4-one (26k):

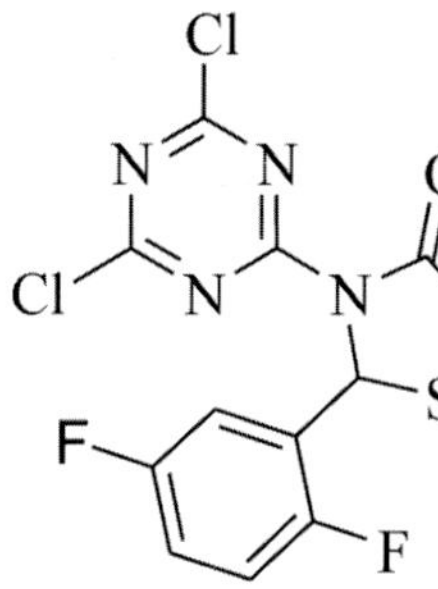

Cl
N N O
Cl N N
S
F
F

Yield: 75%		M.P. 178–180°C
Mol. Formula:	$C_{12}H_6Cl_2F_2N_4OS$	Mol. Weight: 361.10

IR (KBr cm^{-1}): 3120–2825 (Ar-C-H str.), 1650–1555 (Ar-C=C- str.), 1755 (>C=O), 1130–1005 (-C-F-, str.), 8690–820 (Ar-C-H def.), 710–738 (Ar-C-Cl str.), 810 (C-N- str. s-triazine).

^{1}H NMR (400 MHz, CDCl$_3$): δ 7.72 (dd, *J* =6.3, 2.5 Hz, 1H, Ar-H), 7.39 (dd, *J* = 8.6, 4.5, 2.5 Hz, 1H, Ar-H), 6.85 (dd, *J* = 9.5, 8.6 Hz, 1H, Ar-H), 5.80 (s, 1H, -CH$_2$-), 4.15 (m, 2H, -CH$_{2-}$).

MS: (m/z: RA%): 361.1[M+1, 74%]$^+$., 363.67[M+2, 24%], 365.20 [M+4, 9%].

Elemental analysis:	%C	%H	%N
Calculated:	39.69	1.67	15.43
Found:	39.50	1.44	15.33

Synthesis of 3-(3-(4,6-dichloro-1,3,5-triazin-2-yl)-4-oxothiazolidin-2-yl) benzonitrile (26l):

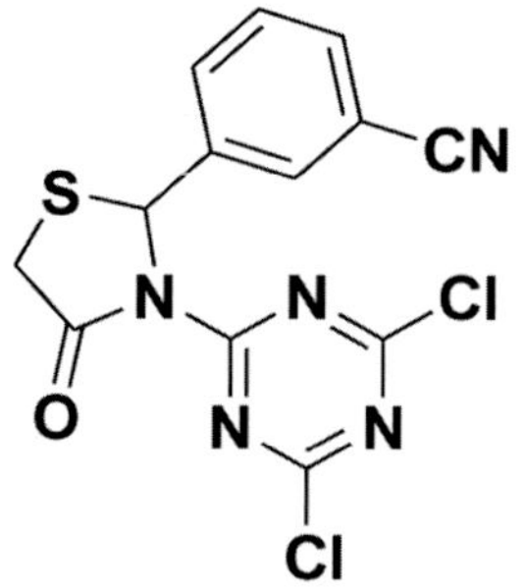

CN
S
N N Cl
O N N
Cl

Yield: 78%		M.P. 145–147°C
Mol. Formula:	$C_{13}H_7Cl_2N_5OS$	Mol. Weight: 350.13

IR(KBr cm^{-1}): 2866 (Ar-C-H str.), 1660 (Ar-C=C- str.), 1755 (>C=O), 1530–1550 (-C=N-, str.), 6670–850 (Ar-C-H def.), 610–770(Ar-C-Cl str.), 802(C-N- str. s-triazine).

^{1}H NMR (400 MHz, $CDCl_3$): δ 7.94 (s, 1H, Ar-H), 7.59 (s, 1H, Ar-H), 7.34–7.21 (m, 1H, Ar-H), 7.04–7.02 (m, 1H, Ar-H), 5.78 (s, 1H, -CH-), 3.40–3.36(m, 2H, $-CH_2-$).

MS: (m/z: RA%): 350.8 [M+1, 70%].$^+$, 352.67 [M+2, 23%], 354.20 [M+4, 7%].

Elemental analysis:	%C	%H	%N
Calculated:	44.33	2.00	19.88
Found:	44.10	1.90	19.56

3.5 RESULTS AND DISCUSSION

The synthesis of the target compounds were accomplished via one-pot MCR. The synthetic pathway used to obtain compounds thiazolidinone triazine derivatives (26a-l) was achieved by the reaction between aromatic aldehyde, 4,6-dichloro-1,3,5-triazin-2-amine and ethyl 2-mercaptoacetate in dry ethanol shown in Scheme 3.1. Their structures were confirmed by analytical and spectral data. The C, H, N, and S elements of the synthesized products were reliable with their predicted structures.

1. **IR Spectra:** The IR (KBr, cm^{-1}) spectrum showed the characteristic bands at The IR spectrum of product showed the characteristic bands in the region 1680–1700 cm^{-1} for carbonyl group of 4-thiazolidinone ring.
2. **^{1}H NMR Spectra:** The ^{1}HNMR spectrum (400 MHz, $CDCl_3$) of compound showed the characteristic signal at 5.50–6.10 ppm for $-CH_2$ protons at position-5 in the 4-thiazolidinone ring and a signal at 4.30–3.50 ppm for -CH protons at position-2 of the 4-thiazolidinone ring. All other signals are at their respective positions for the respective protons in the NMR spectra. The CMR spectrum also show the signal at 160–169 for C=O and 30–40 ppm for $-CH_2$ of thiazolidinone. All other signals appeared at their respective positions. The results of elemental analysis confirmed the product.
3. **Mass Spectra:** The mass spectra of synthesized products are in agreement with their molecular weight. The Triazine thiazolidinone compound (26a) was analyzed for molecular formula as $C_{12}H_7$ F Cl_2N_4OS, m.p. 145–147°C supported by a [M+H]+ ion at m/z 345.
4. **SAR:** All newly synthesized compounds (26a-l) were subjected to preliminary testing for anti-inflammatory activity according to the

method of Hwang et al. in 1993 [45]. Also, ensure anti-microbial activity against various gram-positive, gram-negative bacteria and fungal strains by using an agar well diffusion method [46].

3.6 BIOLOGICAL EVALUATION: ANTI-INFLAMMATORY, ANTIBACTERIAL, AND ANTIFUNGAL ASSAY

The same protocols and procedures that have been followed as in Chapter 2 were used to study the anti-inflammatory, antibacterial, and antifungal activity of the newly synthesized compounds (26a-l). The results were represented in Tables 3.1–3.3.

TABLE 3.1 Anti-Inflammatory Activity Data of 3-(4,6-Dichloro-1,3,5-Triazin-2-yl)-2-Phenylthiazolidin-4-One Derivatives (26a-l)

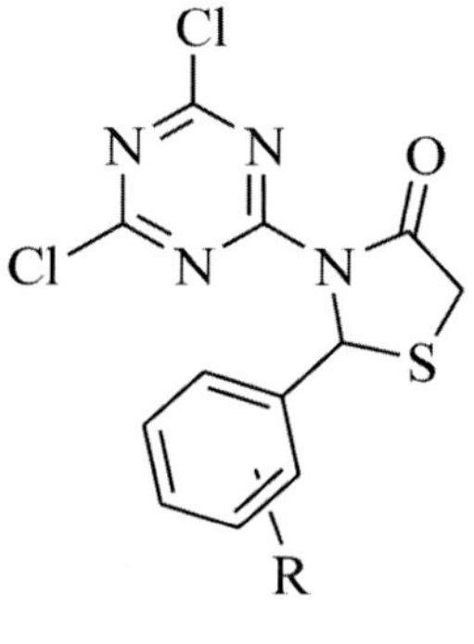

Compounds (26a-l)	(R)	% Inhibition at 10 μM	
		TNF-α	IL-6
26a	Phenyl	12	20
26b	4-OMe	32	29
26c	2-F	70	76
26d	2-F, 4-Br	54	57
26e	4-Br	22	23
26f	2-F, 5-Cl	60	67
26g	3-Me	10	18
26h	4-NO_2	15	20
26i	2-OH	22	29
26j	3-OMe	45	54
26k	2-F, 5-F	72	79
26l	5-CN	30	29
Ref.	Dexamethasone (1μM)	75	81

TABLE 3.2 Antibacterial Activity of 3-(4,6-Dichloro-1,3,5-Triazin-2-yl)-2-Phenylthiazolidin-4-One Derivatives (26a-l) (MIC [a] Values μg/mL)

Compounds (26a-l)	Gram-Positive		Gram-Negative	
	Staphylococcus aureus	***Bacillus subtilis***	***Escherichia coli***	***Salmonella typhimurium***
26a	65	80	70	75
26b	15	15	10	20
26c	80	85	90	85
26d	90	00	85	80
26e	65	70	80	80
26f	00	90	95	00
26g	70	65	75	80
26h	20	20	15	10
26i	35	45	30	40
26j	75	60	80	85
26k	90	00	00	85
26l	15	25	15	20
Ciprofloxacin (Ref.)	20	25	20	20

[a] - Values are the average of three readings.

TABLE 3.3 Antifungal Activity of 3-(4,6-Dichloro-1,3,5-Triazin-2-yl)-2-Phenylthiazolidin-4-on Derivatives (26a-l) (MIC [a] Values μg/mL)

Compounds (26a-l)	***Candida albicans***	***Aspergillus niger***	***Fusarium solani***	***Aspergillus flavus***
26a	80	75	70	80
26b	15	15	10	20
26c	00	90	85	90
26d	80	65	70	85
26e	65	75	70	80
26f	80	70	85	75
26g	00	00	90	95
26h	20	20	15	10
26i	35	45	30	40
26j	90	80	85	00
26k	80	75	80	85
26l	45	35	55	40
Miconazole (Ref.)	20	25	25	15

[a] - Values are the average of three readings.

3.6.1 SAR OF SYNTHESIZED COMPOUNDS

In Tables 3.1, the anti-inflammatory activity of all the compounds has been recorded on the basis of reference standard drug Dexamethasone. Some of the compounds showed highest (26c, 26f, and 26k) anti-inflammatory activity. Thus, from the activity data (Tables 3.1) it is observed that the compounds 26c, 26f, and 26k were found to have been comparatively active as TNF-α and IL-6 inhibitor (up to 63–74% TNF-α and 67–79% IL-6 inhibitory activity). While compounds 26k (74% and 79%) exhibiting the highest inhibition against TNF-α and IL-6 respectively at 10 μM. It is to be noted that these entire compounds found to be equally potent that of the standard Dexamethasone at 1μM concentration. The compounds 26d and 26j exhibited moderate activity (45–57% inhibition) while other compounds exhibited low (26b, 26e, 26i, 26l) to very low (26a, 26g, 26h) activity at the same level of concentration. The anti-inflammatory activity data shows that the presence of fluorine group at C-2 and C-5 of phenyl ring plays an important role in the activity. The presence of 'F' group at C-5 enhances the anti-inflammatory activity (26k). Because of the presence of chlorine instead of fluorine at C-5 compound (26f) little beat reduces the activity. Further, it is observed that the presence of fluorine at position-2, bromine at position-4 and methoxy group at position-3 on phenyl ring shows moderate activity. Interestingly, the presence of other functional groups on different position such as Phenyl, 4-OMe, 4-Br, 3-Me, 2-OH, 4-NO_2, and 5-CN has exhibits low to very low activity.

It reveals from our SAR studies that, the presence of halide (electron-donating deactivating groups) F and Cl at C-5 position tolerates the procytokine activity. It is found that fluorine imparts the special characteristics that enhance therapeutic efficiency and improved pharmacological properties in bioactive molecules. Thus, the compound 26k and 26c were found to be potential anti-inflammatory agents amongst the series of compounds studied. It is found wise to go for the antimicrobial activity of these compounds to further assist for SAR study. The final compounds were screened against a panel of human disease-causing pathogens consisting of four gram-positive and four gram-negative strains as well as antifungal strain.

From antimicrobial activity data shown in Tables 3.2 and 3.3, it is revealed that some analogs of this series have more potency than the standard drug Ciprofloxacin and Miconazole while some of them have comparable potency. Interestingly none of the compounds with high anti-inflammatory activity found to be potent antibacterial or antifungal agents. Thus, the compounds 26b, 26h, and 26l have higher potency against the tested antimicrobial strain. It is cleared from our results that the 4th position of substituent on the terminal benzene ring is the favorite site for high antimicrobial activity. The high potency of these compounds may be attributed to the presence of nonhalogenated electron-donating group (EDG) or EDG-type group's placement at 4-positions. Meanwhile the presence CN (EDG) at C-5 and OH (EDG) at C-2 on the terminal benzene ring (26i and 26l) shows moderately potent antimicrobial activity with respect to standard drug. It is noteworthy to mention that compounds containing halogenated electron-withdrawing group (-Br) at 4th position have no activity against the pathogenic bacteria and fungal strain. Any activity has not been observed in the case of remaining compounds up to a concentration of 200 μg/mL against some bacterial and fungal strains.

It is cleared from results i.e., Tables 3.2 and 3.3, the SAR of antibacterial activity partially correlates with their SAR of antifungal activity as there is some divergence is observed. The same aryl 4th position as observed already is a favorable site of high activity. The compounds 26b and 26h have been found to be two-fold more potent than the standard drug Micanazole as similar to the antibacterial activity trend. While the compounds (26a, 26c, 26d, 26e, 26f, 26g, 26j, and 26k) have no major effect on the antifungal activity also.

3.7 CONCLUSION

In this chapter, we have elaborated on the efforts made toward the potentially active novel series of hybrid 3-(4,6-dichloro-1,3,5-triazin-2-yl)-2-phenylthiazolidin-4-one derivatives as potent anti-inflammatory and antimicrobial agents via efficient synthetic methodology. The present study suggests the role of halogenated electron-withdrawing groups in generating highly potent anti-inflammatory agents from the title hybrid skeleton. Thus the presence of lipophilic -Cl and -F at 5th position tolerates the procytokine activity. Also the presence of nonhalogenated electron-donating group (EDG) or EDG-type of (4-OMe and 4-NO_2) group's present at 4-positions of terminal benzene ring found to be effective potent antimicrobial agents. Our optimization studies on the hybrid derivatives of 3-(4,6-dichloro-1,3,5-triazin-2-yl)-2-phenylthiazolidin-4-one under way and will be reported consequently in the future.

KEYWORDS

- **anti-inflammatory**
- **antimicrobial activity**
- **characterization**
- **s-triazine**
- **synthesis**
- **thiazolidinone**

REFERENCES

1. (a) Demain, A. L., & Sanchez, S., (2009). *J. Antibiotics*, *62*, 5–16. (b) Levy, S. B., & Marshall, B., (2004). *Natural Medicines*, *10*, S122–S129.
2. (a) Gootz, T. D., (2010). *Crit. Rev. Immunol.*, *30*, 79–93. (b) Overbye, K. M., & Barrett, J. F., (2005). *Drug Discov. Today*, *10*, 45–52. (c) Niccolai, D., Tarsi, L., & Thomas, R., (1997). *J. Chem. Commun.*, pp. 2333–2342.
3. (a) Nathan, C., (2004). *Nature*, *431*, 899–902. (b) Raviglione, M. C., (2003). *Tuberculosis*, *83*, 4.
4. (a) Patel, R. V., Kumari, P., Rajani, D. P., Pannecouque, C., De Clercq, E., & Chikhalia, K. H., (2012). *Future Med. Chem.*, *4*, 1053–1065. (b) Mishra, A. R., & Shailendra, S., (2000). *J. Agri. Food Chem.*, *48*, 5465–5468. (c) Patel, D. H., Chikhalia, K. H., Shah, N. K., Patel, D. P., Kaswala, P. B., & Buha, V. M., (2010). *J. Enzyme Inhibition and Med. Chem.*, *25*, 121–125.
5. (a) Patel, R. V., Kumari, P., Rajani, D. P., & Chikhalia, K. H., (2011). *Eur. J. Med. Chem.*, *46*, 4354–4365. (b) Liu, B., Lee, Y., Zou, J., Petrassi, H. M., Joseph, R. W., Chao, W., Michelotti, E. L., Bukhtiyarova, M., Springman, E. B., & Dorsey, B. D., (2010). *Bioorg. and Med. Chem. Lett.*, *20*, 6592–6596. (c) Dianzani, C., Collino, M., Gallicchio, M., Fantozzi, R., Samaritani, S., Signore, G., & Menicagli, R., (2006). *J. Pharmacy and Pharmacology*, *58*, 219–226.
6. (a) Avupati, V. R., Yejella, R. P., Parala, V. R., Killari, K. N., Papasani, V. M. R., Cheepurupalli, P., Gavalapu, V. R., & Boddeda, B., (2013). *Bioorg. and Med. Chem. Lett.*, *23*, 5968–5970. (b) Bhat, H. R., Singh, U. P., Gahtori, P., Ghosh, S. K., Gogoi, K., Prakashe, A., & Singh, R. K., (2013). *New J. Chem.*, *37*, 2654–2662.
7. Shah, D. R., Modh, R. P., & Chikhalia, K. H., (2014). *Future Med. Chem.*, *6*, 463–477.
8. Bhat, H. R., Pandey, P. K., Ghosh, S. K., & Singh, U. P., (2013). *Med. Chem. Res.*, *22*, 5056–5065.
9. Desai, A. D., Mahajan, D. H., & Chikhalia, K. H., (2007). *Ind. J. Chem. Sec.*, *46B*, 1169.
10. Gahtori, P., Ghosh, S. K., Parida, P., Prakash, A., Gogoi, K., Bhat, H. R., & Singh, U. P., (2012). *Exp. Parasitol.*, *130*, 292–299.
11. Baliani, A., Bueno, G. J., Stewart, M. L., Yardley, V., Brun, R., Barrett, M. P., Gilbert, I. H., (2005). *J. Med. Chem.*, *48*, 5570–5579.

12. Singh, U. P., Bhat, H. R., & Gahtori, P., (2012). *J. Mycol. Med.*, *22*, 134–141.
13. Menicagli, R., Samaritani, S., Signore, G., Vaglini, F., & Dalla, V. L., (2004). *J. Med. Chem.*, *47*, 4649–4652.
14. Patel, A. B., Patel, R. V., Kumari, P., Rajani, D. P., & Chikhalia, K. H., (2012). *Med. Chem. Res.*, *22*, 367–381.
15. Chen, X., Zhan, P., Liu, X., Cheng, Z., Meng, C., Shao, S., Pannecouque, C., & De Clercq, E., (2012). *Bioorg. Med. Chem.*, *20*, 3856–3864.
16. Kumar, S., Bhat, H. R., Kumawat, M. K., & Singh, U. P., (2013). *New J. Chem.*, *37*, 581–584.
17. Taranalli, A. D., Bhat, A. R., Srinivas, S., & Saravanan, E., (2008). *Ind. J. Pharmaceutical Sci.*, *70*, 159–164.
18. Karali, N., Gürsoy, A., Kandemirli, F., Shvets, N., Kaynak, F. B., Ozbey, S., Kovalishyn, V., & Dimoglo, A., (2007). *Future Med. Chem.*, *15*, 5888–5904.
19. Kaminskyy, D., Bednarczyk-Cwynar, B., Vasylenko, O., Kazakova, O., Zimenkovsky, B., Zaprutko, L., & Lesyk, R., (2012). *Med. Chem. Res.*, *21*, 3568–3580. (b) Wang, S., Zhao, Y., Zhu, W., Liu, Y., Guo, K., & Gong, P., (2012). *Archiv Der Pharmazie*, *345*, 73–80.
20. Havrylyuk, D., Zimenkovsky, B., Vasylenko, O., Gzella, A., & Lesyk, R., (2012). *J. Med. Chem.*, *55*, 8630–8641.
21. Balzarini, J., Orzeszko-Krzesinska, B., Maurin, J. K., & Orzeszko, A., (2009). *Eur. J. Med. Chem.*, *44*, 303–311.
22. Palekar, V. S., Damle, A. J., & Shukla, S. R., (2009). *Eur. J. Med. Chem.*, *44*, 5112–5116.
23. Omar, K., Geronikaki, A., Zoumpoulakis, P., Camoutsis, C., Sokovic, M., Ciric, A., & Glamoclija, J., (2010). *Future Med. Chem.*, *18*, 426–432.
24. Shih, M. H., & Ke, F. Y., (2004). *Future Med. Chem.*, *12*, 4633–4643.
25. Terzioglu, N., Karali, N., Gürsoy, A., Pannecouque, C., Leysen, P., Paeshuyse, J., Neyts, J., & De Clercq, E., (2006). *Arkivoc*, *1*, 109–118.
26. Ragab, F. A., Eid, N. M., & El-Tawab, H. A., (1997). *Pharmazie*, *52*, 926–929.
27. Raikwar, D., Srivastava, S. K., & Srivastava, S. D., (2008). *J. Ind. Chem. Soc.*, *85*, 78–84.
28. Srinivas, A., Nagaraj, A., & Reddy, C. S., (2008). *J. Heterocycl. Chem.*, *45*, 999–1003.
29. Diurno, M. V., Mazzoni, O., Piscopo, E., Calignano, A., Giordano, F., & Bolognese, A., (1992). *J. Med. Chem.*, *35*, 2910–2912.
30. Patel, H. D., Patel, K. C., Mehta, K. M., & Patel, P. D., (2011). *Elixir Org. Chem.*, *38*, 4122–4126.
31. Hassan, K. H. M., El-Ezbawy, S. R., & Abdel-Wahab, A. A., (1984). *J. Ind. Chem. Soc.*, *56*, 290–292.
32. Pathak, R. B., Chovatia, P. T., & Parekh, H. H., (2012). *Bioorg. and Med. Chem. Lett.*, *22*, 5129–5133.
33. Joshi, S. N., (1991). *J. Ind. Chem. Soc.*, *68*, 627–628.
34. Jadhav, S. A., Shioorkar, M. G., Chavan, O. S., Chavan, R. V., Shinde, D. B., & Pardeshi, R. K., (2015). *Der Pharma Chemica*, *7*(5), 329–334.
35. Sharma, G. V. R., Devi, B., Reddy, K. S., Reddy, M. V., Kondapi, A. K., & Bhaskar, C., (2015). *Heterocycl. Commun.*, *21*(4), 187–190.
36. Bhagwat, V. S., Parvate, J. A., & Joshi, M. N., (1991). *J. Ind. Chem. Soc.*, *68*, 419–420.
37. Zheng, H., Mei, Y. J., Du, K., Shi, Q. Y., & Zhang, P. F., (2013). *Catal. Lett.*, *143*(3), 298–301.
38. Srivastava, T., Haq, W., & Katti, S. B., (2002). *Tetrahedron*, *58*, 7619–7624.
39. Joshi, H. S., (2003). *Ind. J. Chem.*, *42B*, 1544–1548.

40. Fraga-Dubreuil, J., & Bazureau, J. P., (2003). *Tetrahedron*, *59*, 6121–6130.
41. Bhagat, T. M., Swamy, D. K., Badne, S. G., & Kuberkar, S. V., (2011). *Rasayan J. Chem.*, *4*(1), 24–28.
42. Srivastava, S. K., Srivastava, S. L., & Srivastava, S. D., (2000). *J. Ind. Chem. Soc.*, *77*, 104–105.
43. Parrek, D., Chaudhary, M., Pareek, P. K., Kant, R., & Ojha, K. G., (2011). *Der Pharmacia Sinica*, *2*(1), 170–181.
44. Khot, S. S., Kapase, V. S., Kenawade, S., & Dhongade, S. R., (2014). *Inter. J. Pharma. Res. Scholar*, *3*(1), 363–373.
45. Hwang, C., Gatanaga, M., Granger, G. A., & Gatanaga, T., (1993). *J. Immunology*, *151*, 5631.
46. Patal, A. B., Chikhali, K. H., & Kumari, P., (2014). *Eur. J. Med. Chem.*, *79*, 57–65.
47. Sridhar, R., Perumal, P. T., Etti, S., Shanmugam, G., Ponnuswamy, M. N., Prabavathyc, V. R., & Mathivanan, N., (2004). *Bioorg. Med. Chem. Lett.*, *14*, 6035.

CHAPTER 4

Effects of *N*-Acetylcysteine, Glutathione, and Glutathione Disulfide on NIH 3T3, VH10 and MCF-7 Cells Exposed to Ascorbate and Cu(II) Ions

KATARÍNA VALACHOVÁ,[1] IVANA ŠUŠANÍKOVÁ,[2]
DOMINIKA TOPOĽSKÁ,[3] ESZTER BÖGI,[1] and LADISLAV ŠOLTÉS[1]

[1]*CEM, Institute of Experimental Pharmacology and Toxicology, Slovak Academy of Sciences, Dúbravská Cesta 9, SK–84104 Bratislava, Slovakia, E-mail: katarina.valachova@savba.sk (K. Valachová)*

[2]*University of Comenius, Pharmaceutical Faculty, Odbojárov 65/10, SK–83104 Bratislava, Slovakia*

[3]*IQVIA, Vajnorská 100/B, SK–83104 Bratislava, Slovakia*

ABSTRACT

We focused on the treatment of cancer MCF-7 cell line with a high dose of ascorbate, followed by cupric ions themselves and by a mixture of Cu(II) ions and ascorbate. For comparison, we selected normal 3T3 cells. The oxidative system composed of Cu(II) ions and ascorbate applied in both 3T3 and cancer MCF-7 cells did not result in significant differences in (low) viability of these two cell lines. We examined the effect of *N*-acetylcysteine (NAC) on oxidatively damaged MCF-7 and 3T3 cells.

Further, we investigated the effect of the addition of glutathione and glutathione disulfide on 3T3 and VH10 fibroblast cell lines exposed to oxidative damage by Cu(II) ions and ascorbate. The results showed that ascorbate only at the highest examined millimolar concentrations decreased the viability of MCF-7 cells more markedly than did 3T3 cells. The viability of 3T3 cell line was not affected significantly in the presence of NAC and cancer MCF-7 cells were a bit more vital after addition of NAC. Glutathione or glutathione disulfide did not prevent oxidative damage in both 3T3 and VH10 cell lines.

4.1 INTRODUCTION

Reactive oxygen species are generated endogenously and exogenously as a by-product of normal respiration and as a function of biochemical reactions under aerobic conditions. At high levels, they are toxic to cells, but at low levels, they have physiological functions, including activation and modulation of signal transduction pathways and regulation of mitochondrial enzyme activities, as well as modulation of activities of redox-sensitive transcription factors. Mitochondria of normal mammalian cells are the principal endogenous sources of reactive oxygen species. Their main exogenous sources are drugs and several xenobiotics [1–4]. Reactive oxygen species are constantly generated during the metabolic process, and their nature of existence in the form of free radicals, ions, and molecules with a single unpaired electron confers them high reactivity [5]. Growth factors enhance production of hydrogen peroxide that results in mitogen-activated protein kinase activation and DNA synthesis, a phenomenon inhibited by antioxidants. Several observations suppose that reactive oxygen also participates in carcinogenesis: first, oxidative stress can induce DNA damages that lead to genomic instability and possibly stimulates cancer progression; second, elevated reactive oxygen species levels are responsible for constant activation of transcription factors, such as nuclear factor κB and activator protein 1, during tumor progression [2, 6].

Cancer is one of the severe diseases with high mortality despite extensive research and considerable efforts to develop targeted therapies. Numerous studies have provided evidence that changes in redox balance and deregulation of redox signaling is common hallmarks of cancer progression and resistance to treatment. It was demonstrated that cancer cells have permanently high levels of reactive oxygen species due to oncogenic transformation including alteration in metabolic, genetic, and tumor microenvironment. Recent studies have shown that cancer cells are adaptable to elevated levels of reactive oxygen species by activating antioxidant pathways. Thus, targeting the signaling pathways of reactive oxygen species and redox mechanisms involved in cancer development are new potential strategies to prevent cancer [5].

Reactive oxygen species exert a dual effect on cancer cells. On one hand, reactive oxygen species can promote cancer by transforming normal cells through the activation of transcription factors or inhibition of tumor suppressor genes. On the other hand, increased reactive oxygen species levels inhibit the progression of tumor growth through stimulation of proapoptotic-signals, leading to the death of cancer cells. Indeed, the redox metabolism that maintains the homeostasis of reactive oxygen species (ratio between production and detoxification), is critical in cell signaling and in the regulation of cell death [2]. Tumors rapidly exhaust the local oxygen supply creating a hypoxic environment. This hypoxic microenvironment around cancer cells can promote invasion and metastasis as well as resistance to radiation therapy and the effectiveness of anti-cancer drugs. Laurent et al. [2] first demonstrated that tumor cells differ from nontransformed cells not only in the level but also in the origin of reactive oxygen species produced and in the pathways involved in the control of these species.

Vitamin C (ascorbic acid) plays an important role in many biological processes. This large spectrum of biological functions is explained by the fact that it is a specific co-factor for a large family of enzymes, collectively known as Fe and 2-oxoglutarate-dependent dioxygenases. Recent studies have shown that a high dose of ascorbate/ascorbic acid (1–2 mmol/l) inhibits glycolysis in breast cancer stem cells and induces their death [7]. Whether vitamin C has a role in the development and regulation of tumor growth has been a topic of investigation and discussion for decades. There are numerous papers documenting potential anti-tumorigenic effects of ascorbate in- *in vitro* and *in vivo* settings, with many reports on the cytotoxicity toward cancer cells and slowing down tumor growth in animal models [8–12]. Human clinical studies have been only in a recent phase I/II studies with the aim to determine the tolerability of pharmacological doses of ascorbate for patients with advanced cancer [13, 14]. Some of these studies have suggested that high dose ascorbate treatment may have a clinical benefit for patients with pancreatic cancer and other advanced cancers [14–16] but just few patients have been analyzed in these studies to date. Results of Sant et al. [17] suggest a potential role even of physiological doses of vitamin C in breast cancer prevention and treatment.

Copper is one of the most abundant metals in humans. It plays several important roles in cellular metabolism. Copper is an essential micronutrient in human diet, it is a component of metalloenzymes such as cytochrome c oxidase, lysyl oxidase, Cu-Zn superoxide dismutase, tyrosinase, and ceruloplasmin, which are important in fundamental biological pathways. Copper ions also act as electron donors or electron acceptors. However, high levels of these transition metal ions in the form of Cu^{2+} are not only toxic, but those are responsible for the production of reactive oxygen species, which lead to oxidative stress in cells, causing cellular damage and diseases. High levels of copper in the body, either by ingestion or genetic predisposition (Menkes syndrome and Wilson's disease), may contribute to mitochondrial dysfunction, subsequently leading to the production of reactive oxygen species and induction of apoptosis [18–20].

There is broad evidence that copper is an important angiogenic factoring in tumors [19]. The studies showed that while the zinc, iron, and selenium ion concentrations were significantly lower in patients with cancer, the copper ion concentrations in serum were almost always found to be elevated up to 2–3-times compared to samples from normal tissue. Reduction of copper level as an anti-cancer strategy is currently under intense investigation. Copper chelators such as D-penicillamine, tetrathiomolybdenate, clioquinol, and trientine have been shown to inhibit angiogenesis both *in vitro* and *in vivo* [18, 19, 21, 22].

N-Acetylcysteine (NAC) is a derivative of cysteine with an acetyl group attached to the nitrogen atom and like most thiols can be oxidized by several types of free radicals and also serves as an electron-pair donor: NAC reacts fast with $^{\cdot}OH$, $^{\cdot}NO_2$, $CO_3^{\cdot-}$ and thiyl radicals, which eventually lead to the formation of $O_2^{\cdot-}$. It reacts also with nitroxyl, the reduced and protonated form of nitric oxide, which has been demonstrated as a unique species with potentially important pharmacological activities [23]. NAC is administered orally, intravenously or by inhalation. It is relatively low toxic and is associated with mild side effects such as vomiting, nausea, and tachycardia. The half-life of NAC in blood after a single intravenous administration is 5.6 hours [24]. NAC can also reduce disulfide bonds in proteins and bind metal ions yielding metal complexes. *In vitro* and *in vivo* studies have shown that NAC acts like cysteine prodrug: *in vivo,* this thiol compound functions as a precursor of cysteine, which is essentially necessary for glutathione synthesis and replenishment. Oral administration of NAC at doses up to 8000 mg/day does not cause clinically significant adverse reactions [25, 26]. NAC is also a stimulator of the cytosolic enzymes involved in glutathione regeneration, and when it is administered systemically it had been shown to significantly increase levels of glutathione in rat brains [25, 27–29]. Furthermore, NAC can directly protect cells and tissues against oxidative damage by interacting with reactive oxygen species [30]. NAC also has anti-inflammatory effects, alters the levels of neurotransmitters, inhibits proliferation of fibroblasts and keratinocytes, and causes vasodilatation. It is used in therapies to treat neurodegenerative and mental health diseases [25, 31, 32]. Further, it reduced endothelial dysfunction, fibrosis, invasion, and cartilage erosion [33]. NAC also possesses important chemopreventive properties [34, 35]. *It* is known that this compound is involved in several biochemical pathways. It can prevent apoptosis and enhance cell survival by activating an extracellular signal-regulated kinase pathway, a phenomenon useful for the treatment of certain degenerative diseases. NAC directly modifies the activity of several proteins by its reducing activity and has been shown to be a mitochondrial protectant. Its administration was found to protect age-related stress-induced mitochondrial dysfunction in a rat model as well as in a different stress-induced model of mitochondrial dysfunction in mice [29, 36]. Studies *in vivo* showed that dietary intake of NAC can decrease the severity of many diseases including cancer, pancreatitis, cardiovascular diseases, HIV infections, acetaminophen-induced liver toxicity and metal-induced toxicity [25, 31, 37]. Besides the antioxidative effect, NAC can be also pro-oxidative, when it promotes the DNA damage by increasing the production of reactive oxygen species. It was found that NAC is cytotoxic only when administered with copper [38]. Zheng et al. [38] also reported that NAC, which is a membrane-permeable metal-binding compound, might have anti-cancer activity in the presence of metals. The authors postulated that NAC/Cu(II) significantly alters growth and induces apoptosis in human cancer lines, while NAC/Zn(II) and NAC/Fe(III) not demonstrated these effects. Cancer prevention by NAC has been shown to be effective in several animal experiments, e.g., its oral administration completely prevented the induction of various DNA alternations in rat lung cells. Furthermore, it prevented the *in vivo* formation of carcinogen-DNA adducts, and suppressed the development and growth of tumors in rodents [35]. NAC exhibits antitumoral activity by increasing tumor necrosis factor a-dependent T-cell cytotoxicity [39]. The results suggest that systemic NAC therapy promotes anti-angiogenesis through angiostatin production, resulting in endothelial apoptosis and vascular collapse in the tumor [40]. NAC could reduce markers of stromal-cancer metabolic heterogeneity and markers of cancer cell aggressiveness in human breast cancer [15].

Glutathione (γ-L-glutamyl-L-cysteinylglycine) is a tripeptide that is synthesized in the cytosol from glutamic acid, cysteine, and glycine in two adenosine triphosphate-dependent steps. Intracellular levels of glutathione (0.5–10 mmol/l) are controlled by γ-glutamyltranspetidase. Glutathione is a continuous source of cysteine because cysteine can auto-oxidize to cystine and generate potentially unstable reactive oxygen species. Glutathione is present in many cellular compartments, including the endoplasmic reticulum, the nucleus, and mitochondria and is found at higher concentrations in the liver. Glutathione is also found in the extracellular space, such as in bile and plasma. The key function of glutathione is to reduce hydrogen peroxide and other organic peroxides with participation of glutathione peroxidase. Glutathione co-localized in the nucleus has a profound impact on cellular redox homeostasis and nuclear gene expression. Low levels of glutathione are correlated with many autoimmune diseases [41, 42]. Glutathione has an important role in maintaining intracellular redox homeostasis inevitable during hypoxia and production of reactive oxygen species and $^{\cdot}NO$. Glutathione exists in cells in reduced (GSH) and oxidized (glutathione disulfide, GSSG) states. About 90% of total glutathione is present in a reduced form and less than 10% in disulfide form. The decreases in the GSH:GSSG ratios indicate the status classified as oxidative stress [5, 42–44]. Glutathione directly reacts with reactive oxygen and nitrogen species such as $HO^{\cdot}$, HOCl, $RO^{\cdot}$, $RO_2^{\cdot}$, $^1O_2$ and $ONOO^-$, whose reaction often might result in the formation of thiyl radicals ($GS^{\cdot}$) [45]. Glutathione is also involved in several metabolic processes, including the synthesis of proteins and DNA, signal transduction, enzyme activity, gene expression, metabolism, and the intensification of cytoplasmic and transmembrane transport. It is also used to moderate chronic obstructive pulmonary disease, prevent kidney damage, attenuate influenza, and treat pulmonary fibrosis [34, 46]. Another relevant function of glutathione is the detoxification of xenobiotics. Considering the complex physiological function of the glutathione system, its disequilibrium is involved in several pathological pathways and thus plays an important role in cancer and regulation of the progression through the cell cycle and cell survival, growth, and death. The levels of glutathione and enzymes that are related to biosynthesis of glutathione are high in cancer cells, thus leading cancer cells to become resistant to cell death through oxidative stress mechanisms [42, 47, 48].

The aim of the study was to examine a/the effect of a high dose of ascorbate on viability of tumor MCF-7 cell lines and compare them with normal 3T3 cells, b/to examine the effect of Cu(II) ions themselves and a combination of Cu(II) ions with ascorbate as a source of reactive oxygen species on these cell lines in the absence and presence of NAC. The third aim was to examine effects of the addition of glutathione and glutathione disulfide on viability of 3T3 and VH10 cell lines.

4.2 EXPERIMENTAL PART

4.2.1 MATERIALS

Dulbecco's modified eagle's medium (DMEM), fetal bovine serum (FBS), penicillin/streptomycin, trypsin-EDTA solution, phosphate-buffered saline, trolox (6-hydroxy-2,5,7,8-tetramethylchroman-2-carboxylic acid), 3-(4,5-dimethyl-2-thiazolyl)-2,5-diphenyl-2H-tetrazolium bromide (MTT), dimethylsulfoxide (DMSO), glutathione, glutathione disulfide and NAC were obtained from Sigma-Aldrich, Slovakia. Ascorbic acid was obtained from Fluka, Germany, and $CuCl_2 \cdot 2H_2O$ was purchased from Slavus, Slovakia.

4.2.2 CELL CULTURES

Mice NIH 3T3 fibroblasts were obtained from the Central Tissue Bank, University Hospital in Bratislava and from Dr. Diana Vavrincová (Department of Pharmacology and Toxicology, Faculty of Pharmacy, Comenius University in Bratislava, Slovakia). Normal human fibroblasts, type VH10, was a gift from Dr. Horváthová from the Cancer Research Institute, Slovak Academy of Sciences in Bratislava. MCF-7 (ER-positive human breast adenocarcinoma, ECACC, Porton Down, Salisbury, UK) were used. Cells were grown at 37°C in humidified atmosphere with 5% CO_2 in DMEM supplemented with 10% FBS, 100 IU/ml penicillin and 100 µg/ml streptomycin.

4.3 METHODS

1. **Initiation of Oxidative Stress:** Fibroblasts were cultured in 24-well culture plates (initial inoculum 4×10^4 3T3 cells/well and 8×10^4 VH10 cells/well). When the cells were firmly attached to the substratum, the culture medium was replaced with 1 ml of the fresh medium containing Cu(II) ions (final Cu(II) concentration in the well was 5 µmol/l). Then, 15 min later, the solution of ascorbic acid was added to produce reactive oxygen species (final ascorbate concentration in the well was 200 µmol/l). Then, the medium was discarded after 90 min from 3T3 cells (2 h in total) and after 210 min from VH10 cells (4 h in total) and replaced with 1 ml of the fresh medium. Cells were subjected to morphological and biochemical evaluation after 24-h treatment.
2. **Application of Trolox, Glutathione, or Glutathione Disulfide:** Each component was added to the system 15 min after the addition of ascorbic acid. The final concentrations of the agents in the well were in the concentration range 1–200 µmol/l.
3. **Cell Viability Assay:** 24 hours after the initiation of oxidative stress, viability of cells was determined by counting in the Burker chamber. Experiments were performed in triplicate.

The effect of the compounds tested on the activity of mitochondrial dehydrogenases of both NIH 3T3 and MCF-7 cells was assessed as reduction of tetrazolium salt MTT. The cells were seeded (9000 cells/100 μl/well) in the 96-well-plate in a complete medium. The examined agents (cupric ions and ascorbate) at various concentrations dissolved in a complete medium were added 24 h after seeding. In the case of examining the three-component system, the agents were added as follows: cupric ions, NAC, and 1 h later ascorbate. Following the next 24 h incubation, the MTT salt (0.45 mg/ml final concentration) was added and the cells were incubated for 3 hours. After removing the medium with MTT, 100% DMSO was added to lyse the cells and the absorbance was measured (λ = 570 nm) in Infinite M200 spectrofluorometer (Tecan, Switzerland). The amount of created formazan (correlating to the number of viable and metabolically active cells) was calculated as a percentage of control cells and was set to 100%.

4.4 RESULTS AND DISCUSSION

Results in Figure 4.1 illustrate MTT reduction *vs.* concentration of Cu(II) ions in both 3T3 and MCF-7 cells and are compared with untreated cells (control). As seen in left panel 3T3 cell viability dropped from 92% to 34.6% in a concentration range 1.5–300 μmol/l. (The exception are cells exposed to Cu(II) ions at concentration 5 μmol/l).

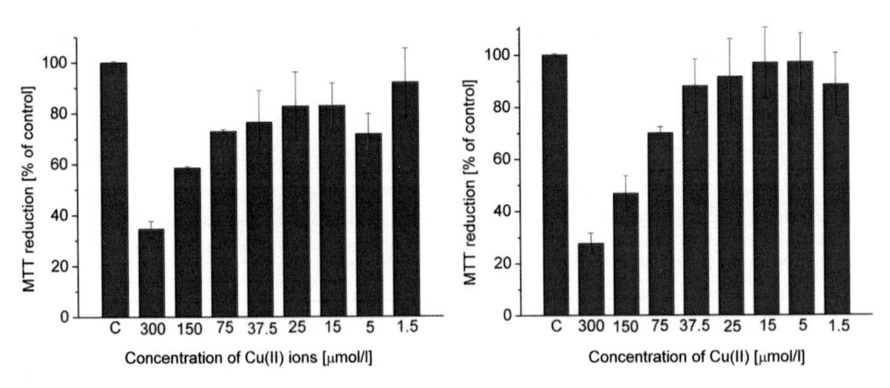

FIGURE 4.1 Viability of 3T3 cells (left panel) and MCF-7 cells (right panel) after exposition to Cu(II) ions determined by MTT test, C-control, p £ 0.05.

Similar results were observed for MCF-7 cells (right panel). The highest cell viability, i.e., 97.1% was reported using 5 μmol/l concentration of Cu(II) ions and was lowered with their increasing concentrations. The amount of vital cells after application of Cu(II) ions at the highest concentration 300 μmol/l was 27.6%.

Further, we investigated the addition of ascorbate at high concentrations (0.375–6 mmol/l) to cancer MCF-7 cells (right panel) and compared it with normal 3T3 cells (left panel). As evident in Figure 4.2 left panel, 3T3 cell viability decreased from 93.7% using ascorbate at concentration 0.375 mmol/l to 13.2% after addition of 6 mmol/l ascorbate. A more pronounced positive effect of ascorbate on decreasing MCF-7 cancer cell viability (right panel) was reported when applying ascorbate at concentrations 1.5 and 6 mmol/l. The cell viability reached 59.2 and 5.3%, respectively. In contrast, viabilities of 3T3 cells after treatment with 1.5 and 6 mmol/l ascorbate were 79.3 and 13.2%, respectively.

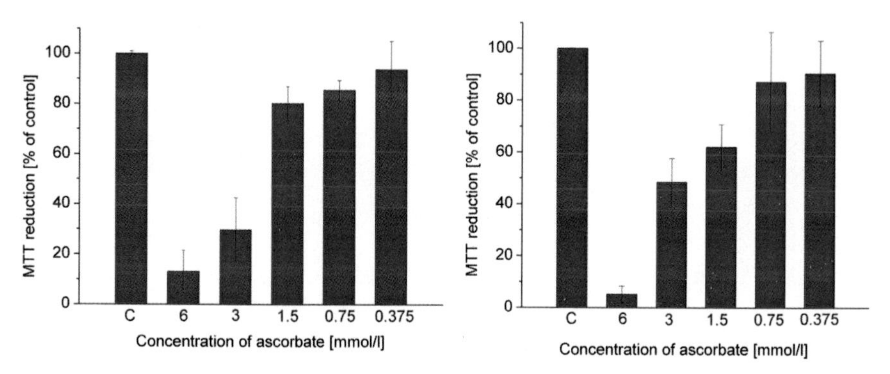

FIGURE 4.2 Viability of 3T3 cells (left panel) and MCF-7 cells (right panel) after exposition to ascorbate determined by MTT test, C-control, p £ 0.01.

High-dose vitamin C has been shown to have significant anticancer effects *in vitro* and *in vivo* [16, 49]. To confirm the effects of vitamin C on the survival of MCF-77 cells, Uetaki et al. [50] examined cell viability using MTT assays. The authors found that cell viability was decreased after exposure of cells to high concentrations of vitamin C (3 or 10 mmol/l), which corresponds to our results. Verrax and Calderon [51] observed that ascorbate at pharmacologic concentrations killed efficiently various cancer cell lines, including MCF-7 reaching EC_{50} from 3 to 7 mmol/l).

Oxidation of ascorbate under catalytic action of Cu(II) ions generates hydrogen peroxide, the so-called Weissberger biogenic oxidative system (W), followed by production of ·OH radicals [52, 53]. For this reason, we examined a combination of ascorbate (1 mmol/l) with different concentrations of Cu(II) ions (1.5, 5 or 15 μmol/l) in both 3T3 and MCF-7 cell lines (Figure 4.3). In 3T3 cells the viability decreased from 76.8% to 9.5% dependent on increasing concentrations of Cu(II) ions (left panel). The addition of ascorbate at higher concentration 1.5 mmol/l to Cu(II) ions (1.5, 5 or 15 μmol/l) resulted in exacerbation of cell viability induced by reactive oxygen species. The results showed 49.5, 4.8, and 3.7% of vital cells, respectively.

In MCF-7 cells (right panel) treated with 1 mmol/l ascorbate in the presence of increasing concentrations of Cu(II) ions (1.5, 5 or 15 μmol/l) a dose-dependent cell death was reported. The cell viability reached 58.2, 26, or 8.3%, respectively. By increasing of the concentration of ascorbate to 1.5 mmol/l we observed greater cell mortality only in the presence of 1.5 μmol/l Cu(II) ions and the valiability of cells reached 32.4%.

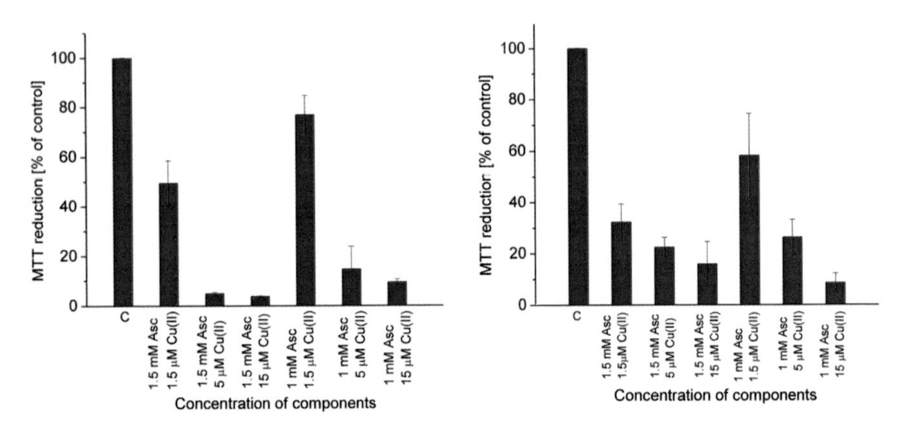

FIGURE 4.3 Viability of 3T3 cells (left panel) and MCF-7 cells (right panel) after exposition to ascorbate (1 or 1.5 mmol/l) and Cu(II) ions (1.5–15 μmol/l) determined by MTT test, C-control, p £ 0.05.

As seen in Figure 4.4 left panel, 3T3 cells oxidatively damaged by ascorbate (1.5 mmol/l), Cu(II) ions (1.5 μmol/l) and NAC (1.5–6 μmol/l) the cell viability varied from 24.7–47.2%. When examining the mixture with less concentrated ascorbate (1 mmol/l), greater levels of Cu(II) ions (5 μmol/l) and NAC (6–20 μmol/l), the percentage of cell viability was lower than when assessing the above mentioned oxidative system. NAC did not prevent cell damage and the cell viability reached values from 17.2 to 29%.

Concerning to MCF-7 cells (right panel), cells treated with all reaction systems reported similar viability.

Uetaki et al. [50] also reported that treatment with NAC (10 mmol/l) significantly inhibited ascorbate (1 mmol/l) and H_2O_2 (1 mmol/l)-induced cytotoxicity of MCF7 cells. NAC also suppressed ascorbate-dependent metabolic changes.

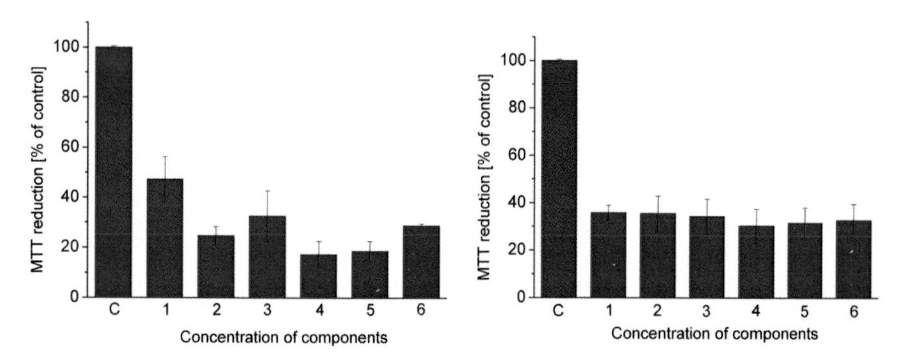

FIGURE 4.4 Viability of 3T3 cells (left panel) and MCF-7 cells (right panel) determined by MTT test after exposition to ascorbate (1.5 mmol/l) with Cu(II) ions (1.5 µmol/l) in the presence of *N*-acetylcysteine (µmol/l): 1.5 (1), 3 (2), 6 (3). Viability of the cells after exposition to ascorbate (1 mmol/l) with Cu(II) ions (5 µmol/l) in the presence of *N*-acetylcysteine (µmol/l): 6 (4), 10 (5), 20 (6), C-control, p £ 0.001.

We selected concentrations of ascorbate 200 µmol/l and of Cu(II) ions 5 µmol/l to reach the 3T3 cell viability at the level ca. 50%. As seen in Figure 4.5, left panel the addition of glutathione resulted in a dose-dependent decrease in cell viability from 55.5% (10 µmol/l) to 14.5% (200 µmol/l). Results were compared with untreated cells (C) and cells treated with trolox (T), where 100% and 98%, respectively were viable. In right panel the results show that the addition of glutathione disulfide at different concentrations did not lead to remarkable differences in cell viability, and varied from 46.5 (200 µmol/l) to 59.5% (1 µmol/l).

As displayed in Figure 4.6, left panel VH10 cells damaged by the Weissberger biogenic oxidative system (W) were vital at the level 54.4%. The addition of extracellular glutathione led to concentration-dependent elevated cell death, whereas the cell viability reached values from 55.5% (10 µmol/l) to 40.1% (200 µmol/l). Results in right panel point to the fact that after the addition of oxidized glutathione the cell viability was comparable to the viability of cells treated with the Weissberger biogenic oxidative system. In contrast, trolox (T) was effective at the level of around 90%.

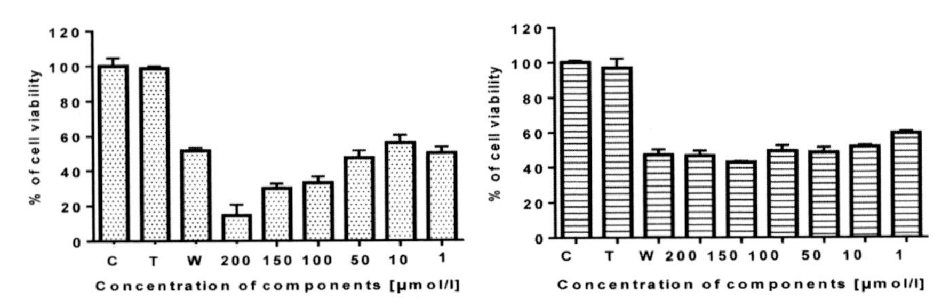

FIGURE 4.5 Viability of 3T3 cells after exposition to ascorbate (200 μmol/l) with Cu(II) ions (5 μmol/l) (W) and after addition of glutathione (left panel) and glutathione disulfide (right panel, 200–1 μmol/l), C-control, T-trolox.

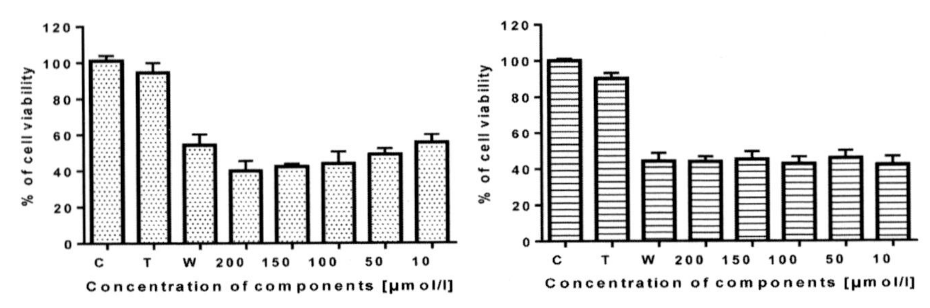

FIGURE 4.6 Viability of VH10 cells after exposition to ascorbate (200 μmol/l) with Cu(II) ions (5 μmol/l) (W) and after addition of glutathione (left panel) or glutathione disulfide (right panel, 200–10 μmol/l), C-control, T-trolox.

A similar oxidative system, particularly Cu(II) ions (0.5 μmol/l) and ascorbate (100 μmol/l) used Campo et al. [54]. As cells, they used human fibroblasts, in which the authors showed a reduction in oxidative stress when pretreated with glycosaminoglycans. Also, similar experiments were performed by Presa et al. [20], who found that the viability and proliferative capacity of 3T3 fibroblasts exposed to $CuSO_4$ (1 μmol/l) or ascorbate individually (2 mmol/l) was not affected. They exposed cells to oxidative stress in the presence of $CuSO_4$ and ascorbate, which caused the activation of caspase-3 and caspase-9, which resulted in the apoptosis of the cells.

4.5 CONCLUSION

It can be concluded that ascorbate added at the highest examined millimolar concentrations affected the viability of MCF-7 cells more effectively than in the case of 3T3 cells. The oxidative system composed of Cu(II) ions and ascorbate did not result in significant differences in the viability of 3T3 and cancer MCF-7 cells. Similarly, the viability of 3T3 cell line was not affected a lot in the presence of NAC. Cancer MCF-7 cells were slightly more vital after the addition of NAC. The addition of glutathione or glutathione disulfide did not have a positive effect to enhance viability on oxidatively damaged 3T3 and VH10 cell lines.

ACKNOWLEDGMENT

The study was supported by the grant VEGA 2/0019/19 and APVV-15-0308.

KEYWORDS

- **ascorbate**
- **cancer cells**
- **fibroblasts**
- ***N*-acetylcysteine**
- **reactive oxygen species**
- **thiol agents**

REFERENCES

1. Fiaschi, A. I., Cozzolino, A., Ruggiero, G., & Giorgi, G., (2005). Glutathione, ascorbic acid, and antioxidant enzymes in the tumor tissue and blood of patients with oral squamous cell carcinoma. *Eur. Rev. Med. Pharmacol. Sci., 9*, 361–367.
2. Laurent, A., Nicco, C., Chéreau, C., Goulvestre, C., Alexandr, E. J., Alves, A., Lévy, E., Goldwasser, F., Panis, Y., Soubrane, O., Weill, B., & Batteux, F., (2005). Controlling tumor growth by modulating endogenous production of reactive oxygen species. *Cancer Res., 65*(3), 948–955.
3. Sharma, P., Jha, A. B., Dubey, R. S., Pessarakli, M. (2012). Reactive oxygen species, oxidative damage, and antioxidative defense mechanism in plants under stressful conditions. *J. Botany*, Article ID 217037, 26 pages.
4. Azimi, I., Petersen, R. M., Thompson, E. W., Roberts-Thomson, S. J., & Monteith, G. R., (2017). Hypoxia-induced reactive oxygen species mediate N-cadherin and Serpine1

expression, EGFR signaling and motility in MDA-MB-468 breast cancer cells. *Sci. Rep., 7*(1), 15140.
5. Kumari, S., Badana, A. K., Mohan, M. G., Shailender, G., & Malla, R. R., (2018). Reactive oxygen species: A key constituent in cancer survival. *Biomark. Insights, 13*, 1–9.
6. Roy, J., Galano, J. M., Durand, T., Le Guennec, J. Y., & Lee, J. C. Y., (2017). Physiological role of reactive oxygen species as promoters of natural defenses. *FASEB J., 31*, 3729–3745.
7. Mastrangelo, D., Pelosi, E., Castelli, G., Lo-Coco, F., & Testa, U., (2018). Mechanisms of anti-cancer effects of ascorbate: Cytotoxic activity and epigenetic modulation. *Blood Cells Mol. Dis., 69*, 57–64.
8. Du, J., Cullen, J. J., & Buettner, G. R., (2012). Ascorbic acid: Chemistry, biology and the treatment of cancer. *Biochim. Biophys. Acta, 1826*, 443–457.
9. Park, S., (2013). The effects of high concentrations of vitamin C on cancer cells. *Nutrients,* **5,** 3496–3505.
10. Wilson, M. K., Baguley, B. C., Wall, C., Jameson, M. B., & Findlay, M. P., (2014). Review of high-dose intravenous vitamin C as an anticancer agent. *Asia Pac. J. Clin. Oncol., 10*, 22–37.
11. Mata, A. M., Carvalho, R. M., Alencar, M. V., Cavalcante, A. A., & Silva, B. B., (2016). Ascorbic acid in the prevention and treatment of cancer. *Rev. Assoc. Med. Bras., 62*(7), 680–686.
12. Cimmino, L., Dolgalev, I., Wang, Y., Yoshimi, A., Martin, G. H., Wang, J., Ng, V·, Xia, B·, Witkowski, M.T., Mitchell-Flack, M·, Grillo, I·, Bakogianni, S. Ndiaye-Lobry, D·, Martín, M.T., Guillamot, M·, Banh, R.S., Xu, M., Figueroa, M.E·, Dickins, R.A., Abdel-Wahab, O·, Park, C.Y., Tsirigos, A·, Neel, B.G., Aifantis, I. (2017). Restoration of TET2 function blocks aberrant self-renewal and leukemia progression. *Cell*, 170, 1079, e20–1095.
13. Stephenson, C. M., Levin, R. D., Spector, T., & Lis, C. G., (2013). Phase I clinical trial to evaluate the safety, tolerability, and pharmacokinetics of high-dose intravenous ascorbic acid in patients with advanced cancer. *Cancer Chemother. Pharmacol., 72*, 139–146.
14. Hoffer, L. J., Robitaille, L., Zakarian, R., Melnychuk, D., Kavan, P., Agulnik, J., Cohen, V., Small, D., & Miller, W. H. Jr., (2015). High-dose intravenous vitamin C combined with cytotoxic chemotherapy in patients with advanced cancer: A phase I-II clinical trial. *PLoS One*, *10*, e0120228.
15. Monti, D. A., Mitchell, E., Bazzan, A. J., Littman, S., Zabrecky, G., Yeo, C. J., Pillai, M. V., Newberg, A. B., Deshmukh, S., & Levine, M., (2012). Phase I evaluation of intravenous ascorbic acid in combination with gemcitabine and erlotinib in patients with metastatic pancreatic cancer. *PLoS One*, *7*(1), e29794.
16. Cieslak, J. A., & Cullen, J. J., (2015). Treatment of pancreatic cancer with pharmacological ascorbate. *Curr. Pharm. Biotechnol., 16*, 759–770.
17. Sant, D. W., Mustaf, S., Gustafson, C. B., Chen, J., Slingerland, J. M., & Wang, G., (2018). Vitamin C promotes apoptosis in breast cancer cells by increasing TRAIL expression. *Sci. Reports, 8*, 5306.
18. Gupte, A., & Mumper, R. J. (2009). Elevated copper and oxidative stress in cancer cells as a target for cancer treatment. *Cancer Treat Rev. 35*(1), 32–46.
19. Wang, F., Jiao, P., Qi, M., Frezza, M., Dou, Q. P., & Yan, B., (2010). Turning tumor-promoting copper into an anti-cancer weapon via high-throughput chemistry. *Curr. Med. Chem., 17*(25), 2685–2698.

20. Presa, F. B., Marques, M. L. M., Viana, R. L. S., Nobre, L. T. D. B., Costa, L. S., & Rocha, H. A. O., (2018). The protective role of sulfated polysaccharides from green seaweed *Udotea flabellum* in cells exposed to oxidative damage. *Marine Drugs, 16*(4), 135.
21. Coates, R. J., Weiss, N. S., Daling, J. R., Rettmer, R. L., & Warnick, G. R., (1989). Cancer risk in relation to serum copper levels. *Cancer Res., 49*, 4353–4356.
22. Yucel, I., Arpaci, F., Ozet, A., Doner, B., Karayilanoglu, T., Sayar, A., & Berk, O., (1994). Serum copper and zinc levels and copper/zinc ratio in patients with breast cancer. *Biol. Trace Elem. Res., 40*(1), 31–38.
23. Samuni, Y., Goldstein, S., Dean, O. M., & Berk, M., (2013). The chemistry and biological activities of *N*-acetylcysteine. *BBA-General Subjects, 1830*(8), 4117–4129.
24. Sansone, R. A., & Sansone, L. A., (2011). Getting a knack for NAC. *Innov. Clin. Neurosci., 8*(1), 10–14.
25. Atkuri, K. R., Mantovani, J. J., Herzenberg, L. A., & Herzenberg, L. A., (2007). N-Acetylcysteine: A safe antidote for cysteine/glutathione deficiency. *Curr. Opin. Pharmacol., 7*(4), 355–359.
26. Aldini, G., Altomare, A., Baron, G., Vistoli, G., Carini, M., Borsani, L., & Sergio, F., (2018). N-acetylcysteine as an antioxidant and disulphide breaking agent: The reasons why. *Free Rad. Res., 52*(7), 751–762.
27. Martinez, B. M., (2000). *N*-Acetylcysteine elicited increase in complex I activity in synaptic mitochondria from aged mice: Implications for treatment of Parkinson's disease. *J. Brain. Res., 859*, 173–175.
28. Sun, S. Y., (2010). *N*-acetylcysteine, reactive oxygen species and beyond. *Cancer Biol. Ther., 9*(2), 109–110.
29. Douiev, L., Soiferman, D., Alban, C., & Saada, A., (2017). The effects of ascorbate, *N*-acetylcysteine, and resveratrol on fibroblasts from patients with mitochondrial disorders. *J. Clin. Med., 6*, 1.
30. Cocco, T., Sgobbo, P., Clemente, M., Lopriore, B., Grattagliano, I., Di Paola, M., & Villani, G., (2005). Tissue-specific changes of mitochondrial functions in aged rats: Effect of a long-term dietary treatment with *N*-acetylcysteine. *J. Free Radic. Biol. Med., 38*(6), 796–805.
31. Adil, M., Suhail Amin, S., & Mohtashim, M. (1995). *N*-Acetylcysteine in dermatology. *Indian J. Dermatol. Venereol. Leprol., 84*(6), 652–659.
32. Tardiolo, G., Bramanti, P., & Mazzon, E., (2018). Overview on the effects of *N*-acetylcysteine in neurodegenerative diseases. *Molecules, 23*, 3305.
33. Zafarullah, M., Li, W. Q., Sylvester, J., & Ahmad, M., (2003). Molecular mechanisms of *N*-acetylcysteine actions. *CMLS, Cell. Mol. Life Sci., 60*, 6–20.
34. Van Zandwijk, N., (1995). *N-Acetylcysteine for Lung Cancer Prevention, 107*(5), 1437–1441.
35. Oikawa, S., Yamada, K., Yamashita, N., Tada-Oikawa, S., & Kawanishi, S., (1999). *N*-acetylcysteine, a cancer chemopreventive agent, causes oxidative damage to cellular and isolated DNA. *Carcinogenesis, 20*(8), 1485–1490.
36. Grattagliano, I., Portincasa, P., Cocco, T., Moschetta, A., Di Paola, M., Palmieri, V. O., & Palasciano, G., (2004). Effect of dietary restriction and *N*-acetylcysteine supplementation on intestinal mucosa and liver mitochondrial redox status and function in aged rats. *J. Exp. Gerontol., 39*, 1323–1332.
37. Mokhtari, V., Afsharian, P., Shahhoseini, M., Kalantar, S. M., & Moini, A., (2017). A review on various uses of N-acetylcysteine. *Cell Journal, 19*(1), 11–17.

38. Zheng, J., Loua, J. R., Zhang, X. X., Benbrook, D. M., Hanigan, M. H., Lind, S. E., & Ding, W. Q., (2010). *N*-acetylcysteine interacts with copper to generate hydrogen peroxide and selectively induce cancer cell death. *Cancer Lett., 298*(2), 186–194.
39. Delneste, Y., Jeannin, P., Potier, L., Romero, P., & Bonnefoy, J. Y., (1997). *N*-Acetyl-L-cysteine exhibits antitumoral activity by increasing tumor necrosis factor dependent T-cell cytotoxicity. *Blood, 90*(3), 1124–1132.
40. Agarwal, A., Muñoz-Nájar, U., Klueh, U., Jaminet, S. C., & Claffey, K. P., (2004). *N*-Acetyl-cysteine promotes angiostatin production and vascular collapse in an orthotopic model of breast cancer. *Am. J. Pathol., 164*(5), 1683–1696.
41. Vergauwen, B., Pauwels, F., & Van, B. J. J., (2003). Glutathione and catalase provide overlapping defenses for protection against respiration-generated hydrogen peroxide in *Haemophilus influenzae. J. Bacteriol., 185*(18), 5555–5562.
42. Corso, C. R., & Acco, A., (2018). Glutathione system in animal model of solid tumors: From regulation to therapeutic target. *Crit. Rev. Oncol./Hematol., 128*, 43–57.
43. Armstrong, J. S., Steinauer, K. K., Hornung, B., Irish, J. M., Lecane, R., Birrell, G. W., Peehl, D. M., & Knox, S. J., (2002). Role of glutathione depletion and reactive oxygen species generation in apoptotic signaling in a human B lymphoma cell line. *Cell Death Differ., 9*, 252263.
44. Weschawalit, S., Thongtlip, S., Phutrakool, P., & Asawanonda, P., (2017). Glutathione and its antiaging and antimelanogenic effects. *Clin. Cosmet. Investig. Dermatol., 10*, 147–153.
45. Lushchak, V. I., (2012). Glutathione homeostasis and functions: Potential targets for medical interventions. *J. Amino Acids, 2012*, 26. Article ID 736837.
46. Qanungo, S., Uysd, J. D., Manevich, Y., Distlerg, A. M., Shaner, B., Hille, E. G., Mieyalg, J. J., Lemastersa, J. J., Townsend, D. M., & Nieminena, A. L., (2014). *N*-acetyl-L-cysteine sensitizes pancreatic cancers to gemcitabine by targeting the NFκB pathway. *Biomed. Pharmacother., 68*(7), 855–864.
47. Allen, J., & Bradley, R. D., (2011). Effects of oral glutathione supplementation on systemic oxidative stress biomarkers in human volunteers. *J. Altern. Complement. Med., 17*(9), 827–833.
48. Fukushima, A., Iwasa, M., Nakabayashi, R., Kobayashi, M., Nishizawa, T., Okazaki, Y., Saito, K., & Kusano, M., (2017). Effects of combined low glutathione with mild oxidative and low phosphorus stress on the metabolism of Arabidopsis thaliana. *Front. Plant Sci., 8*, 1464.
49. Cameron, E., & Pauling, L., (1976). Supplemental ascorbate in the supportive treatment of cancer: Prolongation of survival times in terminal human cancer. *Proc. Natl. Acad. Sci. USA, 73*, 3685–3689.
50. Uetaki, M., Tabata, S., Nakasuka, F., Soga, T., & Tomita, M. (2015). Metabolomic alterations in human cancer cells by vitamin C-induced oxidative stress. *Sci. Rep., 5*, 13896.
51. Verrax, J., & Calderon, P. B., (2009). Pharmacologic concentrations of ascorbate are achieved by parenteral administration and exhibit antitumoral effects. *Free Rad. Biol. Med., 47*, 32–40.
52. Valachová, K., Kogan, G., Gemeiner, P., & Šoltés, L., (2008). Hyaluronan degradation by ascorbate: Protective effects of manganese(II) chloride. *Cellulose Chem. Technol., 42*(9/10), 473–483.

53. Baňasová, M., Valachová, K., Rychlý, J., Janigová, I., Csomorová, K., Mendichi, R., Mislovičová, D., Juránek, I., & Šoltés, L., (2014). Effect of bucillamine on free-radical-mediated degradation of high-molar-mass hyaluronan induced *in vitro* by ascorbic acid and Cu(II) ions. *Polymers, 6*, 2625–2644.
54. Campo, G. M., D'Ascola, A., Avenoso, A., Campo, S., Ferlazzo, A. M., Micali, C., Zanghi, L., & Calatroni, A., (2004). Glycosaminoglycans reduce oxidative damage induced by copper (Cu^{+2}), iron (Fe^{+2}) and hydrogen peroxide (H_2O_2) in human fibroblast cultures. *Glycoconjugate J., 20*, 133–141.

CHAPTER 5

Environmentally Friendly Approach: Synthesis and Biological Evaluation of α-Aminophosphonate Derivatives

SATISH A. DAKE

Department of Engineering Chemistry, Deogiri Institute of Engineering and Management Studies, Station Road, Aurangabad–431005, Maharashtra, India, E-mail: satish_dake57@yahoo.com

ABSTRACT

Synthesis of α-aminophosphonates and their derivatives existing vital role in medicinal and pharmaceutical fields. Green chemistry has become a powerful tool in research field of chemistry and recent areas of interest are the synthesis of bioactive organic molecules *via* multi-component reactions with environmental friendly approach. Multi-component reactions are contribute remarkable advantages such as nontoxic and environmental friendly, atom economy and cost effective. The present chapter expresses various synthetic methodology of α-aminophosphonate derivatives and bioactivity studies of α-aminophosphonates such as antiproliferative, antifungal, anti-allergy agents, anti-inflammatory, antibacterial, inhibitors of HIV protease, potent antibiotics, anti-cancer etc.

5.1 INTRODUCTION

Synthesis of α-aminophosphonates and their derivatives offered important biological and pharmaceutical applications. In the last decade, Green chemistry has become a powerful tool in research field of chemistry and recent areas of interest are the synthesis of bioactive organic molecules *via* multi-component reactions (MCRs) with environmental friendly approach. MCRs methodologies contribute remarkable advantages such as short reaction time, relatively nontoxic and environmentally friendly, atom economy, easy workup procedure, and cost-effective, the catalyst offers other advantages such as greater substrate compatibility, high reaction yields, recyclable, reusable, and the ability to tolerate functional groups, etc. The present chapter expresses various synthetic methodologies of α-aminophosphonate derivatives and bioactivity studies of α-aminophosphonates.

Their synthesis has been received considerable interest because of structural similarity to α-amino acids [1]. They exhibit a versatile range of biological activities such as peptide mimetics, [2] enzyme inhibitors, [3] inhibitors of HIV protease, PTPases, and EPSP synthase, [4] herbicides, [5] fungicides or plant growth regulators, [6] potent antibiotics [7]. The applications of α-aminophosphonates have a wide range from agricultural to medicinal fields as anti-cancer agents [8].

Thus, the efforts are being made in the development of new methods for their easy synthesis [9]. There has been an increasing influence of green approach to medicinal chemistry and research-based chemistry organizations [10]. It is necessary to maintain eco-friendly synthetic pathways by preventing waste generation, avoiding the use of auxiliary substances (e.g., solvents, additional reagents) and minimizing the energy requirements [11]. This needs for the development of convenient and high-yielding synthetic methods [12].

Although some significant advances have been made in reported synthesis, still it has certain limitations such as the use of hazardous solvents, expensive, and toxic catalyst, longer reaction time, and elevated temperature, etc. In many cases, two steps protocol was employed; which increases the reaction times and reduced yield with the purity of the product.

In recent years, solvent-free reactions have gained considerable attention because; these methods are not only valuable for ecological and economical reasons but also for simplicity in procedures with a high yield of products. It emphasized towards the development of clean and green chemical processes and investigations of new and less hazardous catalysts. The reaction modifications concentrate on finding recyclable, reusable solvents such as using ethyl ammonium nitrate ($EtNH_3$) ionic liquid (IL).

We previously reported the use of $EtNH_3$ IL as an excellent catalyst and solvent for the three-component one-pot reaction of an aldehydes, amines, and diethylphosphite to form novel α-aminophosphonates at room temperature (RT). Among the various catalysts, $EtNH_3$ is an environmental friendly, cost-effective, and recyclable catalyst.

ILs are used as green reaction media due to their unique chemical and physical properties such as non-volatility, non-inflammability, thermal stability, and controlled miscibility. They are playing a significant role in reactions as a solvent/catalyst.

5.2 REVIEW OF LITERATURE

5.2.1 BIOLOGICAL IMPORTANCE OF A-AMINOPHOSPHONATE DERIVATIVES

Awad et al. [13] was described molecular docking, molecular modeling, vibrational, and biological studies of some new heterocyclic α-aminophosphonates. These kinds of interactions are probably responsible for the stabilization of a protein inhibitor complex.

Zhu et al. [14] have been designed and synthesized by introducing bioactive quinoline scaffold to *α*-aminophosphonate. The *in vitro* cytotoxic activities of target compounds were first investigated against two human cancer cell lines including Eca109 and Huh7 by MTT assay. Two series of *α*-aminophosphonate derivatives containing a quinoline moiety among them, compounds (1) and (2) which are containing methyl-substituted aniline group were found to be more active against both of two cancer cell lines, with IC_{50} in the range of 2.26 μmol/L–7.46 μmol/L.

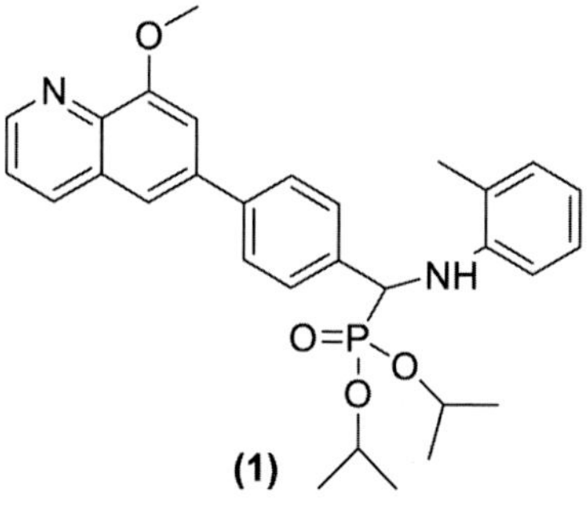

N O NH O=P O O (1)

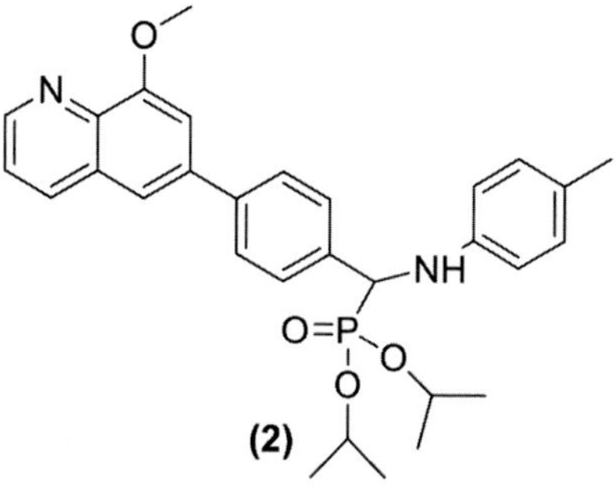

N O NH O=P O O (2)

Chen et al. [15] was described antifungal activity of dehydroabietic acid-based thiadiazole-phosphonate compounds. The target compounds was exhibited excellent antifungal activity against the five fungi tested, in which several compounds displayed even better antifungal effects than the commercial antifungal drug azoxystrobin.

El-Refaie S. Kenawy et al. [16] was described synthesis of α-aminophosphonates containing chitosan moiety. Some compounds are showed high antimicrobial activities against *Escherichia coli* (NCIM2065), *Serratia marcescens*, *Enterobacter cloacae*, *Shigella dysenteriae*, *Salmonella enterica*, and *Proteus vulgaris* as Gram-negative bacteria, *Bacillus subtilis* (PC1219) and *Staphylococcus aureus* (ATCC25292) as Gram-positive bacteria and Candida albicans as a fungus, at low concentrations (2.5–10 mg/mL).

Yang, Jiaqiang et al. [17] were designed and synthesized of phosphonate derivatives bearing quinolinone moiety. The antibacterial activities of the products against *S. aureus*, *E. coli*, drug-resistant *S. aureus* were evaluated by the agar dilution method. The results demonstrated that some compounds exhibited superior activities against drug-resistant *S. aureus* compared with norfloxacin.

Bhagat et al. [18] have been synthesized twenty six structurally diverse α-aminophosphonates (3) and evaluated for in vitro anti-leishmanial activity and cytotoxicity using the MTT assay. Among them, some compounds exhibited anti-leishmanial potency against the L. donovani promastigote with IC_{50} values in the low micromolar range. The structure-activity relationships were quantitatively evaluated by a statistically reliable CoMFA model with high predictive abilities.

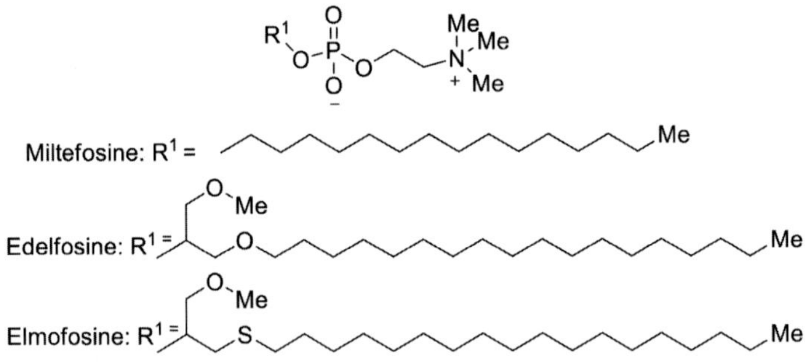

Miltefosine: R^1 =

Edelfosine: R^1 =

Elmofosine: R^1 =

(3)

Nature has chosen phosphates as essential chemicals in many critically important biological processes and materials. They are so versatile and fundamentally important in the chemistry of living systems, in many ways, that it would be difficult to imagine any other chemical types would be able to meet the manifold demands of living systems [19]. The phosphate group is present in (i) intermediates of important biochemical pathways (e.g., sugar phosphates, isopentenyl pyrophosphate), in (ii) structural elements of the cell (e.g., DNA, phospholipids, and protein phosphates), in (iii) the energy management of the cell (e.g., ATP, phosphoenol pyruvate (PEP)) and in (iv) messenger molecules (e.g., myoinositol triphosphate, cAMP). Their phosphonate counterparts are found far less widespread in living organisms [20].

More recently, phosphonates have also found application as antigens for the preparation of catalytic antibodies, because of their excellent transition state analogy with the hydrolysis of amino acids. The mammalian immune system is capable of synthesizing large folded polypeptides (immunoglobulins or antibodies) that bind virtually any natural or synthetic molecule with high affinity and exquisite selectivity. Being carboxylate hydrolysis transition state analogues, phosphonates (4) can be used as antigens to generate specific antibodies that catalyze the hydrolysis of the parent carboxylate (5) [21].

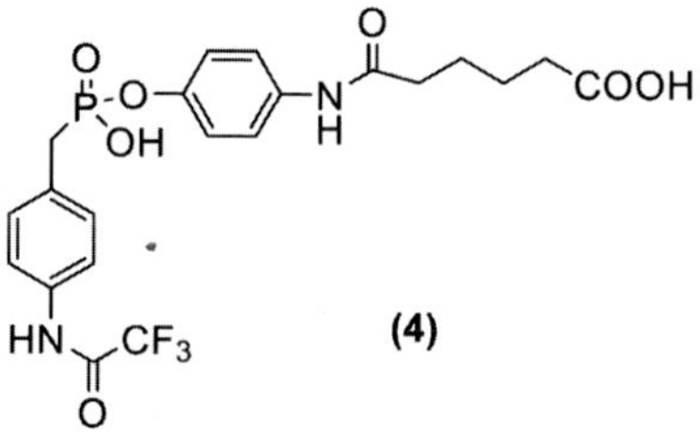

(4)

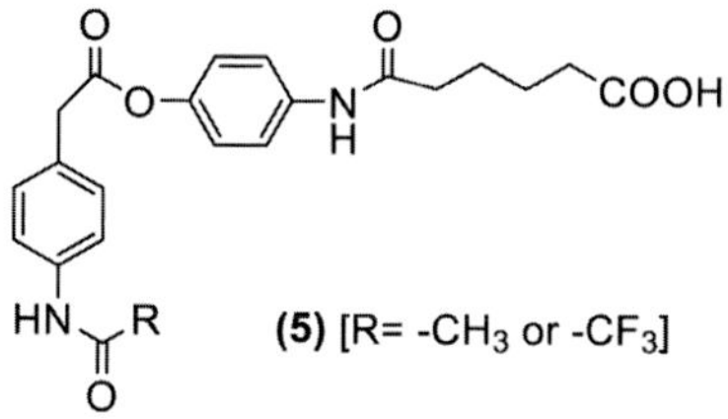

(5) [R= $-CH_3$ or $-CF_3$]

Ciliatine or 2-aminoethane phosphonic acid (6) was the first one to be discovered [22]. Even more exciting was the discovery of naturally occurring phosphonates possessing remarkable biological activity despite their relatively simple structural features, as for instance the antibacterial fosfomycin, [23] (7) and the antimalarial fosmidomycin, [24] (8), isolated from fermentation broths of Streptomyces species. These and other examples caused organophosphorus chemistry to achieve an important and well-recognized place in the search for new drugs [25]. The biological potential of phosphonic acid derivatives may arise from several rationales. Firstly, the methylene group attached to phosphorus is isosteric with oxygen of a phosphate. However, the high stability of the C-P bond would block any natural process involving hydrolysis of a phosphate group. This can be applied in the development of antiviral agents based on naturally occurring nucleotides [26].

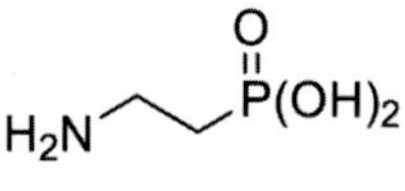

Ciliatine **(6)** 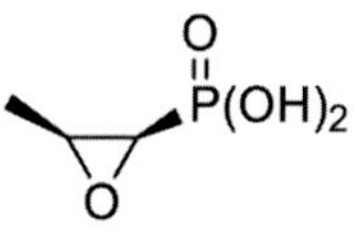 Fosfomycin **(7)** 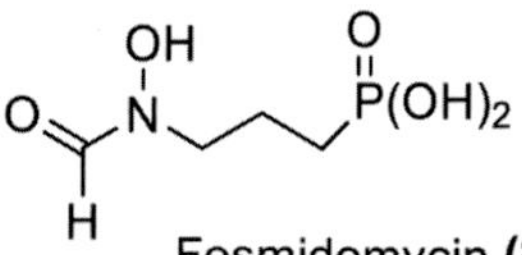 Fosmidomycin **(8)**

Application of the same principle resulted in a major breakthrough in the treatment of bone diseases such as osteoporosis and Paget's disease [27]. Their activity additionally results from the second interesting property of the phosphonate group, namely its ability to complex divalent cations such as Ca^{2+}. Amino acids (amino phosphonates), since enzymes act as catalysts because they are more closely complementary to the transition states than to the substrates or products of a reaction. Phosphono peptides emerge as inhibitors of a wide range of enzymes, [28] such as elastase (9), [29] dipeptidyl dipeptidase IV (10), [30] thrombin (11), [31]. HIV-protease (12), [32] etc. Good results are also obtained with metalloproteases (such ascarboxypeptidase A (13), [33] because of the better recognition of the tetrahedral transition state analog together with the Zn^{2+} complexing properties of the phosphonic acid [34].

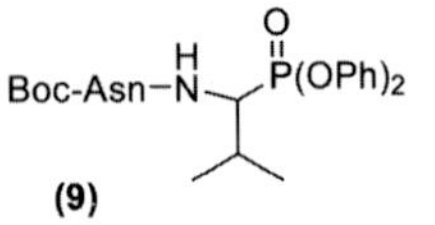

Boc-Asn–NH–P(O)(OPh)₂ (9)

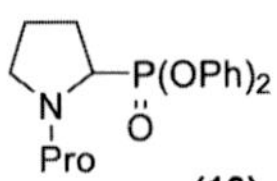

N-Pro P(O)(OPh)₂ (10)

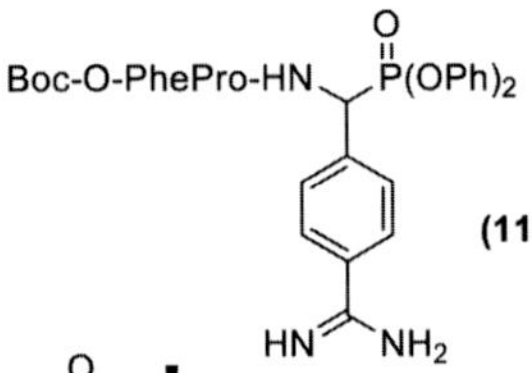

Boc-O-PhePro-HN P(O)(OPh)₂ HN NH₂ (11)

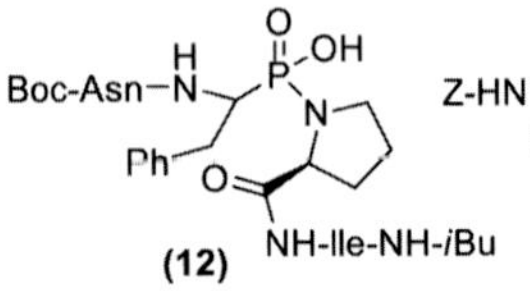

Boc-Asn–NH P(O)OH N Ph O NH-Ile-NH-*i*Bu (12)

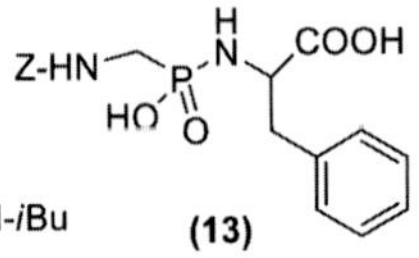

Z-HN P HO O NH COOH (13)

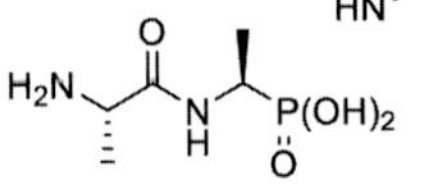

H₂N N H P(OH)₂ O (14)

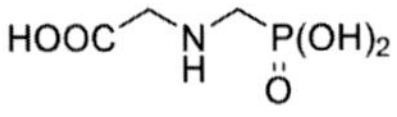

HOOC N H P(OH)₂ O (15)

Phosphonic acid-containing enzyme inhibitors such as alafosfalin (14) which is readily transported into the bacterial cell and hydrolyzed subsequently. The resulting phosphono alanine is a good inhibitor of the alanine racemase, which is essential to the bacterial cell wall synthesis [35]. Phosphonomethyl glycine (glyphosate) (15) on the other hand was developed by Monsanto as an environment-friendly total herbicide with very low mammalian toxicity. Glyphosate selectively inhibits 5-enolpyruvoylshikimate 3-phosphate synthase (EPSPS), blocking the biosynthesis of aromatic plant metabolites, including the aromatic amino acids [36]. Apart from their enzyme inhibitory activity, also neuroactive aminophosphonic acid derivatives have been found [37]. Phosphonylated analogs of glutamate, the most important excitatory neurotransmitter in the mammalian CNS, act as specific agonists or antagonists of certain glutamate receptor subtypes. Furthermore, a considerable enhancement of antagonist potency was achieved by synthesizing conformationally restricted analogs of AP5 (16). This can be performed by incorporating the amino group into a heterocyclic ring system [CPP (17), CGS19755 (18)] [38]. This illustrates the potential of azaheterocyclic phosphonates. Because of the success of constrained amino acids in drug design and biomechanistic investigation of receptor-bound ligand conformations, [39] the corresponding amino phosphonates deserve appropriate attention.

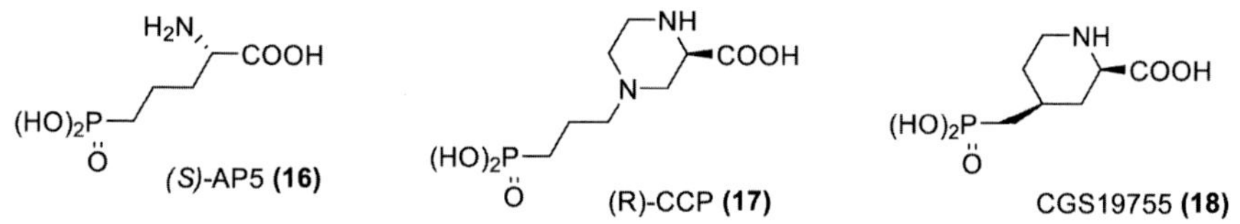

(S)-AP5 **(16)** (R)-CCP **(17)** CGS19755 **(18)**

5.2.2 SYNTHESIS OF A-AMINOPHOSPHONATES USING ENVIRONMENTAL FRIENDLY APPROACH

In recent years, the pollution is serious issue in front of human being therefore, scientist are tried to developed synthesis of drug molecules and various synthetic methodology using Green Chemistry approach. Such as solvent-free reactions, using aqueous medium, ILs, etc. It emphasized towards the development of clean and green chemical processes and investigations of new and less hazardous catalysts.

Recently, various synthetic protocols have been described using green protocol. Rasal's approach [40] was synthesized α-aminophosphonates using zinc oxide nanoflowers (ZnO NFs) as catalyst under solvent-free ultrasonication condition enhances the yield of α-aminophosphonate derivatives. ZnO NFs catalyst was recycled and reused without any significant loss of catalytic activity. This methodology becomes attractive advantages such as eco-friendly fast, cost-effective, clean, easy procedure (Scheme 5.1).

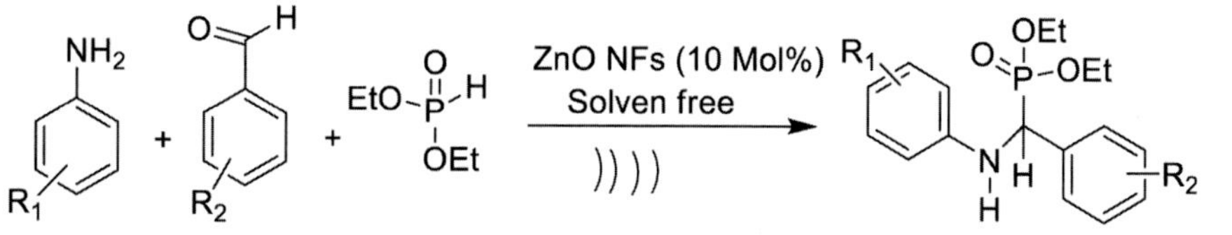

ZnO NFs (10 Mol%) Solven free))))

SCHEME 5.1 Synthesis of alpha-aminophosphonates using zinc oxide nanoflowers (ZnO NFs).

Deshmukh's approach [41] was reported synthesis of novel fluorine containing α-aminophosphonate derivatives (Scheme 5.2). Novel Fluorinated α-aminophosphonate compounds were screened for antiproliferative and apoptosis activity on human nonsmall cell lung carcinoma cells (A549) and human skin melanoma cells (SK-MEL-2).

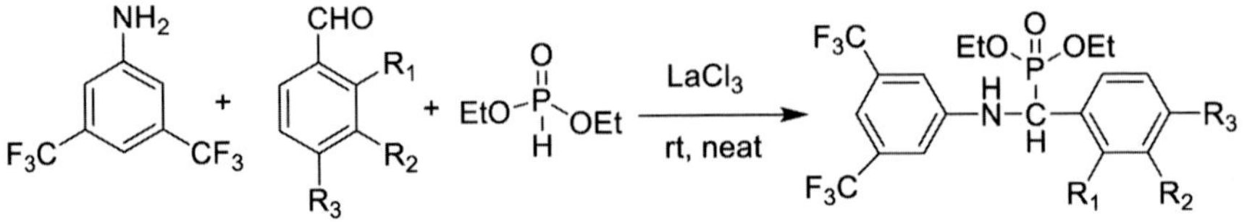

SCHEME 5.2 Synthesis of alpha-aminophosphonates using catalyst $LaCl_3$.

Isca's approach [42] was described the conversion of xylose and xylan into large variety of α-aminophosphonates with excellent yields and good chemoselectivity catalyzed by $HReO_4$ at 140°C (Scheme 5.3).

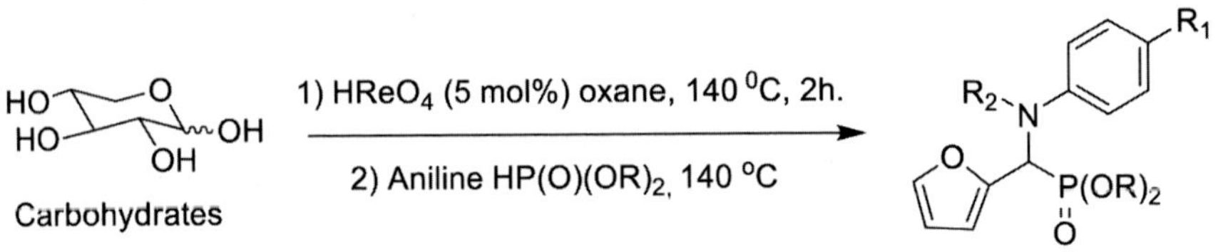

SCHEME 5.3 Synthesis of alpha-aminophosphonates using catalyst $HReO_4$.

Balint's approach [43] were reported the synthesis and utilization of optically active α-aminophosphonate derivatives by Kabachnik-Fields reaction in the microwave (MW)-assisted condensation with paraformaldehyde and >P(O)H species, such as dialkyl phosphites, ethyl phenyl-H-phosphinate and diphenylphosphine oxide to provide optically active α-aminophosphonates, α-aminophosphinate and α-amino phosphine oxide, or using a two equivalent quantity of paraformaldehyde and $(RO)_2P(O)H$ or $Ph_2P(O)H$, bis (phosphonomethyl)amines and bis(phosphinoylmethyl) amine, respectively. The bis(phosphinoylmethyl)amine served as a precursor for an optically active bidentate P-ligand in the synthesis of a chiral platinum complex (Scheme 5.4).

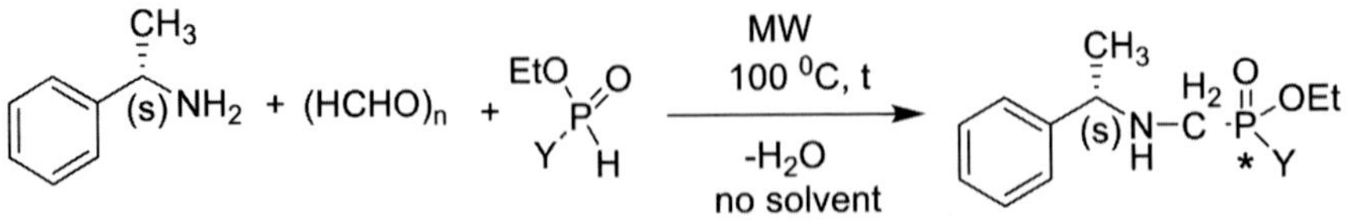

SCHEME 5.4 Synthesis of alpha-aminophosphonates in the MW-assisted condensation.

Mohammadiyan's approach [44] was synthesized α-aminophosphonates in one-pot, the three-component reaction by using aqueous formic acid (37%) as a green organocatalyst at 65°C under solvent-free condition (Scheme 5.5).

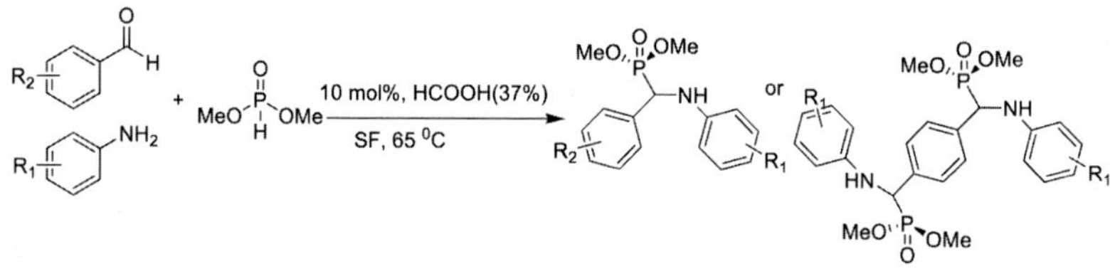

SCHEME 5.5 Synthesis of alpha-aminophosphonates under solvent free condition.

Taheri-Torbati's approach [45] was synthesis of α-aminophosphonates by using 2-aminobenzothiazoles, aromatic aldehydes and trimethyl phosphite in the presence of NS-CSs as microorganocatalyst nitrogen-sulfur-doped carbon spheres (NSCSs) at 50°C under solvent-free condition (Scheme 5.6).

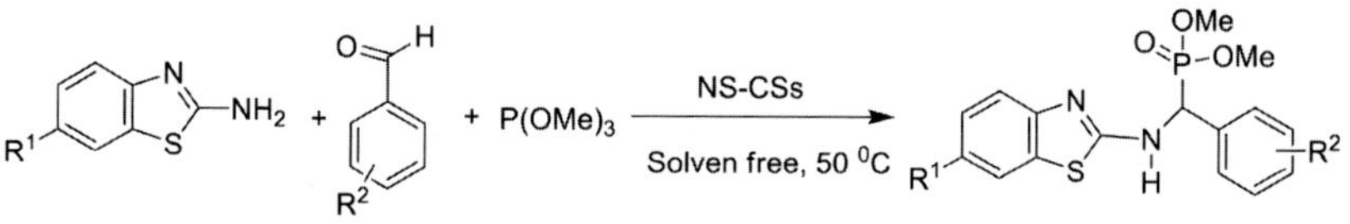

SCHEME 5.6 Synthesis of alpha-aminophosphonates by using nitrogen-sulfur-doped carbon spheres.

Ghafuri's approach [46] were reported the preparation of nanomagnetic sulfated zirconia ($Fe_3O_4@ZrO_2/SO_4^{2-}$) and characterized by various instrumental methods. It was applied as an efficient nanocatalyst in the synthesis of α-aminophosphonate derivatives in Kabachnik-Fields reaction. Nano ($Fe_3O_4@ZrO_2/SO_4^{2-}$) is heterogeneous acidic catalyst. This synthetic method has several advantages including high yields, short reaction times, easy workup, and environmentally benign reaction conditions (Scheme 5.7).

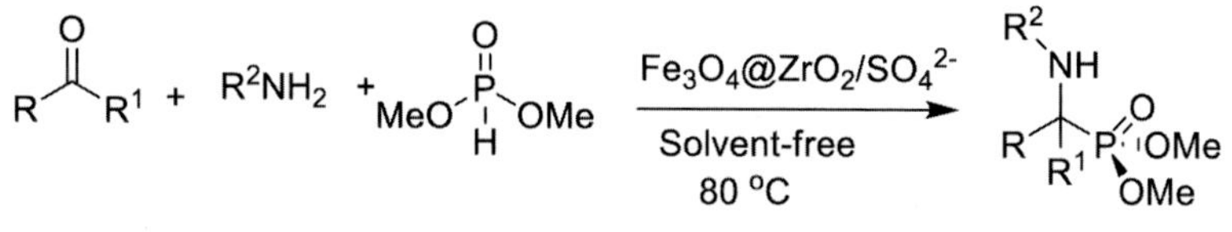

SCHEME 5.7 Synthesis of alpha-aminophosphonates by using $Fe_3O_4@ZrO_2/SO_4^{2-}$.

Radai's approach [47] has been accomplished in a new, environmentally friendly way to afford the corresponding α-hydroxyphosphonates by using

Pudovik reaction between substituted benzaldehydes and dialkyl phosphites (Scheme 5.8).

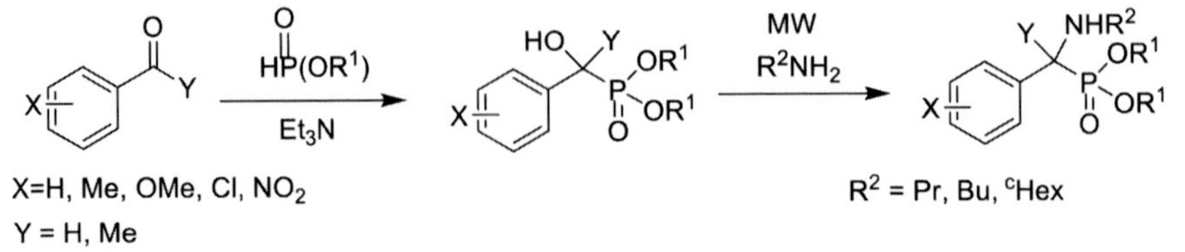

X=H, Me, OMe, Cl, NO_2
Y = H, Me
R^2 = Pr, Bu, ^{c}Hex

SCHEME 5.8 Synthesis of alpha-hydroxyphosphonates by using Pudovik reaction.

Hellal's approach [48] was described convenient and high yielding procedure for the synthesis of diethyl α-aminophosphonates in water by using aromatic aldehydes, aminophenols, and dialkyl phosphites in the presence of a low catalytic amount (10 mol%) of citric, malic, tartaric, and oxalic acids as a catalyst (Scheme 5.9).

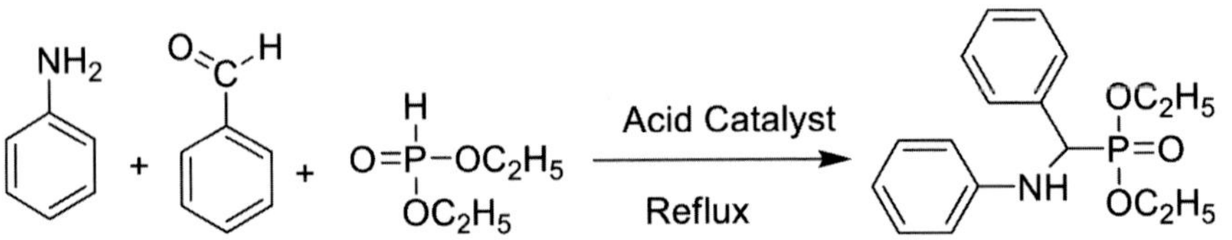

SCHEME 5.9 Synthesis of diethyl alpha-aminophosphonates in water by acids catalyst.

Shaikh's approach [49] were reported series of dimethyl(phenyl (phenylamino) methyl)-phosphonates and novel dimethyl ((phenylamino)(2-(prop-2-yn-1-yloxy) phenyl)methyl)-phosphonates by using $[Et_3NH][HSO_4]$ as an efficient, eco-friendly, and reusable catalyst with aldehydes, amines, and trimethyl phosphite in solvent-free conditions (Scheme 5.10).

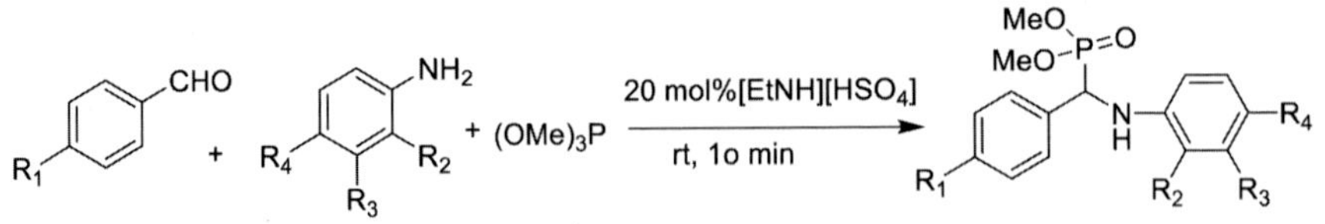

SCHEME 5.10 Synthesis of alpha-aminophosphonates by using $[Et_3NH][HSO_4]$ catalyst.

Mirzaei's approach [50] was designed, synthesized sulfonic acid functionalized IL and used as a Bronsted acid catalyst for the synthesis of

α-aminophosphonates containing benzothiazole at RT under solvent-free conditions (Scheme 5.11).

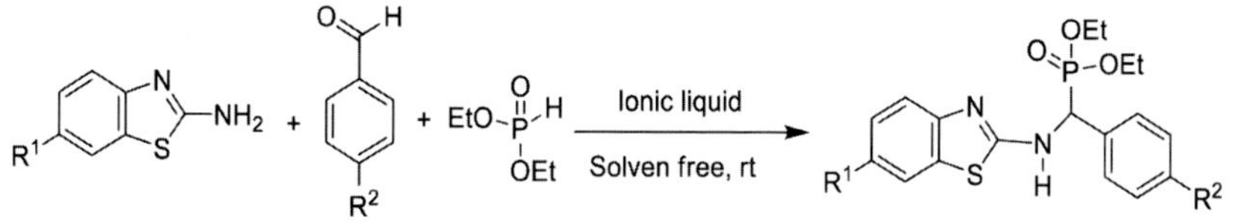

O H
N
R^1 S NH_2 + + EtO-P(O)-H OEt
R^2
Ionic liquid
Solven free, rt
OEt
O=P-OEt
R^1 N S N H R^2

SCHEME 5.11 Synthesis of alpha-aminophosphonates using Bronsted acid catalyst at RT under solvent-free conditions.

Mirzaei's approach [51] were synthesized of α-aminophosphonates through condensation of aryl aldehydes, aryl amines and diethyl phosphite in the absence of catalyst and organic solvents at 50°C temperature (Scheme 5.12).

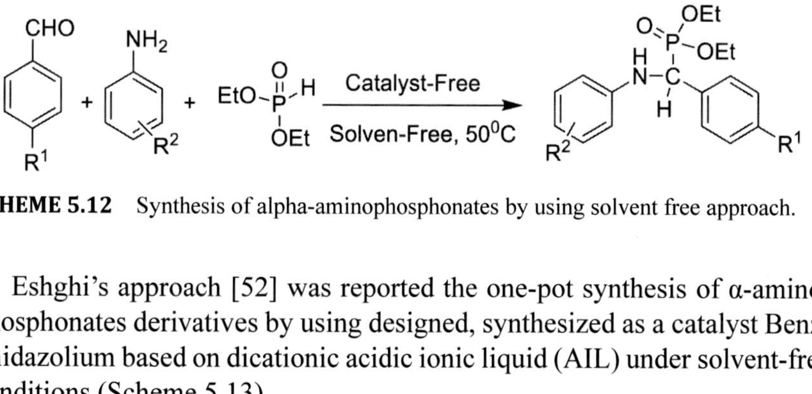

CHO
R^1
+ NH_2 R^2
+ EtO-P(O)-H OEt
Catalyst-Free
Solven-Free, 50^0C
O OEt
P-OEt
H N-C H
R^2 R^1

SCHEME 5.12 Synthesis of alpha-aminophosphonates by using solvent free approach.

Eshghi's approach [52] was reported the one-pot synthesis of α-amino-phosphonates derivatives by using designed, synthesized as a catalyst Benz-imidazolium based on dicationic acidic ionic liquid (AIL) under solvent-free conditions (Scheme 5.13).

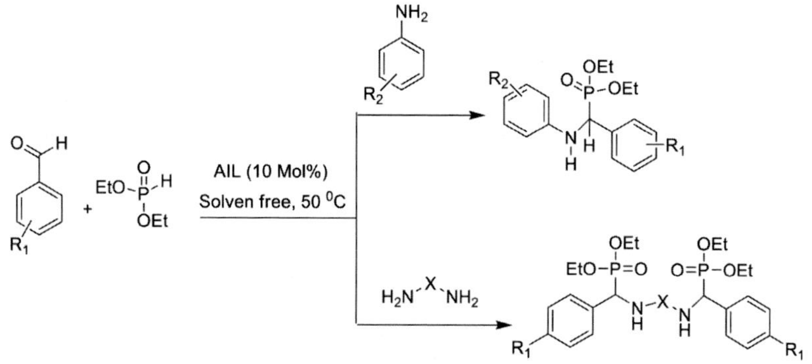

O H
R_1
+ EtO-P(O)-H OEt
AIL (10 Mol%)
Solven free, 50 ^0C
NH_2
R_2
R_2 OEt O=P-OEt
N H H R_1
H_2N-X-NH_2
OEt OEt
EtO-P=O O=P-OEt
R_1 N-X-N H H R_1

SCHEME 5.13 Synthesis of alpha-aminophosphonates by benzimidazolium based on dicationic acidic ionic liquid (AIL).

Yu's approach [53] has been developed a multi-component one-pot reaction of aldehydes or ketones with amines and diethyl or triethyl phosphite to give the α-aminophosphonates in the presence of quaternary ammonium salts which are environmental friendly, inexpensive, and recyclable catalyst under RT (Scheme 5.14).

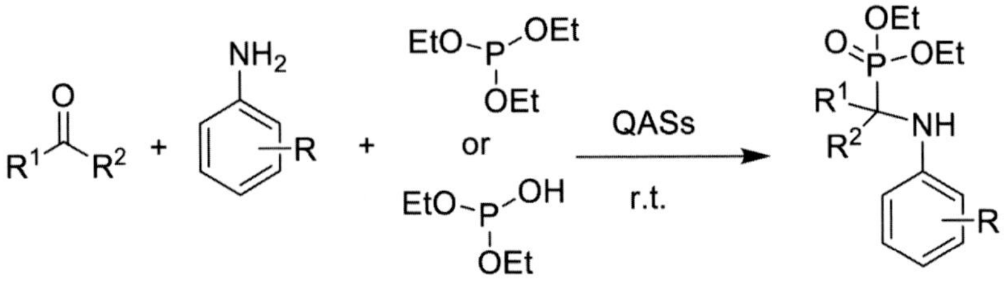

SCHEME 5.14 Synthesis of alpha-aminophosphonates in the presence of quaternary ammonium salts.

Peng's approach [54] were reported the synthesis of α-aminophosphonates *via* a one pot three-component reaction with aldehydes, amines, and triethyl phosphite/diethyl phosphite at RT under solvent-free conditions or in aqueous media by using prepared sulfated choline-based IL [Ch-OSO_3H] (Scheme 5.15).

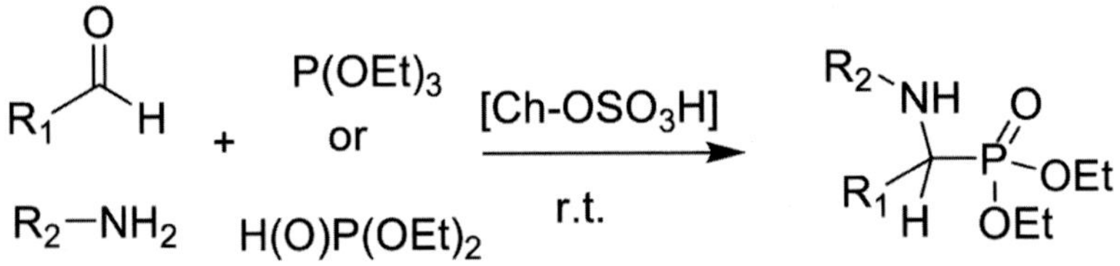

SCHEME 5.15 Synthesis of alpha-aminophosphonates using [Ch-OSO_3H] ionic liquid.

Shakourian-Fard's approach [55] were described a highly efficient magnetic Brønsted acid catalyst was synthesized based on immobilization of 2-methylimidazole functionalized by chlorosulfonic acid on the surface of silica-coated magnetic nanoparticles (NPs) (Fe_3O_4@SiO_2-2mimSO_3H). The catalyst has an excellent activity and recyclable for the synthesis of α-aminophosphonates in a green approach (Scheme 5.16).

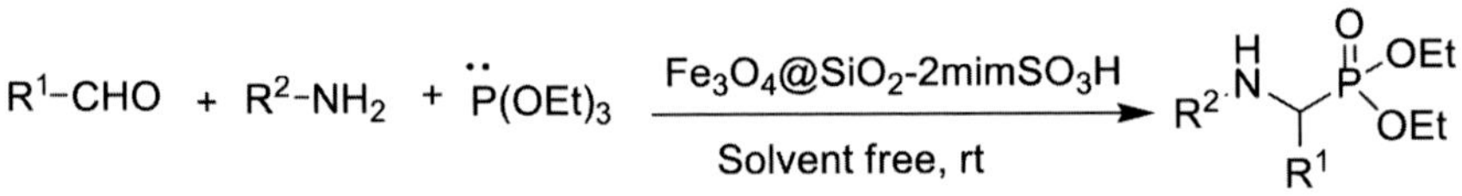

SCHEME 5.16 Synthesis of alpha-aminophosphonates using silica-coated magnetic nanoparticles (Fe_3O_4@SiO_2-2mimSO_3H).

Guna Subba Reddy's approach [56] has been developed environmentally benign method to afford α-aminophosphonates using MW irradiation reaction with an amine, an aldehyde and diethyl phosphite catalyzed by Amberlyst-15 under solvent-free conditions (Scheme 5.17).

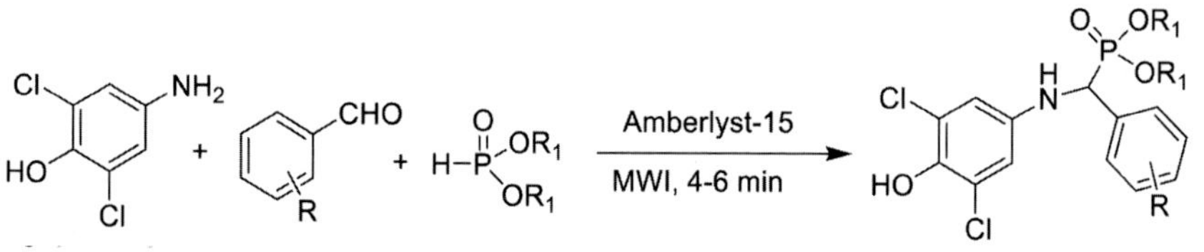

SCHEME 5.17 Synthesis of alpha-aminophosphonates catalyzed by using Amberlyst-15.

Sundar's approach [57] have been synthesized new series of α-aminophosphonates under solvent-free conditions by using 2,3-dihydrobenzo[b][1,4] dioxine-6-carbaldehyde, amines, and dimethyl phosphite in the presence of nano-TiO_2 as a catalyst at 50°C (Scheme 5.18).

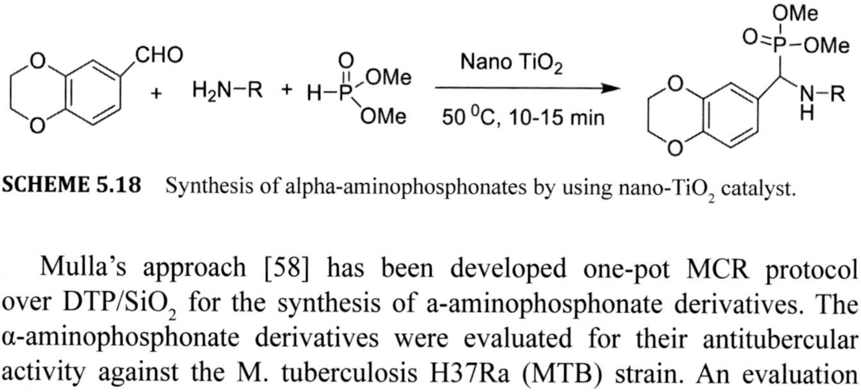

SCHEME 5.18 Synthesis of alpha-aminophosphonates by using nano-TiO_2 catalyst.

Mulla's approach [58] has been developed one-pot MCR protocol over DTP/SiO_2 for the synthesis of a-aminophosphonate derivatives. The α-aminophosphonate derivatives were evaluated for their antitubercular activity against the M. tuberculosis H37Ra (MTB) strain. An evaluation of the data on the cytotoxicity and antimicrobial activity shows promising antitubercular agents (Scheme 5.19).

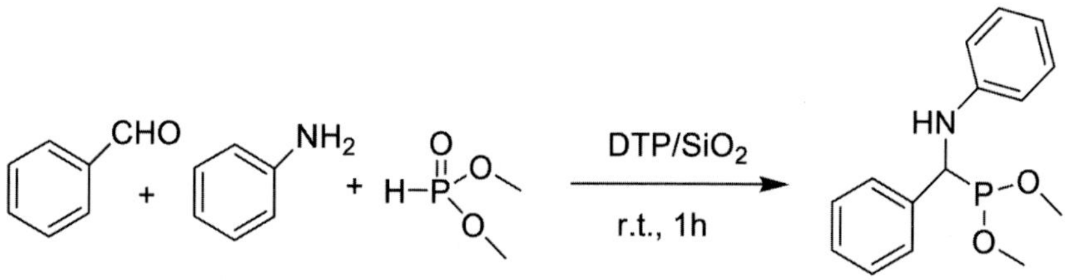

SCHEME 5.19 Synthesis of alpha-aminophosphonates by using DTP/SiO_2 catalyst.

Tekale's approach [59] was reported green and atom-efficient one-pot protocol for synthesis of some α-aminophosphonates using micron-particulate AlN/Al as a new reusable heterogeneous catalyst by the Kabachnik-Fields reaction under solvent-free conditions (Scheme 5.20).

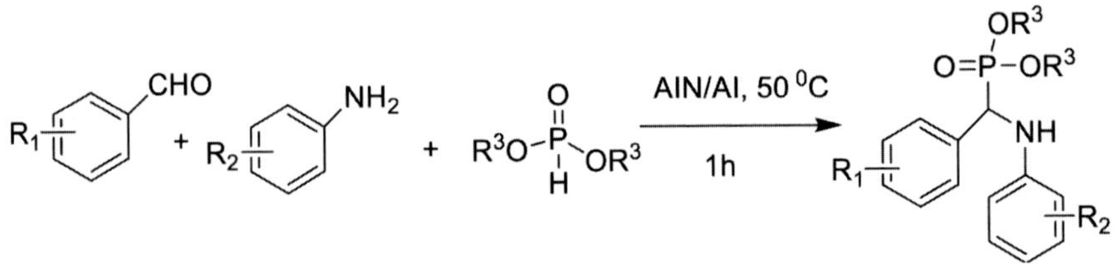

SCHEME 5.20 Synthesis of alpha-aminophosphonates using micron-particulate AIN/AI catalyst.

Ya-Qin Yu's approach [60] has been developed a novel and simple approach to the multi-component one-pot reaction of aldehydes, diethyl phosphite, and azides to form α-aminophosphonates under solvent-free conditions at RT. In the presence of iodine and iron, aryl azides were, for the first time, used as substrates for the synthesis of α-aminophosphonates (Scheme 5.21).

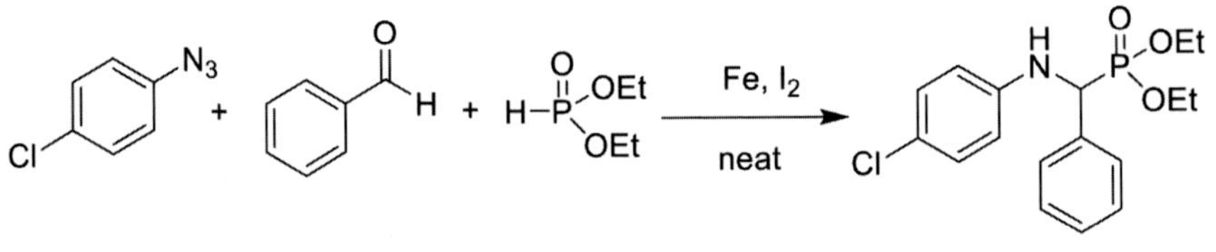

SCHEME 5.21 Synthesis of alpha-aminophosphonates under solvent-free conditions.

Shaterian's approach [61] has been found to be an excellent Nano-TiO_2 catalyst for the green synthesis of α-aminophosphonates under ambient and solvent-free conditions. The synthesis of α-aminophosphonates from arylaldehydes, anilines, and triethylphosphite by employing TiO_2 NPs as a catalyst is described. This green and mild nano catalytic procedure showed good recyclability and provide cleaner conversion in a short reaction time. These advantages make the protocol feasible and economically attractive for researchers (Scheme 5.22).

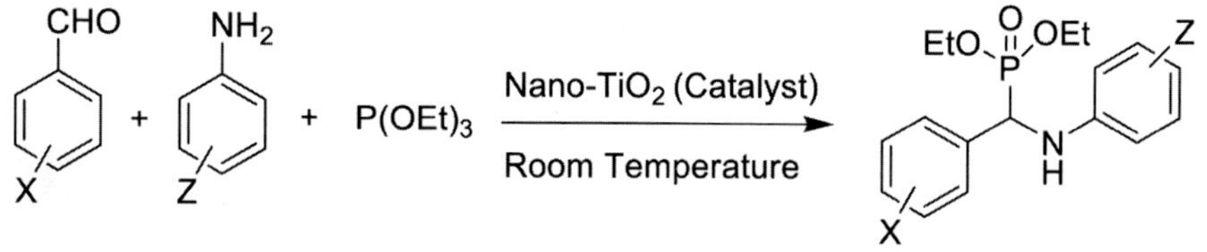

SCHEME 5.22 Synthesis of alpha-aminophosphonates using nano-TiO_2 catalyst.

Fang's approach [62] was prepared a biodegradable SO_3H-functionalized IL and used as the catalyst for the synthesis of α-aminophosphonates from aldehydes, amines, and triethylphosphite/diethylphosphite at RT under solvent-free conditions or in aqueous media. The products could be simply separated from the reaction mixture and the catalyst could be recycled and reused for several times without noticeably reducing catalytic activity (Scheme 5.23).

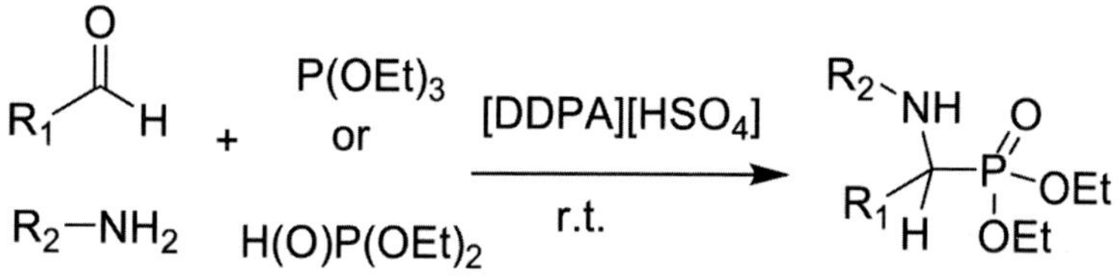

SCHEME 5.23 Synthesis of alpha-aminophosphonates using SO_3H-functionalized IL catalyst.

Sundar's approach [63] has been developed green one-pot three-component synthesis for α-aminophosphonates by condensation of aldehydes, amines, and diethylphosphite by using nonionic surfactant Tween-20 as catalyst in aqueous media. The major advantages of this novel method are green reaction conditions with water as solvent, simple workup, less reaction times, and high to moderate yields (Scheme 5.24).

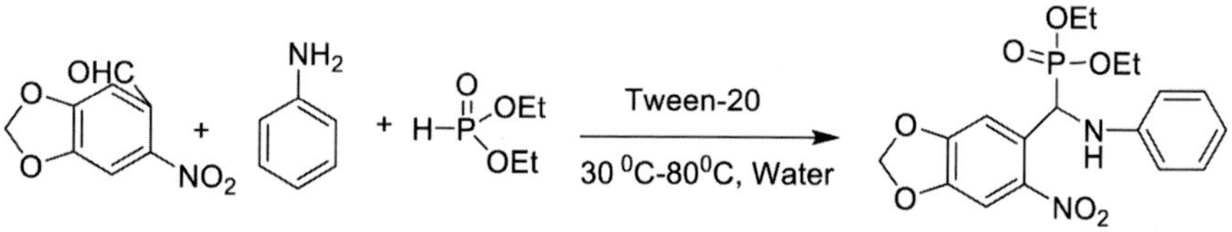

SCHEME 5.24 Synthesis of alpha-aminophosphonates using tween-20 as catalyst in aqueous media.

Dake, approach [64] were reported IL $EtNH_3$ is used as an excellent catalyst and solvent for three-component one-pot reaction of an aldehydes, amines, and diethylphosphite to form novel α-aminophosphonates at RT. Among the various catalysts, the preparation of $EtNH_3$ is an environmental friendly, cost effective and recyclable catalyst (Scheme 5.25).

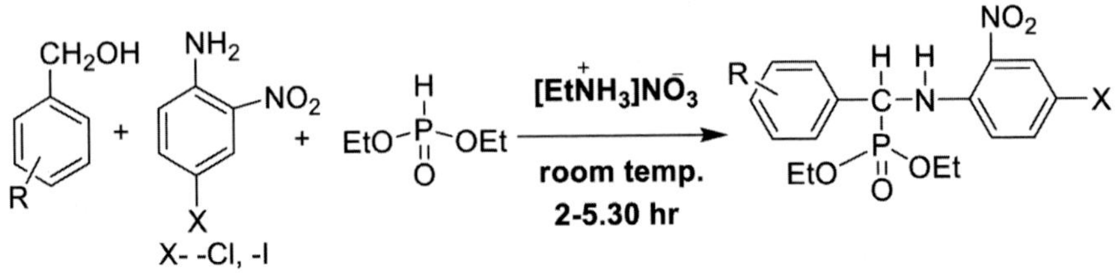

SCHEME 5.25 Synthesis of alpha-aminophosphonates using ionic liquid ethyl ammonium nitrate.

Possible mechanism of α-aminophosphonates: IL $EtNH_3$ involved in the formation of activated imines; so that the phosphite is facilitated to give a phosphonium intermediate which further reacts with water generated during imine formation to givc α-amino phosphonates (Scheme 5.26). In this process the $EtNH_3$ provides a proton for the formation of activated imine and behave as a protic solvent. The mechanistic details of these processes have been outlined as follows:

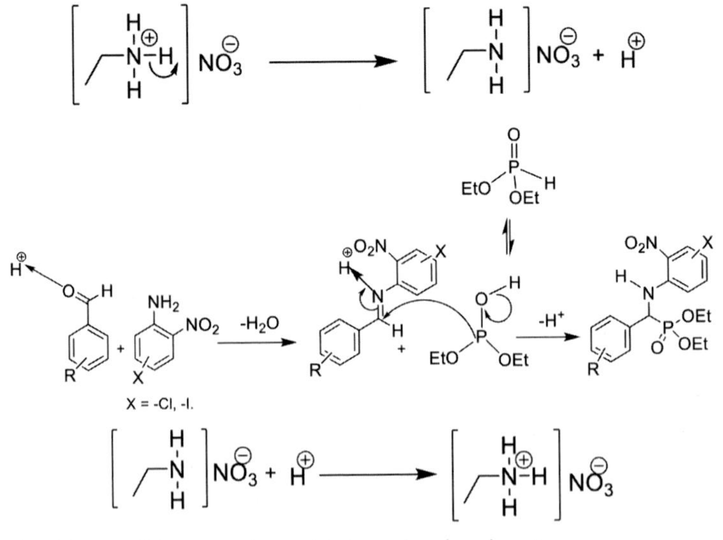

SCHEME 5.26 Possible mechanism of alpha-aminophosphonates.

Sobhani's approach [65] was achieved one-pot synthesis of α-aminophosphonates from aldehydes/ketones, HMDS, and diethyl phosphite using molecular iodine as a catalyst under solvent free conditions (Scheme 5.27).

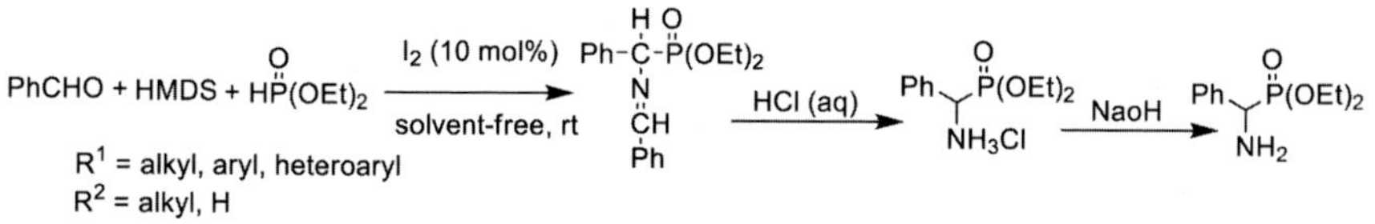

SCHEME 5.27 Synthesis of alpha-aminophosphonates using molecular iodine catalyst.

Jafari's approach [66]: $CeCl_3.7H_2O$ was used as a Lewis acid catalyst for the synthesis of α-aminophosphonates by three-component coupling of aldehydes, aromatic amines, and diethylphosphite under solvent-free conditions (Scheme 5.28).

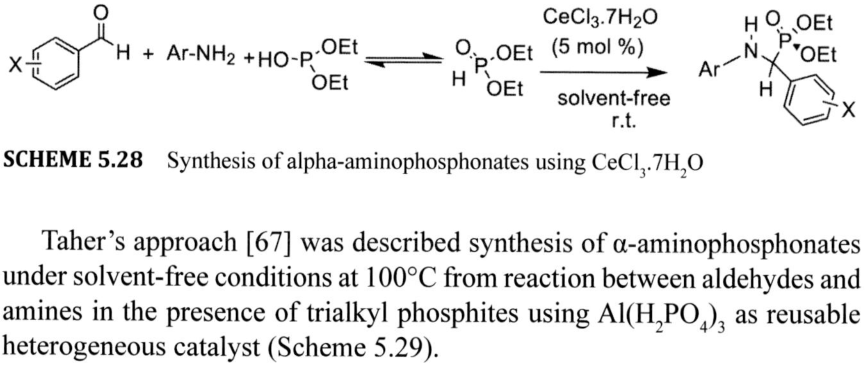

SCHEME 5.28 Synthesis of alpha-aminophosphonates using $CeCl_3.7H_2O$

Taher's approach [67] was described synthesis of α-aminophosphonates under solvent-free conditions at 100°C from reaction between aldehydes and amines in the presence of trialkyl phosphites using $Al(H_2PO_4)_3$ as reusable heterogeneous catalyst (Scheme 5.29).

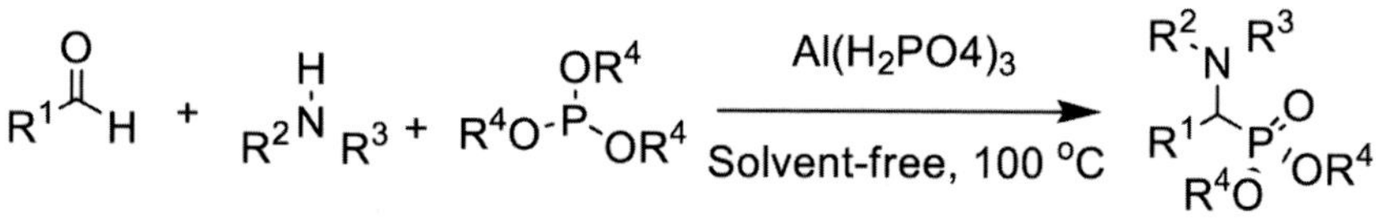

SCHEME 5.29 Synthesis of alpha-aminophosphonates using $Al(H_2PO_4)_3$ catalyst.

Akbari's approach [68] were described the one-pot, three-component synthesis of α-aminophosphonates from aldehydes and ketones at RT in water by using sulfonic acid functionalized IL as a Bronsted acid catalyst (Scheme 5.30).

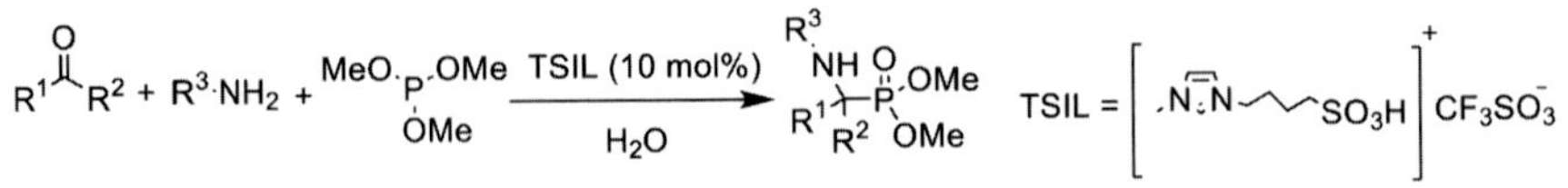

SCHEME 5.30 Synthesis of alpha-aminophosphonates using sulfonic acid functionalized ionic liquid.

Bhanushali's approach [69] was reported synthesis of α-aminophosphonates from the three-component reaction of an amine, an aldehyde, and diethyl phosphite catalyzed by $ZrOCl_2.8H_2O$ (Scheme 5.31).

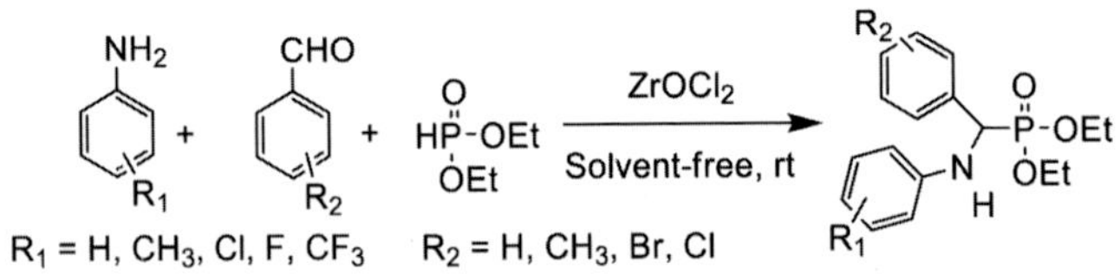

SCHEME 5.31 Synthesis of alpha-aminophosphonates by using catalyzed by $ZrOCl_2.8H_2O$

Kassaee's approach [70] were reported Zinc oxide nanoparticles (ZnO NPs, ca. 22 nm) was used as a catalyst in the solvent free, three component couplings of aldehydes, aromatic amines, dialkyl phosphites at RT to α-amino phosphonates.

Keglevich's approach [71] were expressed the condensation of the oxo-component, the amine and the >P(O)H species to afford α-aminophosphonates or phosphine oxides involves MW irradiation transformations (Scheme 5.32).

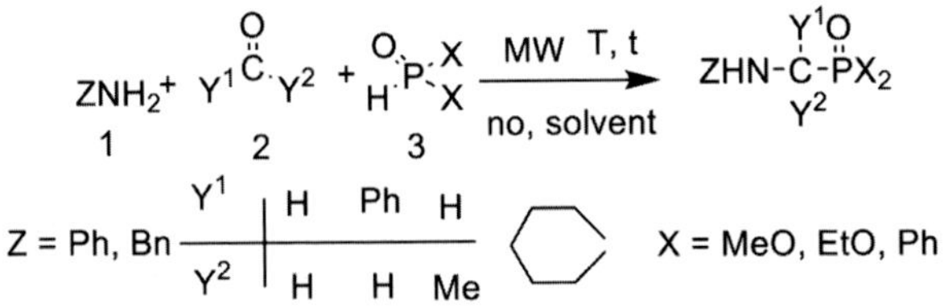

SCHEME 5.32 Synthesis of alpha-aminophosphonates by using MW irradiation transformations.

Mitragotri's approach [72] were reported sulfamic acid-catalyzed solvent-free protocol for the synthesis of α-aminophosphonates by three component condensation between aldehydes, amines, and diethyl phosphite at ambient temperature (Scheme 5.33).

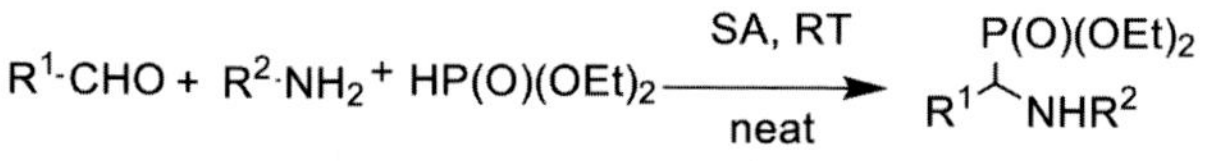

$$R^1\text{-CHO} + R^2\text{-NH}_2 + \text{HP(O)(OEt)}_2 \xrightarrow[\text{neat}]{\text{SA, RT}} R^1\text{CH(NHR}^2\text{)P(O)(OEt)}_2$$

SCHEME 5.33 Sulfamic acid catalyzed solvent-free synthesis of alpha-aminophosphonates.

Mohammad's approach [73] were described the synthesis of α-amino- and α-hydroxy phosphonates by using new and highly flexible procedure under solvent free condition (Scheme 5.34).

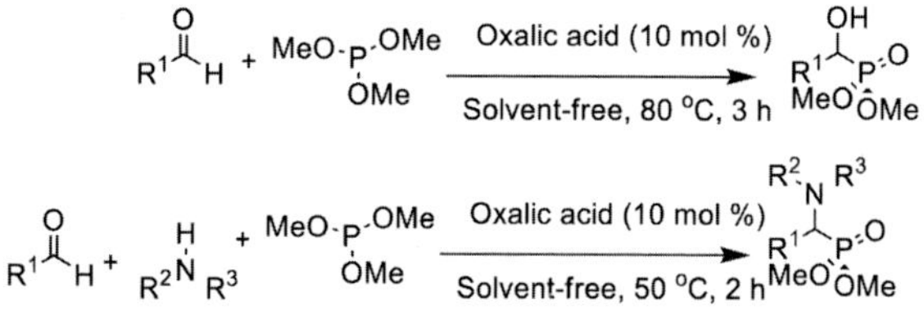

$$R^1\text{CHO} + \text{MeO-P(OMe)}_2 \xrightarrow[\text{Solvent-free, 80 °C, 3 h}]{\text{Oxalic acid (10 mol \%)}} R^1\text{CH(OH)P(O)(OMe)}_2$$

$$R^1\text{CHO} + R^2\text{NHR}^3 + \text{MeO-P(OMe)}_2 \xrightarrow[\text{Solvent-free, 50 °C, 2 h}]{\text{Oxalic acid (10 mol \%)}} R^1\text{CH(NR}^2\text{R}^3\text{)P(O)(OMe)}_2$$

SCHEME 5.34 Oxalic acid catalyzed solvent-free synthesis of alpha-aminophosphonates.

Bhattacharya's approach [74] has been reported three-component reaction of amine, aldehyde or a ketone and diethyl phosphite catalyzed by Amberlite-IR 120 resin to afforded α-amino phosphonates (Scheme 5.35).

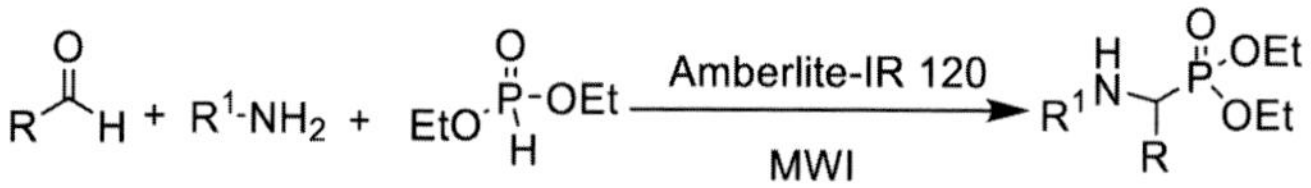

$$R\text{CHO} + R^1\text{-NH}_2 + \text{EtO-P(O)(H)-OEt} \xrightarrow[\text{MWI}]{\text{Amberlite-IR 120}} R^1\text{NH-CH(R)-P(O)(OEt)}_2$$

SCHEME 5.35 Amberlite-IR 120 resin catalyzed synthesis of alpha-aminophosphonates.

Mahmood's approach [75] were reported three-component reaction promoted by Amberlyst-15 of amine, aldehyde, and trimethyl phosphite (Kabachnik-Fields reaction) affords the α-aminophosphonates at ambient temperature.

Bhagat's approach [76] were reported Zirconium(IV) compounds as catalysts for a three-component one-pot reaction of an amine, an aldehyde or a ketone and a di/trialkyl/aryl phosphite to form α-aminophosphonates under solvent-free conditions (Scheme 5.36).

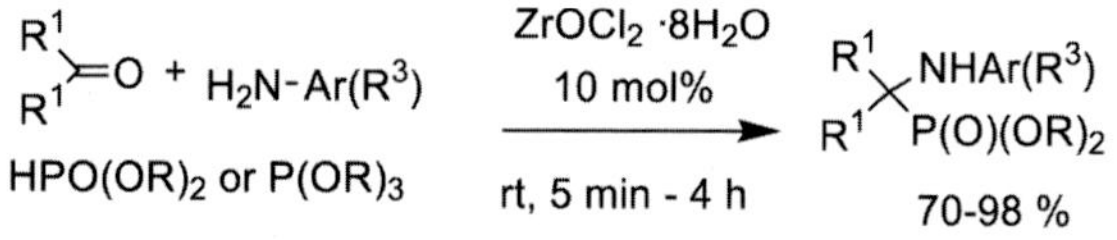

$$R^1R^1\text{C=O} + \text{H}_2\text{N-Ar(R}^3\text{)} + \text{HPO(OR)}_2 \text{ or } \text{P(OR)}_3 \xrightarrow[\text{rt, 5 min - 4 h}]{\text{ZrOCl}_2\cdot 8\text{H}_2\text{O, 10 mol\%}} R^1R^1\text{C(NHAr(R}^3\text{))P(O)(OR)}_2 \quad 70\text{-}98\ \%$$

SCHEME 5.36 $ZrOCl_2.8H_2O$ catalyzed solvent-free synthesis of alpha-aminophosphonates.

Hosseini's approach [77] was reported an innovative route to prepare substituted α-aminophosphonates (Scheme 5.37).

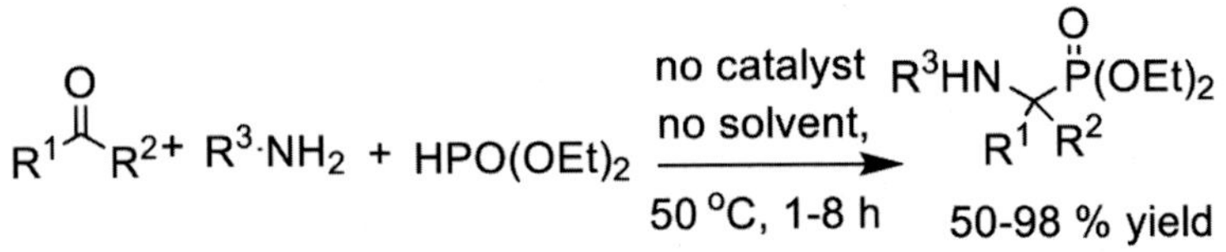

$R^1C(O)R^2$ + R^3NH_2 + $HPO(OEt)_2$ → (no catalyst, no solvent, 50 °C, 1-8 h) R^3HN–$C(R^1)(R^2)$–$P(O)(OEt)_2$, 50-98 % yield

SCHEME 5.37 Synthesis of alpha-aminophosphonates by using solvent-free condition.

Keglevich's approach [78]: ILs are especially suitable to accommodate catalytic reactions. Three-component reaction of aldehyde, amine, and diethyl phosphite catalyzed [bmimPF$_6$] IL at temperature 26°C was reported (Scheme 5.38).

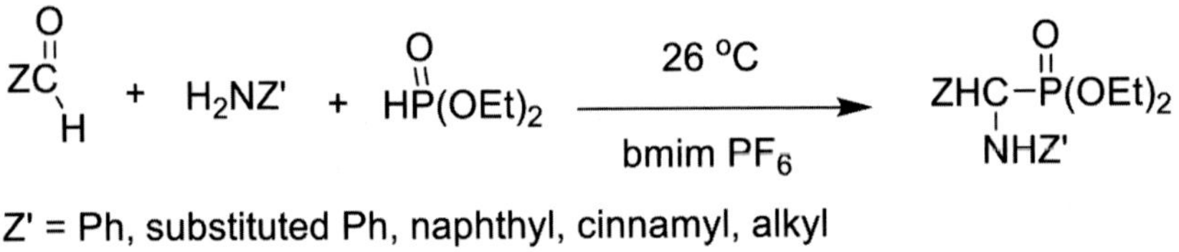

$ZC(O)H$ + H_2NZ' + $HP(O)(OEt)_2$ → (26 °C, bmim PF_6) $ZHC(NHZ')$–$P(O)(OEt)_2$

Z, Z' = Ph, substituted Ph, naphthyl, cinnamyl, alkyl

SCHEME 5.38 Synthesis of alpha-aminophosphonates catalyzed by [bmimPF$_6$] ionic liquid.

Li's approach [79] were synthesized new α-aminophosphonates by the Kabachnik-Fields reaction of 3,4,5-trimethoxybenzaldehyde (TMB) with *p*- or *m*-bromoaniline and a dialkyl phosphite under solvent-free conditions (Scheme 5.39).

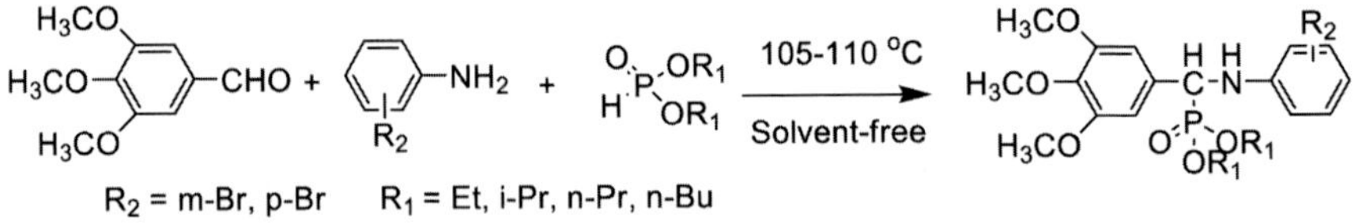

3,4,5-$(H_3CO)_3C_6H_2$–CHO + $R_2C_6H_4$–NH_2 + $HP(O)(OR_1)_2$ → (105-110 °C, Solvent-free) 3,4,5-$(H_3CO)_3C_6H_2$–CH(P(O)(OR$_1$)$_2$)–NH–$C_6H_4R_2$

R_2 = m-Br, p-Br R_1 = Et, i-Pr, n-Pr, n-Bu

SCHEME 5.39 Synthesis of alpha-aminophosphonates under solvent-free condition.

Bhagat's approach [80] was reported three-component reaction of amine, aldehyde or ketone and di-/trialkyl phosphite (Kabachnik-Fields reaction) under solvent-free conditions to afford the γ-aminophosphonates (Scheme 5.40).

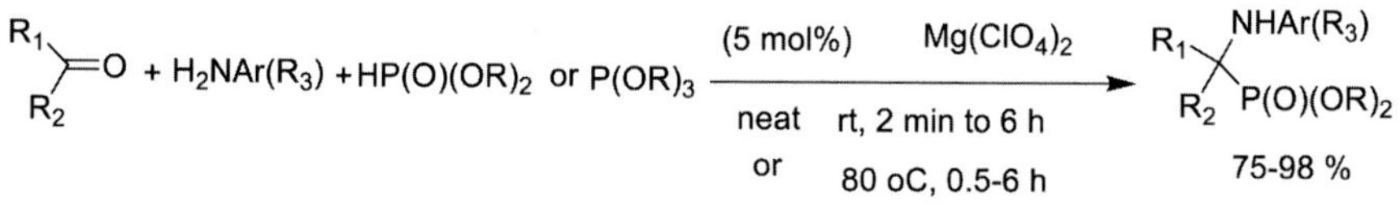

SCHEME 5.40 Synthesis of aminophosphonates catalyzed by $Mg(ClO_4)_2$.

Blaszczyk's approach [81]: Synthesis of 5-substituted-(2-thioxo-imidazolidin-4yl)-phosphonic acid diethyl esters from metallated diethyl isothiocyanatomethyl phosphonates and activated imines has been developed (Scheme 5.41).

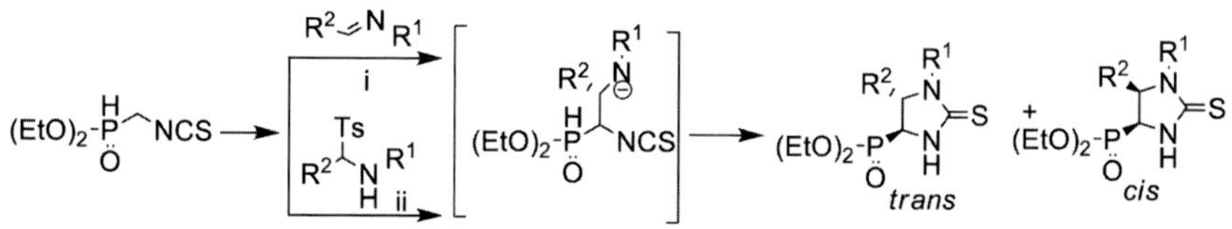

SCHEME 5.41 Synthesis of isothiocyanatomethyl phosphonate derivatives.

Song's approach [82]: Boron trifluoride diethyl etherate catalyzed the Mannich type reaction of 2-trifluoromethyl-4-bromoaniline and *O,O*-dialkylphosphite with 2-fluorobenzoaldehyde under ultrasonic irradiation to product α-aminophosphonates.

Bhanushali et al., approach [83]: has been reported the synthesis of α-aminophosphonates using anhydrous $ZrOCl_2.8H_2O$ as an environment friendly catalyst.

Jin's approach [84] was employed Mannich type reaction to synthesize α-aminophosphonates by reacting substituted benzothiazole, dialkyl phosphite and substituted benzaldehyde in ILs. The Mannich reaction provides a clean and atom economic method to access α-aminophosphonates. The IL [bmim][PF_6], accelerated the Mannich addition reaction to several folds and the products were obtained in excellent yield. The benzothiazole substituted α-aminophosphonate was active against PC3 cell lines (Scheme 5.42).

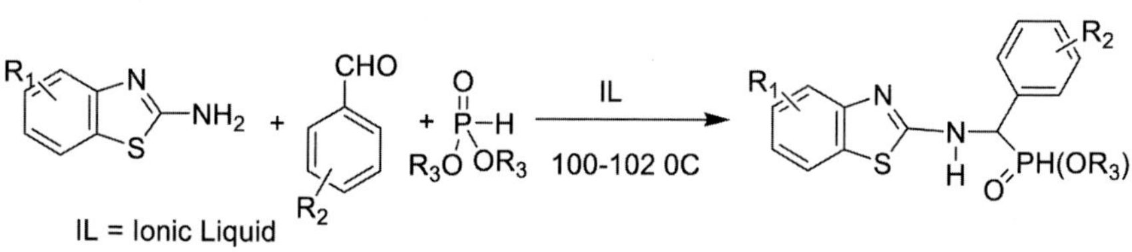

IL = Ionic Liquid

SCHEME 5.42 Synthesis of alpha-aminophosphonates catalyzed by [bmim][PF_6]IL.

Peipei's approach [85] was reported in chloroaluminate-based IL [Bmim] Cl-$AlCl_3$, N= 0.67, the one-pot coupling reaction of carbonyl compounds, amines, and diethyl phosphite to α-amino phosphonates.

Jie's approach [86] were expressed component reactions of aldehydes, amines, and diethyl phosphite catalyzed by $FeCl_3$ in ethanol or under solvent free condition afforded the α-aminophosphonates (Scheme 5.43).

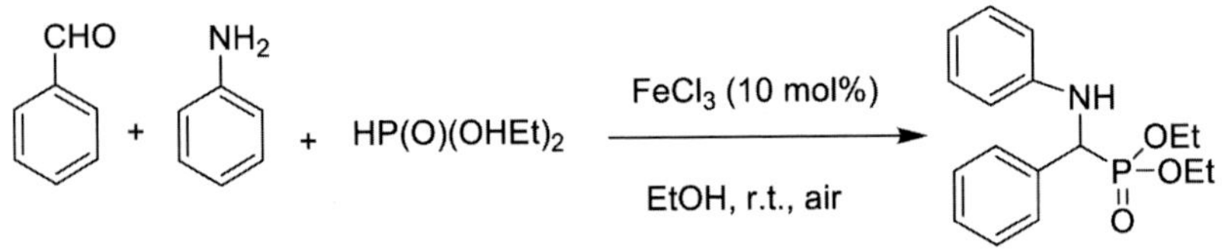

CHO NH$_2$ + + HP(O)(OHEt)$_2$ → FeCl$_3$ (10 mol%), EtOH, r.t., air → NH, OEt, P–OEt, O

SCHEME 5.43 Synthesis of alpha-aminophosphonates catalyzed by $FeCl_3$.

Mu's approach [87] was developed a solvent-free and catalyst-free method for the synthesis of α-aminophosphonates by a MW-assisted three-component Kabachnik-Fields reaction involving aldehyde, amine, and dimethyl phosphite (Scheme 5.44).

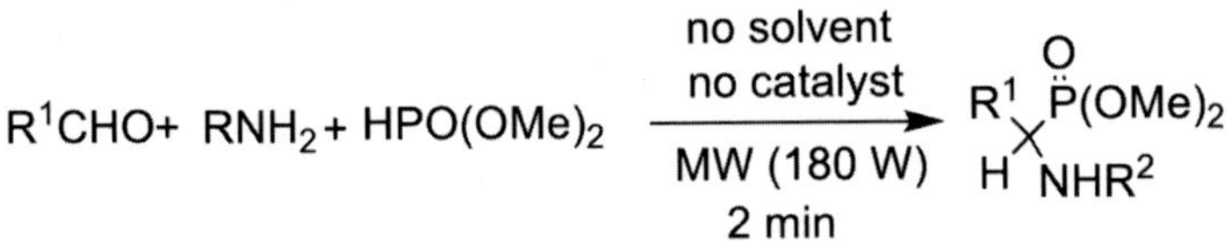

$R^1CHO + RNH_2 + HPO(OMe)_2$ → no solvent, no catalyst, MW (180 W), 2 min → R^1, O, P(OMe)$_2$, H, NHR^2

SCHEME 5.44 Synthesis of alpha-aminophosphonates by solvent-free and catalyst-free method.

Xu's approach [88] has been developed a simple method for the synthesis of various *α*-aminophosphonate derivatives under MW irradiation (Scheme 5.45).

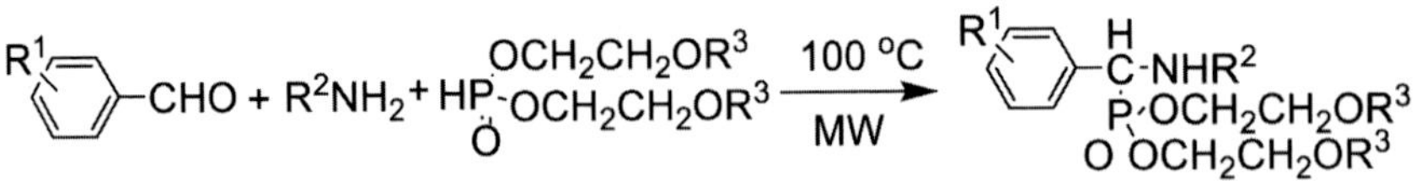

R^1–C$_6$H$_4$–CHO + R^2NH_2 + HP(O)(OCH$_2$CH$_2$OR^3)$_2$ → 100 °C, MW → R^1–C$_6$H$_4$–CH(NHR^2)–P(O)(OCH$_2$CH$_2$OR^3)$_2$

SCHEME 5.45 Synthesis of alpha-aminophosphonates under microwave irradiation.

Davis's approach [89] were prepared 5, 6, and 7-membered cyclic γ-amino phosphonates, amino acid surrogates in pure form *via* the diastereomeric addition of metal phosphonates to masked oxo sulfinimines (Scheme 5.46).

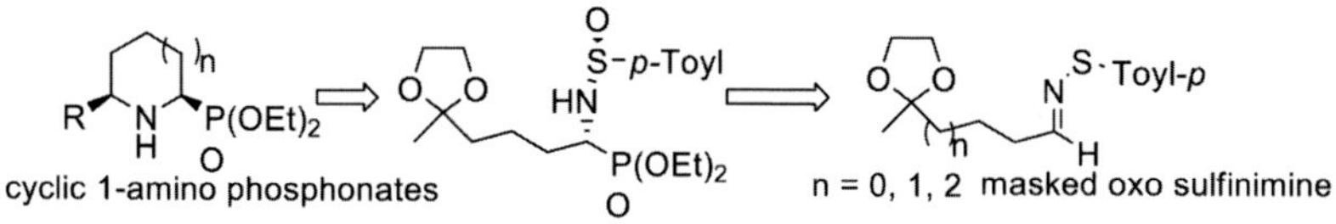

SCHEME 5.46 Synthesis of aminophosphonate derivatives.

Firouzabadi's approach [90] Synthesis of α-aminophosphonates catalyzed by several metal triflates [$M(OTf)_n$, M = Li, Mg, Al, Cu, Ce] in the absence of solvent (Scheme 5.47).

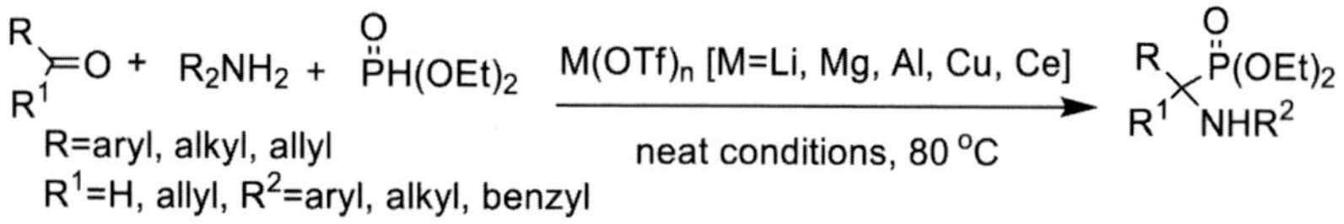

SCHEME 5.47 Synthesis of alpha-aminophosphonates in the absence of solvent.

Xu's approach [91] were synthesized α-hydroxyl aminophosphonates by three-component coupling reactions of aldehydes, hydroxylamines, and diethyl phosphite using 1-butyl-3-methylimidazolium tetrafluoroborate ([bmim]BF_4) or 1-butyl-3-methylimidazolium hexafluorophosphate ([bmim]PF_6) ILs (Scheme 5.48).

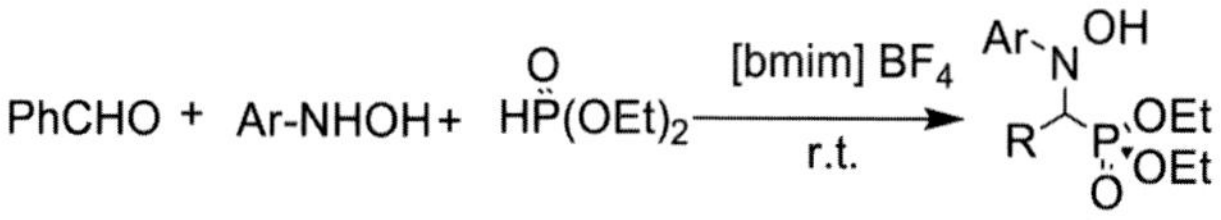

SCHEME 5.48 Synthesis of alpha-hydroxylaminophosphonates by using ([bmim]BF_4)/([bmim]PF_6)ILs.

Chandrasekhar's approach [92] was achieved the solvent and catalyst-free three-component coupling of carbonyl compounds, amines, and triethylphosphite producing α-aminophosphonates at ambient temperature (Scheme 5.49).

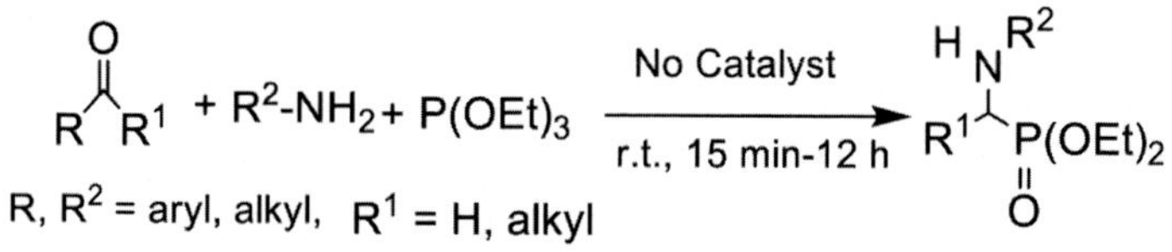

SCHEME 5.49 Synthesis of alpha-aminophosphonates under solvent and catalyst-free conditions.

Kaboudin's approach [93] were synthesized α-aminophosphonates from 1-hydroxyphosphonates under solvent free conditions using MW irradiation (Scheme 5.50).

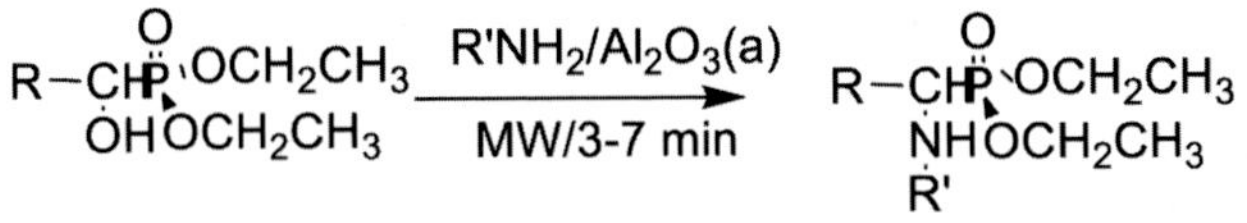

SCHEME 5.50 Synthesis of aminophosphonates using microwave irradiation.

Davis's approach [94] has been developed an aza-Darzens reaction, involving the addition of chloromethylphosphonate anions to enantiopure sulfinimines for the asymmetric synthesis of aziridine-2-phosphonates (Scheme 5.51).

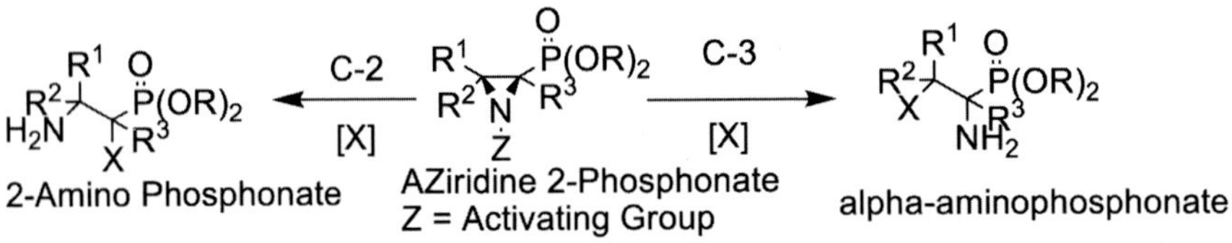

SCHEME 5.51 Asymmetric synthesis of aziridine-2-phosphonate derivatives.

Ranu's approach [95] was synthesized α-aminophosphonates through a solvent-free and catalyst free condensation of carbonyl compound, amine, and diethyl phosphite (Scheme 5.52).

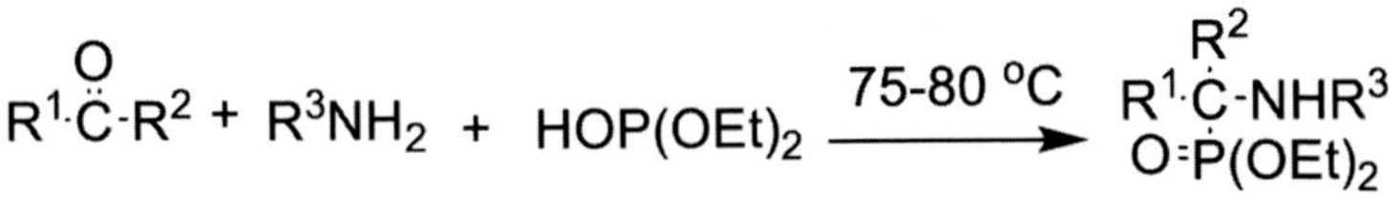

SCHEME 5.52 Synthesis of alpha-aminophosphonates by using solvent-free and catalyst-free conditions.

5.2.3 *SYNTHESIS OF A-AMINOPHOSPHONATES USING VARIOUS TRANSFORMATIONS*

Various synthetic protocols have been described for the synthesis of α-aminophosphonates. The nucleophilic addition of phosphites to imines (Kabachnik-Fields reaction) represents a convenient route for their preparation. A variety of Lewis acids such as $SnCl_4$, [96] $ZrCl_4$, [97] $BF_3.OEt_2$,

[98] BrDMSBr (bromodimethylsulfonium bromide), [99] metal perchlorates, [100] metal triflates, [101] $TaCl_5$-SiO_2, [102] $InCl_3$, [103] $TiCl_4$, [104] and $SbCl_3$-Al_2O_3, [105] have been used in α-aminophosphonate synthesis.

Odinets's approach [106] was achieved synthesis of cyclic α-aminophosphonates by addition of diethyl phosphite to cyclic imines bearing alkyl, aryl, or heteroaryl substituents at the α-position in diethyl ether at RT (Scheme 5.53).

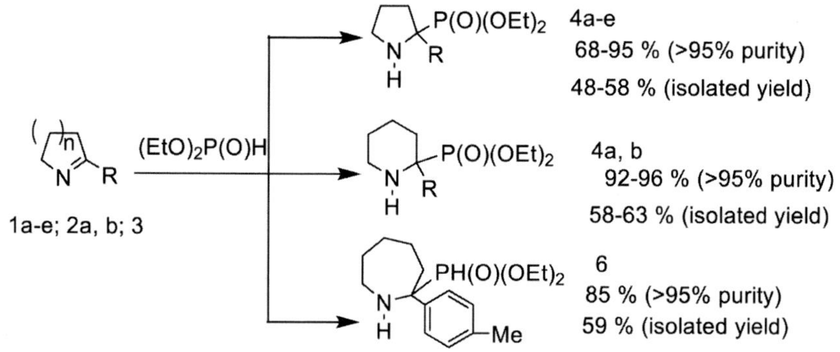

(EtO)2P(O)H

1a-e; 2a, b; 3

P(O)(OEt)2 4a-e 68-95 % (>95% purity) 48-58 % (isolated yield)

P(O)(OEt)2 4a, b 92-96 % (>95% purity) 58-63 % (isolated yield)

PH(O)(OEt)2 6 85 % (>95% purity) 59 % (isolated yield)

SCHEME 5.53 Synthesis of cyclic alpha-aminophosphonate derivatives at Room Temperature.

Tian's approach [107]: Preparation of α-aminophosphonates under one pot reaction of aldehydes with amines and dialkyl phosphites using catalytic amounts of triphenylphosphine was described.

Heydari's approach [108] was used Trifluoroethanol for one-pot, three-component coupling reactions of aldehydes or ketones, amines, and trimethylsilyl cyanide or trimethyl phosphate to afford the corresponding α-amino nitriles or α-amino phosphonates, without use of an acid or base catalyst (Scheme 5.54).

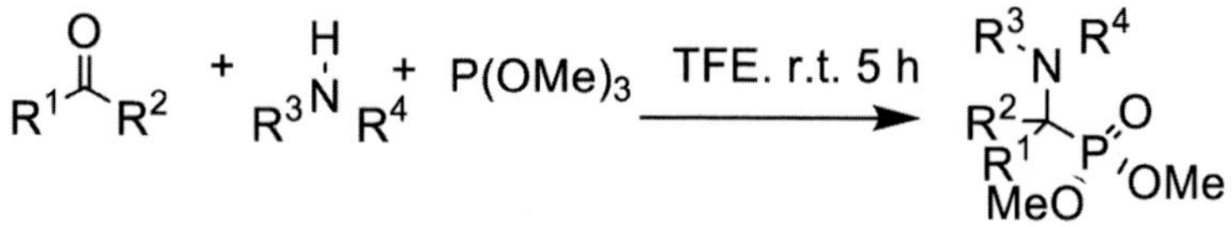

$R^1C(O)R^2 + R^3NHR^4 + P(OMe)_3 \xrightarrow{\text{TFE. r.t. 5 h}} R^1R^2C(NR^3R^4)P(O)(OMe)_2$

SCHEME 5.54 Synthesis of alpha-aminophosphonates by using trifluoroethanol.

Rezaei's approach [109]: Solution of $FeCl_3$ in THF was used for Mannich type reaction of aldehyde; amine and phosphite compounds form α-aminophosphonates (Scheme 5.55).

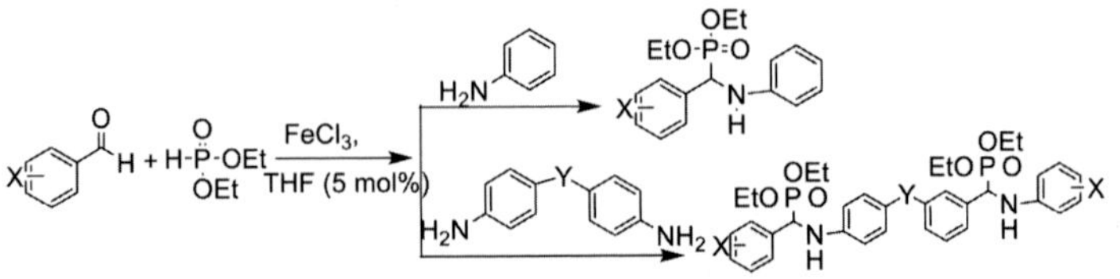

SCHEME 5.55 Synthesis of alpha-aminophosphonates catalyzed by using $FeCl_3$.

Syamala's approach [110] was developed synthesis of α-aminophosphonates *via* gallium triiodide catalyzed coupling of carbonyl compounds, amines, and diethyl phosphite in dichloromethane (Scheme 5.56).

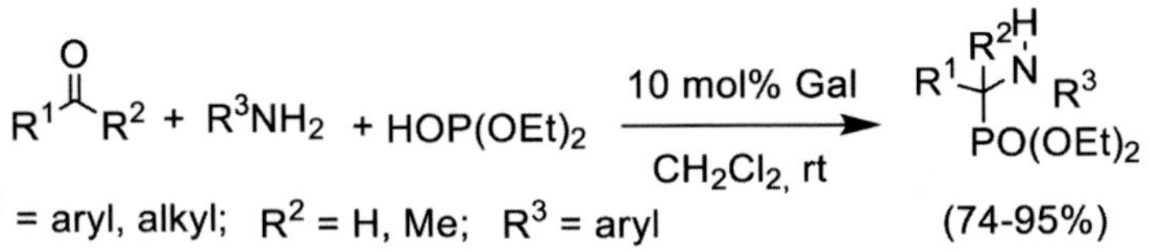

SCHEME 5.56 Synthesis of α-aminophosphonates catalyzed by gallium triiodide.

Kaboudin's approach [111]: A general method has been developed for the synthesis of α-aminophosphonic esters using TsCl as a catalyst (Scheme 5.57).

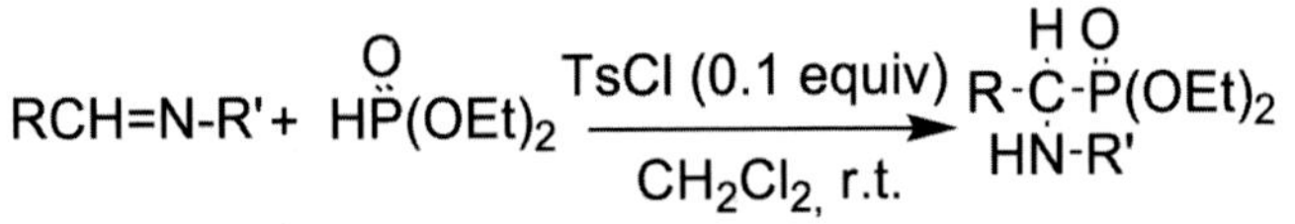

SCHEME 5.57 Synthesis of alpha-aminophosphonates catalyzed by using gallium triiodide.

Wang's approach [112] were synthesized series of naphthoquinone fused cyclic α-aminophosphonates and naphthoquinone fused cyclic α-aminophosphonic monoester (Scheme 5.58).

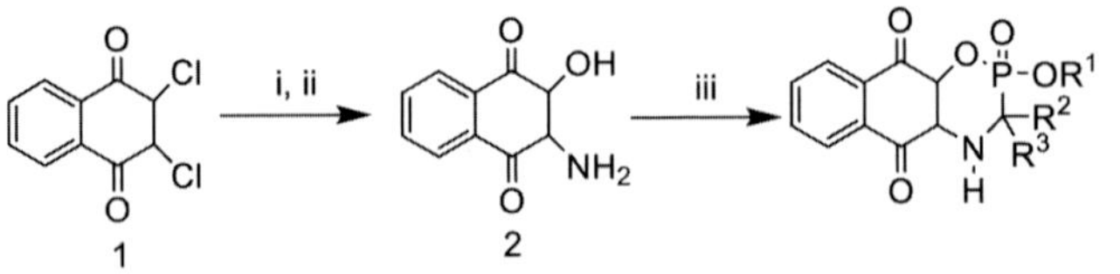

Scheme 1. Reagent and conditions: i) $NaNO_2$, HCl, CH_3OH, H_2O, 80 °C, 3h; ii) $Na_2S_2O_4$, CH_3CH_2OH, H_2O, rt, 30 min, 78% for two steps: iii) R^1OPCl_2, $R_2(CO)R_3$, THF, 0°C to rt, 24 h

SCHEME 5.58 Synthesis of series of naphthoquinone fused cyclic α-aminophosphonate derivatives.

Sayed's approach [113] were synthesized the 1,2,4,3,5-triazadiphosphinanyl and bis-(α-aminophosphonate) derivatives bearing chromone moieties by addition of diethyl phosphite (Scheme 5.59).

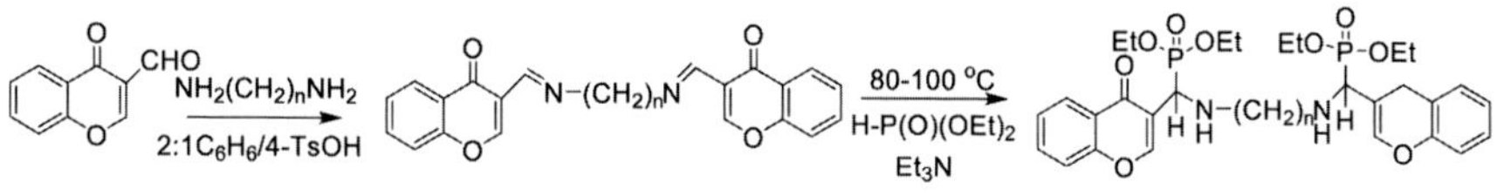

SCHEME 5.59 Synthesis of bis-(alpha-aminophosphonate) derivatives.

Chen's approach [114] were reported nucleophilic addition of dialkylphosphites to *N-tert*-butane sufinimines in presence of potassium carbonate or potassium fluoride, afforded acyclic α-alkyl-α-aminophosphonates and cyclic α-aminophosphonates (Scheme 5.60).

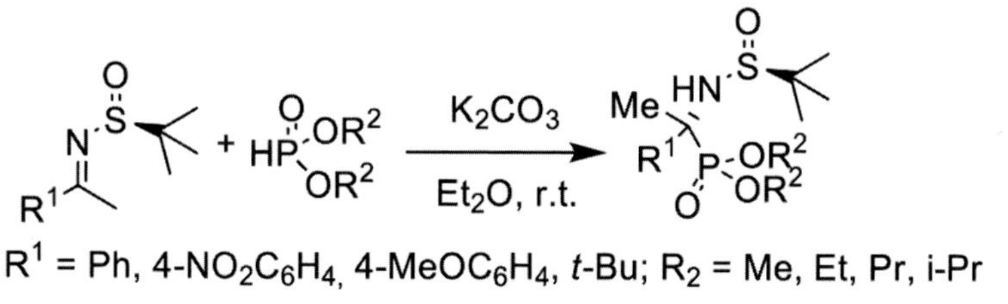

R^1 = Ph, 4-$NO_2C_6H_4$, 4-$MeOC_6H_4$, *t*-Bu; R_2 = Me, Et, Pr, i-Pr

SCHEME 5.60 Synthesis of cyclic alpha-aminophosphonates by using K_2CO_3.

Sobhani's approach [115] was reported the preparation of diethyl α-aminophosphonates in water by one-pot reaction of aldehydes, amines, tri/dialkyl phosphites (Scheme 5.61).

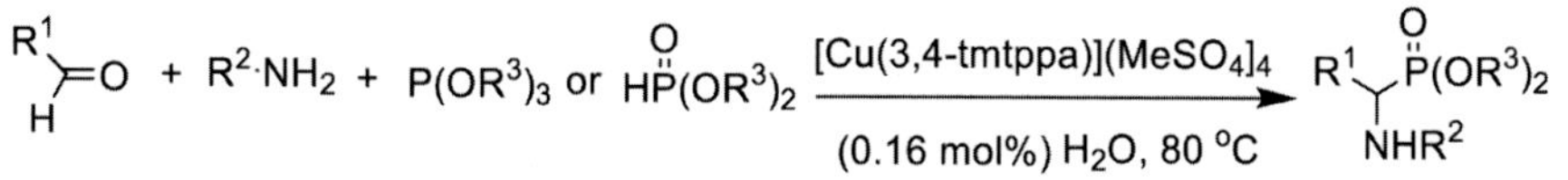

SCHEME 5.61 Synthesis of diethyl alpha-aminophosphonates in water as a solvent.

Matveeva's approach [116] was studied interaction of aliphatic and aromatic aldehydes and ketones with optically active L-α-amino acids or their esters in the three component catalytic one-pot synthesis of α-aminophosphonates (Scheme 5.62).

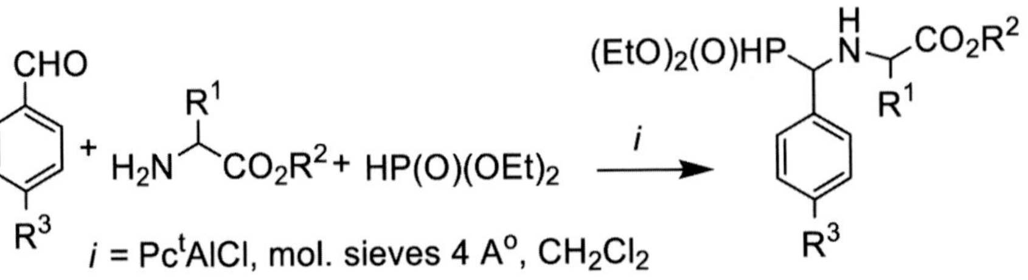

CHO
R[1]
+ H_2N CO_2R^2 + HP(O)(OEt)$_2$ —i→ $(EtO)_2(O)HP$ N CO_2R^2 H R[1]
R[3] R[3]
i = Pc[t]AlCl, mol. sieves 4 A°, CH_2Cl_2

SCHEME5.62 Three-componentcatalyticmethodforsynthesisofalpha-aminophosphonates.

Hua's approach [117] were synthesized dialkylphenyl(4-pyridylcarbonylamino) methylphosphonates *via* the Mannich reaction and peptide coupling (Scheme 5.63).

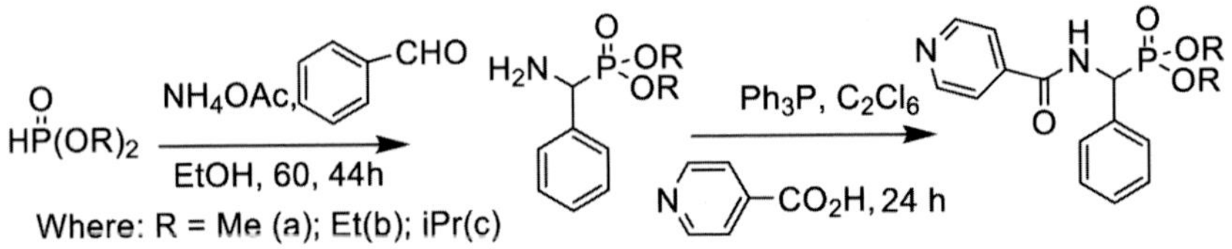

O
HP(OR)$_2$ —NH_4OAc, CHO / EtOH, 60, 44h→ H_2N P-OR OR (O) —Ph_3P, C_2Cl_6 / N CO$_2$H, 24 h→ N H N O P-OR OR O

Where: R = Me (a); Et(b); iPr(c)

SCHEME 5.63 Synthesis of α-aminophosphonate derivatives *via* the Mannich reaction.

Heydari's approach [118]: The synthesis of α-amino-, α-hydrazino- and α-N-hydroxyaminophosphonates were described by the reaction between trimethyl- phosphite and *in situ* generated imines (iminium salts, oximes, hydrazones, nitrones) by using phenyltrimethylammonium chloride as catalyst in dichloromethane at 40°C (Scheme 5.64).

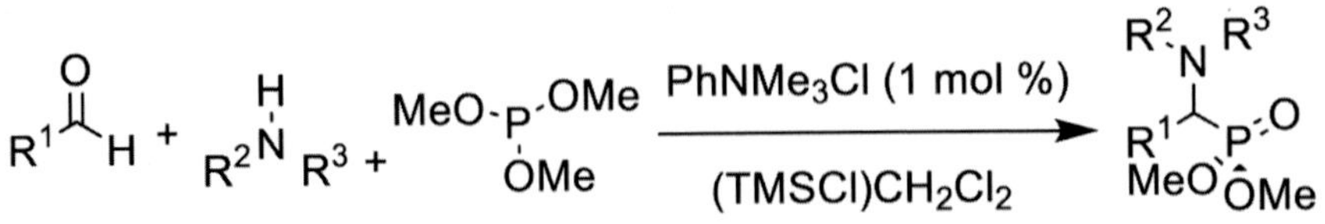

O
R[1] H + R^2 N(H) R^3 + MeO-P-OMe OMe —PhNMe$_3$Cl (1 mol %) / (TMSCl)CH_2Cl_2→ R^2 N R^3 R[1] P=O MeO OMe

SCHEME 5.64 Synthesis of α-aminophosphonates by using phenyltrimethylammonium chloride as catalyst.

Narayana's approach [119]: α-Aminophosphonates were synthesized in high yields by the Kabachnik-Fields reaction (Scheme 5.65).

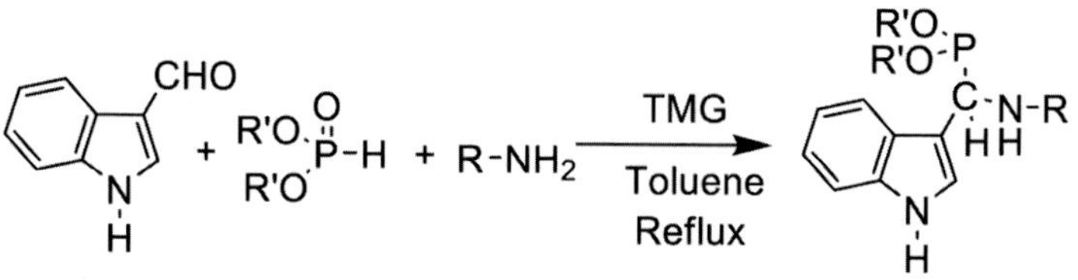

SCHEME 5.65 Synthesis of α-aminophosphonates by the Kabachnik-Fields reaction.

Kaboudin's approach [120]: The synthesis of bis-[1-diethoxyphosphoryl-arylmethyl]amines using acetyl chloride as catalyst in hydrophosphonylation of imine was described (Scheme 5.66).

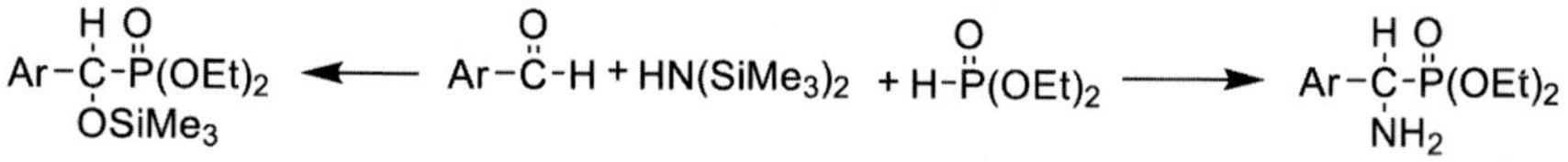

SCHEME 5.66 Synthesis of alpha-aminophosphonates by using acetyl chloride as catalyst.

Kaboudin's approach [121] was reported the synthesis of α-acet-oxyphosphonates from aldehydes with diethylphosphite in the presence of acetic anhydride under solvent free conditions using magnesium oxide (Scheme 5.67).

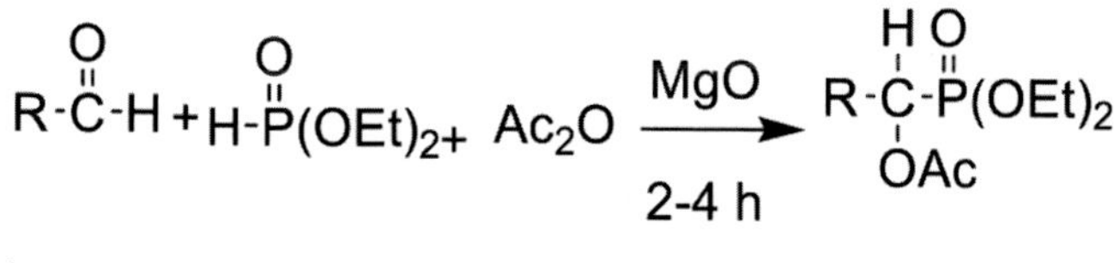

SCHEME 5.67 Synthesis of alpha-aminophosphonates under solvent-free conditions by using MgO.

Xu's approach [122] was reported the preparation of α-aminophosphonates by one-pot condensation of aldehydes, amines, and dialkyl phosphites using catalytic amounts of lanthanide chloride (Scheme 5.68).

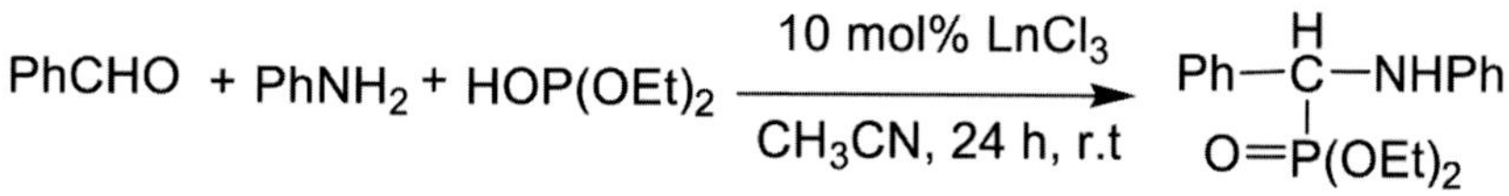

SCHEME 5.68 Synthesis of alpha-aminophosphonates using catalyst lanthanide chloride.

Matveeva's approach [123] has synthesized α-aminophosphonates derivatives by using t-PcAlCl (Scheme 5.69).

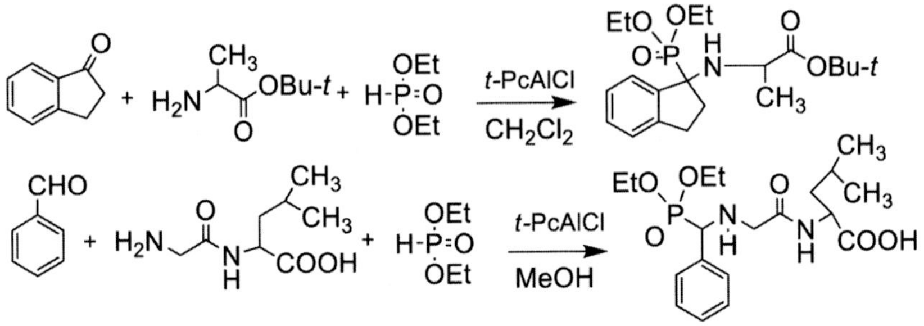

SCHEME 5.69 Synthesis of alpha-aminophosphonates by using t-PcAlCl.

Paraskar's approach [124] were reported Copper(II)triflate catalyzed three-component condensation reaction of aldehyde, amine, and $P(OMe)_3$ in acetonitrile at RT gives the α-aminophosphonates (Scheme 5.70).

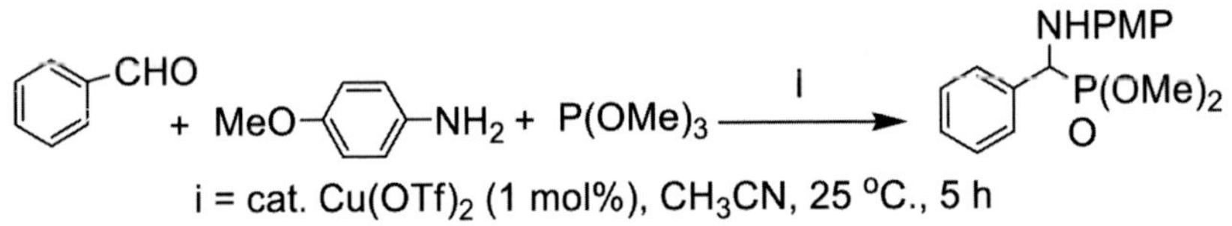

i = cat. $Cu(OTf)_2$ (1 mol%), CH_3CN, 25 °C., 5 h

SCHEME 5.70 Synthesis of alpha-aminophosphonate derivatives at room temperature.

Wu's approach [125] was studied three-component reactions of aldehydes, amines, and diethyl phosphite catalyzed by $Mg(ClO_4)_2$ or molecular iodine afforded the α-aminophosphonates (Scheme 5.71).

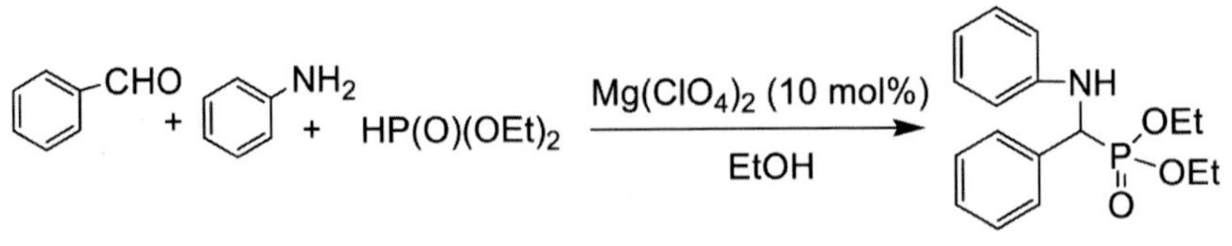

SCHEME 5.71 Synthesis of alpha-aminophosphonates catalyzed by $Mg(ClO_4)_2$ or molecular iodine.

Meenen's approach [126] was reported synthesis of α-aminophosphonates derivatives catalyzed by $HP(O)(OMe)_2$ in methanol as solvent from the reaction of an imine with a dialkyl phosphite (Scheme 5.72).

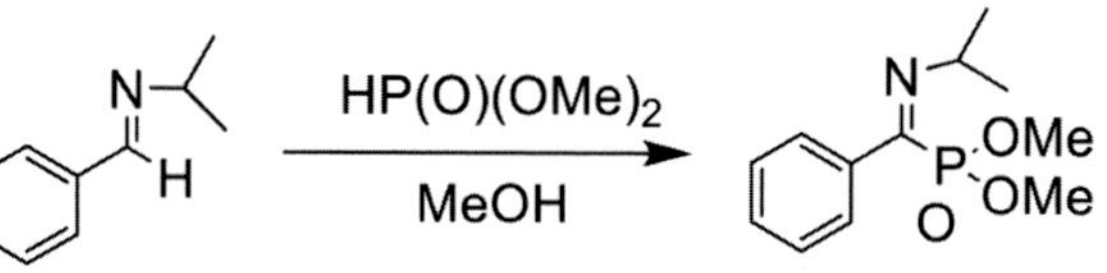

SCHEME 5.72 Synthesis of alpha-aminophosphonates catalyzed by $HP(O)(OMe)_2$.

Kaboudin's approach [127] was studied the treatment of aromatic aldehydes with ammonia and diethylphosphite gives diethyl *N*-(arylmethylene)-1-aminoarylmethyl-phosphonates which hydrolyzed to diethyl α-aminoarylmethylphosphonates (Scheme 5.73).

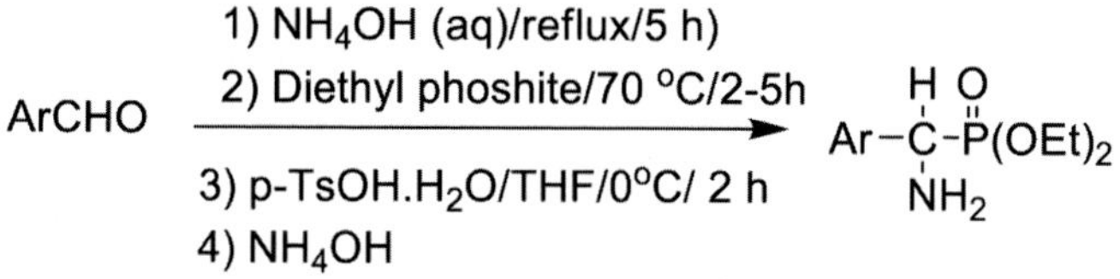

SCHEME 5.73 Synthesis of diethyl alpha-aminoarylmethylphosphonate derivatives.

Dziemidowicz's approach [128] has been studied anhydrous potassium monoalkyl phosphonates prepared from the corresponding dialkyl phosphonates and potassium trimethylsilanolate under non-aqueous conditions (Scheme 5.74).

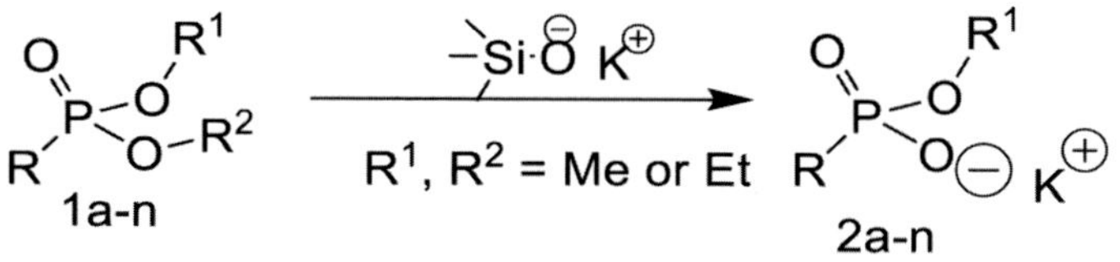

SCHEME 5.74 Synthesis of alpha-aminophosphonates from the corresponding dialkyl phosphonates and potassium trimethylsilanolates.

Ghosh's approach [129]: One-pot synthesis of α-amino phosphonates derived from nitro substituted anilines, aldehydes, and diethyl phosphite has been carried out by employing 5 mol% of $In(OTf)_3$ (Scheme 5.75).

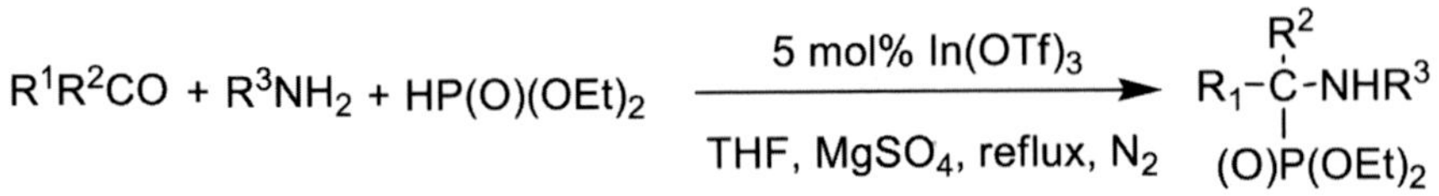

$$R^1R^2CO + R^3NH_2 + HP(O)(OEt)_2 \xrightarrow[\text{THF, MgSO}_4\text{, reflux, N}_2]{\text{5 mol\% In(OTf)}_3} R_1\text{-}C(R^2)(\text{NHR}^3)\text{-}(O)P(OEt)_2$$

SCHEME 5.75 Synthesis of alpha-aminophosphonates catalyzed by $In(OTf)_3$.

Xu'sapproach[130]wasdescribedthepreparationofα-aminophosphonates under one pot reaction of aldehydes with amines and dialkyl phosphites using SmI_2 (Scheme 5.76).

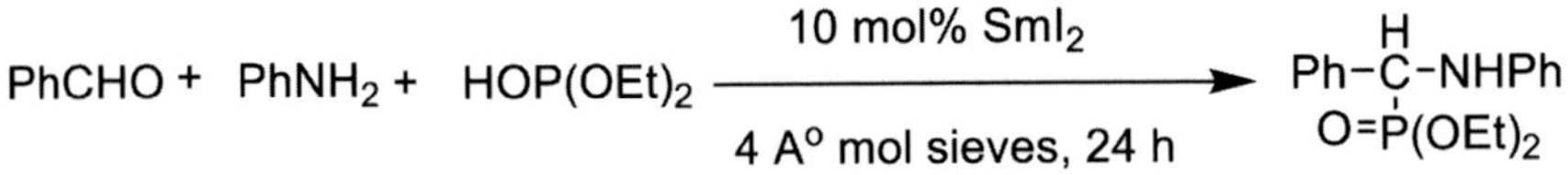

$$PhCHO + PhNH_2 + HOP(OEt)_2 \xrightarrow[\text{4 A}^o\text{ mol sieves, 24 h}]{\text{10 mol\% SmI}_2} Ph\text{-}CH(NHPh)\text{-}P(=O)(OEt)_2$$

SCHEME 5.76 Synthesis of alpha-aminophosphonates using SmI_2.

Matveeva's approach [131] has been developed the synthesis of α-aminophosphonates on basis of the Kabachnik Fields reaction in the presence of tetra-*tert*-butylphthalocyanines (Scheme 5.77).

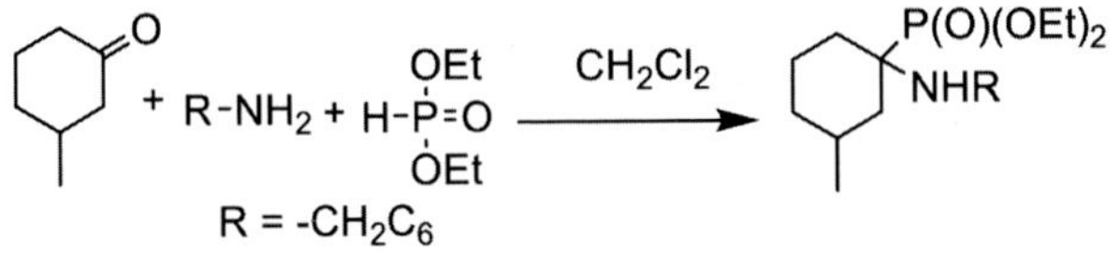

O + R-NH₂ + H-P(=O)(OEt)(OEt) —CH₂Cl₂→ P(O)(OEt)₂ / NHR

R = -CH₂C₆

SCHEME 5.77 Synthesis of alpha-aminophosphonates in the presence of tetra-*tert*-butylphthalocyanines.

Song's approach [132]: α-aminophosphonates containing fluorine was synthesized by Mannich type reactions (Scheme 5.78).

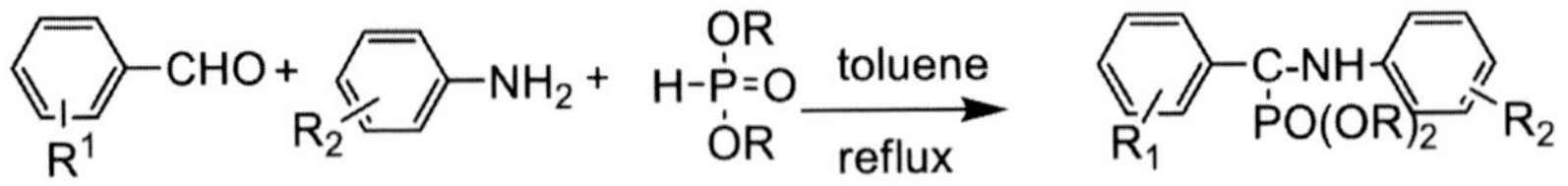

R^1-C₆H₄-CHO + R_2-C₆H₄-NH₂ + H-P(=O)(OR)(OR) —toluene, reflux→ R_1-C₆H₄-CH(PO(OR)₂)-NH-C₆H₄-R_2

R_1=H, o-F, p-F, p-Cl; R_2=CF_3, CH_3, H, NO_2; R=Me, Et, i-Pr, n-Pr, n-Bu

SCHEME 5.78 Synthesis of alpha-aminophosphonates under reflux condition.

Saidi's approach [133] was prepared Imines in situ by reaction of aldehydes and ketones with primary amines in ethereal solution of $LiClO_4$ react readily at ambient temperature with trialkylphosphite to give high yields of α-aminophosphonates (Scheme 5.79).

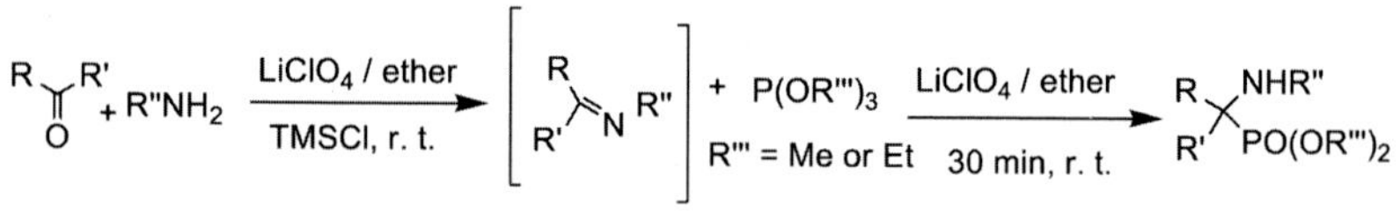

R(C=O)R' + R"NH2 —[LiClO4 / ether; TMSCl, r. t.]→ [R(R')C=N–R"] + P(OR"')3 (R"' = Me or Et) —[LiClO4 / ether; 30 min, r. t.]→ R(R')C(NHR")PO(OR"')2

SCHEME 5.79 Synthesis of alpha-aminophosphonates by using catalyst $LiClO_4$.

Chandrasekhar's approach [134]: $TaCl_5$-SiO_2 has been utilized as Lewis acid catalyst for three component coupling of carbonyl compounds, aromatic amines and diethyl phosphite to produce α-aminophosphonates (Scheme 5.80).

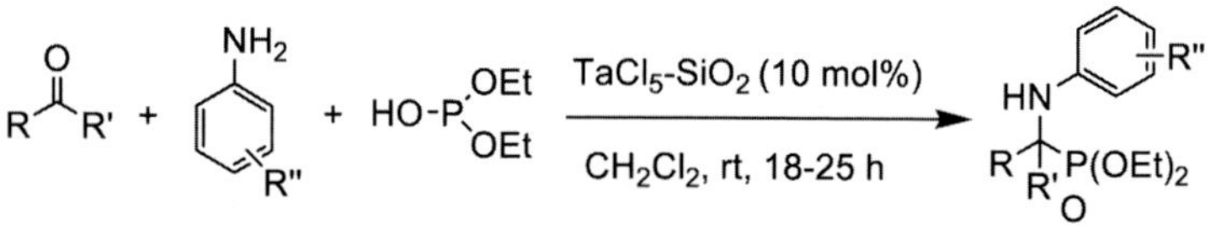

R(C=O)R' + NH2-C6H4-R" + HO-P(OEt)2 —[TaCl5-SiO2 (10 mol%); CH2Cl2, rt, 18-25 h]→ R(R')C(NH-C6H4-R")P(O)(OEt)2

SCHEME 5.80 Synthesis of alpha-aminophosphonates by using catalyst $TaCl_5$-SiO_2.

Green's approach [135]: α-aminophosphonates may be generated by the treatment of α-nitrophosphonates with $LiBH_4/Me_3SiCl$ in THF at RT (Scheme 5.81).

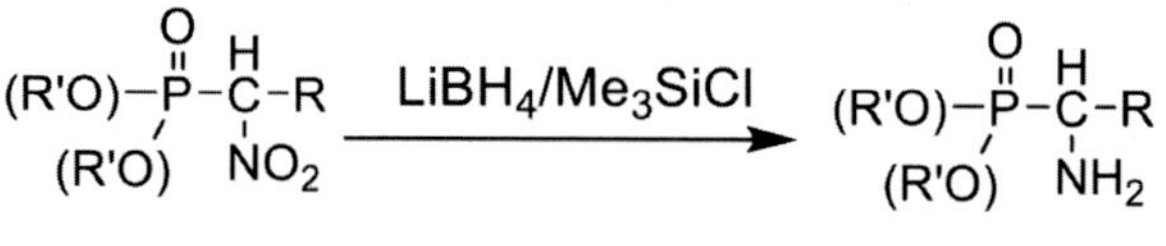

(R'O)2P(O)–CH(NO2)–R —[LiBH4/Me3SiCl]→ (R'O)2P(O)–CH(NH2)–R

SCHEME 5.81 Synthesis of alpha-aminophosphonates by using $LiBH_4/Me_3SiCl$.

Gancarz's approach [136]: Aminophosphonates has been synthesized from aromatic ketones through the formation of hydroxyphosphonates (Scheme 5.82).

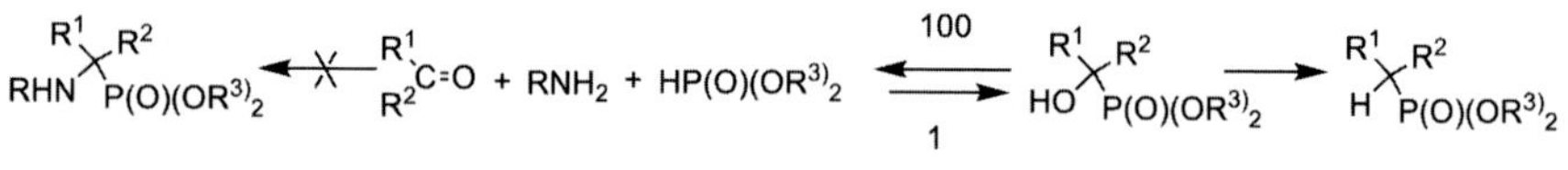

RHN–C(R1)(R2)–P(O)(OR3)2 ←✕— R1R2C=O + RNH2 + HP(O)(OR3)2 ⇄ (100 : 1) HO–C(R1)(R2)–P(O)(OR3)2 → H–C(R1)(R2)–P(O)(OR3)2

SCHEME 5.82 Synthesis of aminophosphonates derived from aromatic ketones.

5.3 RESULTS AND DISCUSSION

In the mid of 20th-century catalysts played a vital role in synthetic Chemistry. At the end of the century, we saw many reforms in these fields towards clean technology and legislations. The industries are also compelled to use green technology avoiding the hazardous solvents. The green chemistry has been utilized to reduce or eliminate the use of hazardous substances. This embraces the principle of clean synthesis involving the improvements in process selectivity, high atom efficiency, and easy separation with the re-use of non-product components.

The comparison study of the different reaction conditions and result for benzaldehydes, substituted anilines, diethylphosphite, and $EtNH_3$ catalyzed at RT. Literature methods shows less yields of products, large reaction time i.e., 24 hrs. The present work is reported in Table 5.1 which is clearly demonstrates the advantages of this methodology.

TABLE 5.1 Solvent and Catalyst Effect for the Synthesis of α-Aminophosphonates

Entries	Catalyst (mol%)	Solvent	Temperature (°C)	Time (h)	Yield (%)	References
1	CF_3CO_2H (25)	-	r.t.	24	96	[137]
2	$ZrCl_4$ (10)	CH_3CN	r.t.	4.5	92	[138]
3	$InCl_3$ (10)	THF	r.t.	11	93	[139]
4	TiO_2 (20)	-	50	2.5	98	[140]
5	$[bimim]BF_6$ (1 mL)	-	r.t.	8	84	[141]
6	$BI(NO_3)_3.5H_2O$		r.t.	10	93	[142]
7	$SbCl_3/Al_2O_3$ (5)	CH_3CN	r.t.	3	91	[143]
8	$TaCl_5$-SiO_2 (10)	CH_2Cl_2	r.t.	20	90	[144]
9	-	-	r.t.	24	N.R.	[141, 145]
10	$CeCl_{3.}7H_2O$ (5)	-	r.t.	5	95	[66]
11	EAN (2 mL)	-	r.t.	5.30	97	[64]

5.4 SUMMARY AND CONCLUSION

In this chapter, we express the synthesis and biological evolution of α-aminophosphonate derivatives which is useful for researcher to further study on α-aminophosphonates. Recently, Scientist are tired to developed the synthesis of biologically active compounds such as novel α-aminophosphonates by using ILs and other environmental friendly as catalyst and solvents. Apart from being relatively nontoxic and environmentally friendly, the catalyst offers other advantages such as greater substrate compatibility, high reaction yields, short reaction times, recyclable, reusable, and the ability to tolerate functional groups, making it an important in addition to the reported methods. Aromatic amines and aldehydes are lipophilic in nature; utilized in therapeutic applications. These methods have been implemented for the synthesis of α-aminophosphonates because:

- Ecofriendly approach;
- Maintain atom economy;
- Easy work up procedure;
- Excellent yield of products.

ACKNOWLEDGMENTS

The authors are thankful to Director Dr. Ulhas Shiurkar and HOD Dr. S. L. Dhondge, Deogiri Institute of Engineering and Management Studies, Station Road, Aurangabad–431005, (MS), India for encouragement during the process of carrying out this work.

KEYWORDS

- **bioactive compounds**
- **environmental friendly**
- **green chemistry**
- **ultrasonication**
- **zinc oxide nanoflowers**
- **α-aminophosphonate derivatives**

REFERENCES

1. (a) Fields, S. C., (1999). *Tetrahedron, 55,* 12237. (b) Yuan, C., & Chen, S., (1992). *Synthesis,* 1124. (c) Hirschmann, R., Smith, A. B., Taylor, C. M., Benkovic, P. A.,

Taylor, S. D., Yager, K. M., Sprengler, P. A., & Venkovic, S. J., (1994). *Science, 265,* 234. (d) Yokomatsu, T., Yoshida, Y., & Shibuya, S., (1994). *J. Org. Chem., 59,* 7930.

2. Kafarski, P., & Lejczak, B., (1991). *Phosphorus, Sulfur, Silicon Relat. Elem., 63,* 1993.
3. (a) Allen, M. C., Fuhrer, W., Tuck, B., Wade, R., & Wood, J. M., (1989). *J. Med. Chem., 32,* 1652. (b) Gannousis, P. P., & Bartlett, A., (1987). *J. Med. Chem., 30,* 1603.
4. (a) Meyer, J. H., & Bartlett, P. A., (1998). *J. Am. Chem. Soc., 120,* 4600. (b) Peyman, K. H., Spanig, J., & Ruppert, D., (1993). *Angew. Chem. Int. Ed. Engl., 32,* 1720.
5. Barder, A., (1988). *Aldrichim Acta, 21,* 15.
6. (a) Maier, L., (1990). *Phosphorus, Sulfur, Silicon, 32,* 1720.
7. Baylis, E. K., Camobell, C. D., & Dingwall, J. G., (1984). *J. Chem. Soc. Perkin Trans., 1,* 2845.
8. Kafarski, P., & Lejczak, B., (2001). *Curr. Med. Chem. Anti-Cancer Agents, 1,* 301.
9. (a) Bhattacharya, A. K., & Kaur, T., (2007). *Synlett,* p. 745. (b) Wu, J., Sun, W., Sun, X., & Xia, H. G., (2006). *Green Chem., 8*, 365. (c) Kaim, L. E., Grimaud, L., & Hadrot, S., (2006). *Tetrahedron Lett., 47*, 3945. (d) Mu, X. J., Lei, M. Y., Zou, J. P., & Zhang, W., (2006). *Tetrahedron Lett., 47*, 1125. (e) Kabachink, M. M., Zobnina, E. V., & Beletskaya, I. P., (2005). *Synlett,* p. 1393. (f) Kaboudin, B., & Moradi, K., (2005). *Tetrahedron Lett., 46*, 2989. (g) Kaboudin, B., & Moradi, K., (2005). *Tetrahedron Lett., 46*, 1209. (h) Zhan, Z. P., Yang, R. F., & Li, J. P., (2005). *Chem. Lett., 34*, 1042. (i) Ghosh, R., Maiti, S., Chakraborty, A., & Maiti, D. K., (2004). *J. Mol. Catal. A, 210*, 53. (j) Firouzabadi, H., Iranpoor, N., & Sobhani, S., (2004). *Synthesis,* p. 2692. (k) Joly, G. D., & Jacobsen, E. N., (2004). *J. Am. Chem. Soc., 126*, 4102. (l) Azizi, N., Rajabi, F., & Saidi, M. R., (2004). *Tetrahedron Lett., 45*, 9233. (m) Azizi, N., & Saidi, M. R., (2003). *Tetrahedron, 59*, 5329. (n) Matveeva, E. D., Podrugina, T. A., Tishkovskaya, E. V., Tomilova, L. G., & Zefirov, N. S., (2003) *Synlett,* p. 2321. (o) Kukhar, V. P., Soloshonok, V. A., & Solodenko, V. A., (1994). *Phosphorus Sulfur Silicon Relat. Elem., 92*, 239.
10. Alfonso, K., Colberg, J., Dunn, P. J., Fevig, T., Jenings, S., Johnson, T. A., Klein, H. P., Knight, C., Nagy, M. A., Perry, D. A., & Stefaniak, M., (2008). *Green Chem., 10*, 31.
11. Dilbeck, T. P., Anastas, P., Black, D. S., Breen, J., Collins, T., Memoli, S., Miyamoto, J., Polyakoff, M., & Tumas, W., (2000). *Pure Appl. Chem., 72,* 1207.
12. Potosky, J., (2005). *Drug Discovery Today, 10*, 115.
13. Awad, M. K., Abdel-Aal, M. F., Atlam, F. M., & Hekal, H. A., (2019). Molecular docking, molecular modeling, vibrational and biological studies of some new heterocyclic α-aminophosphonates. *Spectrochimica Acta-Part A: Molecular and Biomolecular Spectroscopy, 206,* 78–88.
14. Zhu, X. F., Zhang, J., Sun, S., Guo, Y. C., Cao, S. X., & Zhao, Y. F., (2017). Synthesis and structure-activity relationships study of α-aminophosphonate derivatives containing a quinoline moiety. *Chinese Chemical Letters, 28*(7), 1514–1518.
15. Chen, N. Y., Duan, W. G., Liu, L. Z., Li, F. Y., Lu, M. P., & Liu, B. M., (2015). Synthesis and antifungal activity of dehydroabietic acid-based thiadiazole-phosphonates. *Holzforschung, 69*(9), 1069–1075.
16. El-Refaie, S. K., Mohamed, M. A., & Khalil, M. S. A., (2015). Synthesis and antimicrobial activity of α-aminophosphonates containing chitosan moiety. *Arabian Journal of Chemistry, 8*, 427–432.
17. Yang, J., Hu, Y., Gu, Q., Li, M., Li, M., & Song, B., (2014). Synthesis and antibacterial activities of novel phosphonate derivatives containing quinolinone moiety. *Chin. J. Org. Chem., 34*, 829–834.

18. Bhagat, S., Shah, P., Garg, S. K., Mishra, S., Kamal, K. P., Singh, S., & Chakraborti, A. K., (2014). Aminophosphonates as novel anti-leishmanial chemotypes: Synthesis, biological evaluation, and CoMFA studies. *Med. Chem. Comm., 5*(5), pp. 665–670.
19. Westheimer, F. H., (1987). *Science, 235,* 1173.
20. Mastalerz, P., & Kafarski, P., (2000). "Naturally occurring aminophosphonic and aminophosphinic acids." In: Kukhar, V. P., & Hudson, H. R., (eds.), "*Aminophosphonic and Aminophosphinic Acids." John* Wiley and Sons, Ltd., Chichester.
21. Lerner, R. A., Benkovic, S. J., & Schultz, P. G., (1991). *Science, 252,* 659.
22. Horiguchi, M., & Kandatsu, M., (1959). *Nature, 184,* 901.
23. Hendlin, D., (1969). *Science, 166,* 122.
24. (a) Kamiya, T., Hemmi, K., Takeno, H., & Hashimoto, M., (1980). *Tetrahedron Lett., 21,* 95. (b) Wiesner, J., Borrmann, S., & Jomaa, H., (2003). *Parasitol Res., 90,* S71. (c) Lell, B., Ruangweerayut, R., Wiesner, J., Missinou, M. A., Schindles, A., Baranek, T., Hintz, M., Hutchinson, D., Jomaa, H., & Kremsner, P. G., (2003). *Antimicrob. Agents Chemother., 47,* 735.
25. (a) Quin, L. D., (2000). "Organophosphorus chemistry in biology, agriculture, and technology." In: "*A Guide to Organophosphorus Chemistry.*" John Wiley and Sons, Inc., New York. (b) Fields, S. C., (1999). *Tetrahedron, 55,* 12237.
26. (a) De Clercq, E., & Holy, A., (2005). *Nat. Rev. Drug Discov., 4,* 928. (b) De Clercq, E. J., (2001). *Pharmacol. Exp. Ther., 297,* 1.
27. Fleisch, H., (2002). *Breast Cancer Res., 4,* 30.
28. Kafarski, P., & Lejczak, B., (2000). "The biological activity of phosphono- and phosphinopeptides," In: Kukhar, V. P., & Hudson, H. R., (eds.), *"Aminophosphonic and Aminophosphinic Acids."* John Wiley and Sons, Ltd., Chichester.
29. (a) Oleksyszyn, J., & Powers, J. C., (1991). *Biochemistry, 30,* 485. (b) Oleksyszyn, J., & Powers, J. C., (1989). *Biochem. Biophys. Res. Commun., 161,* 143. (c) Boduszek, B., Brown, A. D., & Powers, J. C. J., (1994). *Enzyme Inhib., 8,* 147.
30. (a) De Meester, I., Belyaev, A., Lambeir, A. M., De Meyer, G. R. Y., Van Osselaer, N., Haemers, A., & Scharpe, S., (1997). *Biochem. Pharmacol., 54,* 173. (b) Lambeir, A. M., Borloo, M., De Meester, I., Belyaev, A., Augustyns, K., Hendriks, D., Scharpe, S., & Haemers, A., (1996). *Biochim. Biophys. Acta, 76,* 1290. (c) Oleksyszyn, J., Boduszek, B., Kam, C. M., & Powers, J. C., (1994). *J. Med. Chem., 37,* 226. (d) Wang, C. L., Taylor, T. L., Mical, A. J., Spitz, S., & Reilly, T. M., (1992). *Tetrahedron Lett., 33,* 7667.
31. (a) Green, D., Patel, G., Elgendy, S., Baban, J. A., Skordalakes, E., Husman, W., Goodwin, C. A., Scully, M. F., Kakkar, V. V., & Deadman, J., (1996). *Phosphorus, Sulfur, Silicon Relat. Elem., 110,* 533. (b) Cheng, L., Goodwin, C. A., Scully, M. F., Kakkar, V. V., & Claeson, G., (1991). *Tetrahedron Lett., 32,* 7333.
32. (a) Camp, N. P., Hawkins, P. C. D., Hitchcock, P. B., & Gani, D., (1992). *Bioorg. Med. Chem. Lett., 2,* 1047.
33. Jacobsen, N. E., & Bartlett, P. A., (1981). *J. Am. Chem. Soc., 103,* 654.
34. Breuer, E., (2004). "Carbamoylphosphonates a new class of *in vivo* active matrix metalloproteinase inhibitors." *16th International Conference on Phosphorus Chemistry.* Birmingham, UK.
35. Allen, J. G., Atherton, F. R., Hall, M. J., Hassall, C. H., Holmes, S. W., Lambert, R. W., Nisbet, L. J., & Ringrose, P. S., (1978). *Nature, 272,* 56.

36. Hudson, H. R., (2000). "Aminophosphonic and aminophosphinic acids and their derivatives as agrochemicals." In: Kukhar, V. P., & Hudson, H. R., (eds.), *"Aminophosphonic and Aminophosphinic Acids."* John Wiley and Sons, Ltd., Chichester.
37. Jane, D. E., (2000). "Neuroactive aminophosphonic and aminophosphinic acid derivatives." In: Kukhar, V. P., & Hudson, H. R., (eds.), *"Aminophosphonic and Aminophosphonic Acids."* John Wiley and Sons, Ltd., Chichester.
38. (a) Lehmann, J., Hutchison, A. J., McPherson, S. E., Mondadori, C., Schmutz, M., Sinton, C. M., Tsai, C., Murphy, D. E., Steel, D. J., Williams, M., Cheney, D. L., & Wood, P. L., (1988). *J. Pharmacol. Exp. Ther., 246,* 65. (b) Hutchison, A. J., Williams, M., Angst, C., De Jesus, R., Blanchard, L., Jackson, R. H., Wilusz, E. J., Murphy, D. E., Bernard, P. S., Schneider, J., Campbell, T., Guida, W., & Sills, M. A., (1989). *J. Med. Chem., 32,* 2171.
39. Park, K. H., & Kurth, M. J., (2002). *Tetrahedron, 58,* 8629.
40. Rasal, S., Jain, S., & Shimpi, N. G., (2018). Reusable zinc oxide nanoflowers for the synthesis of α-aminophosphonates under solvent-free ultrasonication. *Synthetic Communications, 48*(18), 2420–2434.
41. Deshmukh, S. U., Kharat, K. R., Yadav, A. R., Shisodia, S. U., Damale, M. G., Sangshetti, J. N., & Pawar, R. P., (2018). Synthesis of novel α-aminophosphonate derivatives, biological evaluation as potent antiproliferative agents and molecular docking. *Chemistry Select, 3*(20), 5552–5558.
42. Vera, M. I. S., & Fernandes, A. C., (2018). Direct synthesis of α-aminophosphonates from biomass resources catalyzed by $HReO_4$. *Green Chemistry*. doi: 10.1039/C8GC01343H.
43. Erika, B., Adam, T., Dorottya, K., Bela, M., Konstantin, K., Matyas, C., & Gyorgy, K., (2017). Synthesis and utilization of optically active α-aminophosphonate derivatives by Kabachnik-fields reaction. *Tetrahedron, 73*, 5659–5667.
44. Esmaeil, M., Hossein, G., & Ali, K., (2017). A new procedure for synthesis of α-aminophosphonates by aqueous formic acid as an effective and environment-friendly organocatalyst, *J. Chem. Sci., 129*(12), 1883–1891.
45. Taheri-Torbati, M., Eshghi, H., Rounaghi, S. A., Shiri, A., & Mirzaei, M., (2017). Synthesis, characterization, and application of nitrogen–sulfur-doped carbon spheres as an efficient catalyst for the preparation of novel α-aminophosphonates. *Journal of the Iranian Chemical Society, 14*(9), 1971–1982.
46. Hossein, G., Afsaneh, R., & Hamid, R. E. Z., (2016). Highly efficient solvent free synthesis of α-aminophosphonates catalyzed by recyclable nano magnetic sulfated zirconia ($Fe_3O_4@ZrO_2/SO_4^{2-}$), *RSC Adv.* doi: 10.1039/C5RA13173A.
47. Zita, R., Nóra, Z. K., Zoltan, M., & Gyorgy, K., (2016). Synthesis of α-hydroxyphosphonates, α-aminophosphonates, phosphorus, sulfur, silicon, and the related elements. *21st International Conference on Phosphorus Chemistry (ICPC-2016), 191,* 11, 12. doi: 10.1080/10426507.2016.1213261.
48. Abdelkader, H. S. C., & Lasnouni, T., (2016). An eco-friendly procedure for the efficient synthesis of diethyl α-aminophosphonates in aqueous media using natural acids as a catalyst. *Korean J. Chem. Eng., 33*(8), 2366–2373.
49. Shaikh, M. H., Subhedar, D. D., Kalam, K. F. A., Sangshetti, J. N., & Shingate, B. B., (2016). $[Et_3NH][HSO_4]$-catalyzed one-pot, solvent-free synthesis and biological evaluation of α-amino phosphonates. *Research on Chemical Intermediates, 42*(5), 5115–5131.
50. Mirzaei, M., Eshghi, H., Hasanpour, M., & Sabbaghzadeh, R., (2016). Synthesis, characterization, and application of [1-methylpyrrolidin-2-one-SO_3H]Cl as an efficient catalyst for the preparation of α-aminophosphonate and docking simulation of ligand

bond complexes of cyclin-dependent kinase 2. *Phosphorus, Sulfur and Silicon and the Related Elements, 191*(10), 1351–1357.

51. Mirzaei, M., Eshghi, H., Rahimizadeh, M., Bakavoli, M., Matin, M. M., Hosseinymehr, M., Rudbari, H. A., & Bruno, G., (2015). An eco-friendly three component manifold for the synthesis of α-aminophosphonates under catalyst and solvent-free conditions, x-ray characterization and their evaluation as anticancer agents. *Journal of the Chinese Chemical Society, 62*(12), 1087–1096.
52. Eshghi, H., Mirzaei, M., Hasanpour, M., & Mokaber-Esfahani, M., (2015). Benzimidazolium dicationic ionic liquid as an efficient and reusable catalyst for the synthesis of α-aminophosphonates and bis (α-aminophosphonates) under solvent-free condition. *Phosphorus, Sulfur and Silicon and the Related Elements, 190*(10), 1606–1620.
53. Yu, Y. Q., & Xu, D. Z., (2015). A simple and green procedure for the one-pot synthesis of α-aminophosphonates with quaternary ammonium salts as efficient and recyclable reaction. *Media Synthesis (Germany), 47*(13), 1869–1876.
54. Peng, H., Sun, S., Hu, Y., Xing, R., & Fang, D., (2015). Clean procedure for the synthesis of A-aminophosphonates catalyzed by choline-based ionic liquid. *Heteroatom Chemistry, 26*(3), 215–223.
55. Mehdi, S. F., Ali, H. R., Somayyeh, K., Ahmad, B., & Mohammad, M. H., (2014). Synthesis of α-aminophosphonates in the presence of a magnetic recyclable Fe_3O_4@SiO_2-2mimSO_3H nanocatalyst. *Bull. Chem. Soc. Jpn., 87*(9), 982–987.
56. Guna, S. R., Kunda, U. M. R., Chereddy, S. S., Sarva, S. S., Buchammagari, H., Sirasanagandla, S., & Cirandur, S. R., (2014). Neat synthesis and antioxidant activity of α-aminophosphonates. *Arabian Journal of Chemistry, 7,* 833–838.
57. Sundar, C. S., Reddy, N. B., Prasad, S. S., Rao, K. U. M., Prakash, S. H. J., & Reddy, C. S., (2014). The synthesis and bioactivity of dimethyl (2,3-dihydrobenzo[b][1,4]dioxin-6-yl)(arylamino)methylphosphonates. *Phosphorus, Sulfur and Silicon and the Related Elements, 189*(4), 551–557.
58. Mulla, S. A. R., Pathan, M. Y., Chavan, S. S., Gample, S. P., & Sarkar, D., (2014). Highly efficient one-pot multi-component synthesis of α-aminophosphonates and bis-α-aminophosphonates catalyzed by heterogeneous reusable silica supported dodecatungstophosphoric acid (DTP/SiO_2) at ambient temperature and their antitubercular evaluation against mycobactrium tuberculosis. *RSC Advances, 4*(15), 7666–7672.
59. Tekale, S. U., Kauthale, S. S., Shaikh, R. U., Marathe, R. P., Nawale, R. B., & Pawar, R. P., (2014). Aluminum nitride catalyzed solvent-free synthesis of some novel biologically active α-aminophosphonates. *Journal of the Iranian Chemical Society, 11*(3), 717–724.
60. Ya-Qin, Y. P., (2013). An efficient and convenient procedure for the one-pot synthesis of α-aminophosphonates from aryl azides under solvent-free conditions. *Synthesis, 45,* 2545–2550.
61. Shaterian, H. R., Farbodeh, J., & Mohammadnia, M., (2013). Nano-TiO_2: An eco-friendly and clean reusable heterogeneous catalyst for preparation of α-aminophosphonates under ambient and solvent-free conditions. *Phosphorus, Sulfur and Silicon and the Related Elements, 188*(7), 850–854.
62. Fang, D., Cao, Y., & Yang, J., (2013). Clean procedure for the synthesis of α-aminophosphonates catalyzed by biodegradable ionic liquid. *Phosphorus, Sulfur and Silicon and the Related Elements, 188*(7), 826–832.

63. Sundar, C. S., Srinivasulu, D., Nayak, S. K., & Reddy, C. S., (2012). Tween-20: An efficient catalyst for one-pot synthesis of α-aminophosphonates in aqueous media. *Phosphorus, Sulfur and Silicon and the Related Elements, 187*(4), 523–534.
64. Dake, S. A., Raut, D. S., Kharat, K. R., Mhaske, R. S., Deshmukh, S. U., & Pawar, R. P., (2011). Ionic liquid promoted synthesis, antibacterial and *in vitro* antiproliferative activity of novel α-aminophosphonate derivatives. *Bioorganic and Medicinal Chemistry Letters, 21*(8), 2527–2532.
65. Sobhani, S., & Vafaee, A., (2010). *J. Iran. Chem. Soc., 7,* 227–236.
66. Jafari, A. A., & Nazarpour, M., & Alibeik. M. A., (2010). *Heteroatom Chemistry, 21*(6).
67. Taher, M. M., Khorassani, H., Sayyed, M. H., Hazeri, R., Sajadikhah, N., Seyed, S., & Mohsen, R., (2010). *Chin. J. Chem., 28,* 285–288.
68. Akbari, J., & Heydari, A., (2009). *Tetrahedron Letters, 50,* 4236–4238.
69. Bhanushali, M. J., Nandurkar, N. S., Jagtap, S. R., & Bhanage, B. M., (2009). $ZrOCl_2 \cdot 8H_2O$: An efficient catalyst for one-pot synthesis of α-aminophosphonates under solvent-free conditions. *Synth. Commun., 39,* 845–859.
70. Kassaee, M. Z., Movahedi, F., & Masrouri, H., (2009). *Synlett., 8,* 1326–1330.
71. Keglevich, G., & Szekrényi, A., (2008). *Letters in Organic Chemistry, 5,* 616–622.
72. Mitragotri, S. D., Pore, D. M., Desai, U. V., & Wadgaonkar, P. P., (2008). *Cat. Commun., 9,* 1822–1826.
73. Mohammad, S., Baharfar, V. R., Tajbakhsh, M., Heydari, A., Baghbanian, S. M., & Khaksar, S., (2008). *Tetrahedron Lett., 49,* 6501–6504.
74. Bhattacharya, A. K., & Rana, K. C., (2008). *Tetrahedron Lett., 49,* 2598–2601.
75. Mahmood, T., Akbar, H., Heshmatollah, A., Mercedeh, G., & Samad, K., (2008). *Synthesis,* 352–354.
76. Bhagat, S., & Chakraborti, A. K., (2008). *J. Org. Chem., 73,* 6029–6032.
77. Hosseini-Sarvari, M., (2008). *J. Iran. Chem. Soc., 5,* 118–124.
78. Keglevich, G., Baan, Z., Hermecz, I., Novak, T., & Odinets, I. L., (2007). *Current Organic Chemistry, 11,* 107–126.
79. Li, C., Song, B., Yan, K., Xu, G., Hu, D., Yang, S., Jin, L., Xue, W., & Lu, P., (2007). *Molecules, 12,* 163–172.
80. Bhagat, S., & Chakraborti, A. K., (2007). *J. Org. Chem., 72,* 1263–1270.
81. Błaszczyk, R., & Gajda, T., (2007). *Tetrahedron Lett., 48,* 5859–5863.
82. Song, B. A., Zhang, G. P., Yang, S., Hu, D. Y., & Jin, L. H., (2007). *Current Organic Chemistry, 11*(1), 107–126.
83. Bhanushali, M. J., Nandurkar, N. S., Jagtap, S. R., & Bhanage, B. M., (2009). *Synthetic Communications, 39,* 845–859.
84. Jin, L., Song, B., Zhang, G., Xu, R., Zhang, S., Gao, X., Hu, D., & Yang, S., (2006). *Bioorg. Med. Chem. Lett., 16,* 1537.
85. Peipei, S., Xin, Y., & Zhixin, H., (2006). *Journal of Chemical Research, 4,* 240–241.
86. Jie, W., Wei, S., Wei, Z. W., & Hong-Guang, X., (2006). *Chinese Journal of Chemistry, 24,* 1054–1057.
87. Mu, X. J., Lei, M. Y., Zoua, J. P., & Zhang, W., (2006). *Tetrahedron Let., 47,* 1125–1127.
88. Xu, Y., Yan, K., Yan, B., Yan, X. G., Yang, S., Xue, W., Hu, D., Lu, P., Ouyang, G., & Zhuo, C. L. J., (2006). *Molecules, 11,* 666–676.
89. Davis, F. A., Lee, S. H., & Xu, H., (2004). *J. Org. Chem., 69,* 3774–3781.
90. Firouzabadi, H., Iranpoor, N., & Sobhani, S., (2004). *Synthesis, 16,* 2692–2696.
91. Yadav, J. S., Reddy, B. V., & Sreedhar, S. P., (2003). *Adv. Synth. Catal., 345,* 564–567.

92. Chandrasekhar, S., Narsihmulu, C., Sultana, S. S., Saritha, B., & Prakash, S. J., (2003). *Synlett., 4,* 505–506.
93. Kaboudin, B., (2003). *Tetrahedron Lett., 44,* 1051–1053.
94. Davis, F. A., Wu, Y., Yan, H., McCoull, W., & Prasad, K. R., (2003). *J. Org. Chem., 68,* 2410–2419.
95. Ranu, B. C., & Hajra, A., (2002). *Green Chemistry, 4,* 551–554.
96. Kunz, H., & Laschat, S., (1992). *Synthesis,* 90.
97. Yadav, J. S., Reddy, B. V. S., Sarita, R. K., Bhaskar, R. K., & Prasad, A. R., (2001). *Synthesis,* p. 2277.
98. Ha, H. J., & Nam, G. S., (1992). *Synth. Commun., 22,* 1143.
99. Kudrimoti, S., & Bommena, V. R., (2005). *Tetrahedron Lett., 46,* 1209.
100. (a) Bhagat, S., & Chakraborti, A. K., (2007). *J. Org. Chem., 72,* 1263. (b) Saidi, M. R., & Azizi, N., (2002). *Synlett.,* 1347.
101. (a) Firouzabadi, H., Iranpoor, N., & Sobhani, S., (2004). *Synthesis,* p. 2692. (b) Ghosh, R., Maiti, S., Chakraborty, A., & Maiti, D. K., (2004). *J. Mol. Catal. A: Chem., 210,* 53. (c) Azizi, N., Rajabi, F., & Saidi, M. R., (2004). *Tetrahedron Lett., 45,* 9233.
102. Chandrasekhar, S., Jaya, P. S., Jagadeshwar, V., & Narsihmulu, C., (2001). *Tetrahedron Lett., 42,* 5561.
103. Ranu, B. C., Hajra, A., & Jana, U., (1999). *Org. Lett., 1,* 1141.
104. Reddy, Y. T., Reddy, P. N., Kumar, B. S., Rajput, P., Sreenivasulu, N., & Rajitha, B., (2007). *Phosphorous, Sulphur Silicon Relat. Elem., 182,* 161.
105. Ambica, K. S., Taneja, S. C., Hundal, M. S., & Kapoor, K. K., (2008). *Tetrahedron Lett., 49,* 2208.
106. Odinets, I. L., Artyushin, O. I., Shevchenko, N., Petrovskii, P. V., Nenajdenko, V. G., & Roschenthaler, G. V., (2009). *Synthesis, 4,* 577–582.
107. Tian, Y. P., Xu, F., Wang, Y., Wang, J. J., & Li, H. L., (2009). *Journal of Chemical Research, 2,* 78–80.
108. Heydari, A., Khaksar, S., & Tajbakhsh, M., (2009). *Tetrahedron Lett., 50,* 77–80.
109. Rezaei, Z., Firouzabadi, H., Iranpoor, N., Ghaderi, A., Jafari, M. R., Jafari, A. A., & Zare, H. R. E., (2009). *J. of Med. Chem., 44,* 4266–4275.
110. Syamala, M., (2009). *Organic Preparations and Procedures International, 41,* 1–68.
111. Kaboudin, B., & Jafari, E., (2008). *Synlett., X,* A–C.
112. Wang, B., Miao, Z. W., Wang, J., Chen, R. Y., & Zhang, X. D., (2008). *Amino Acids, 35,* 463–468.
113. Sayed, A. T. E., (2008). *Arkivoc., II,* 71–79.
114. Chen, Q., Li, J., & Yuan, C., (2008). *Synthesis, 18,* 2986–2990.
115. Sobhani, S., Mozaffar, A. E. S., & Jalili, F., (2008). *Journal of Organometallic Chemistry, 693*(21/22), 3313–3317.
116. Matveeva, E. D., Podrugina, T. A., Prisyazhnoi, M. V., Rusetskaya, I. N., & Zefirov, N. S., (2007). *Russian Chemical Bulletin, International Edition, 56*(4), 798–805.
117. Hua, F., Meijuan, F., Xiaoxia, L., Guo, T., & Yufen, Z., (2007). *Heteroatom Chemistry, 18*(1), 9–15.
118. Heydari, A., & Arefi, A., (2007). *Cat. Commun., 8,* 1023–1026.
119. Narayana, R. M. V., Siva, K. B., Balakrishna, A., Reddy, C. S., Nayak, S. K., & Reddy, C. D., (2007). *Arkivoc., XV,* 246–254.
120. Kaboudin, B., Moradi, K., & Sardarian, A. R., (2007). *Arkivoc., XIII,* 210–217.
121. Kaboudin, B., & Karimi, M., (2006). *Bioorg. and Med. Chem. Lett., 16,* 5324–5327.

122. Xu, F., Luo, Y., Wu, J., Shen, Q., & Chen, H., (2006). *Heteroatom Chemistry, 17*(5), 389–392.
123. Matveeva, E. D., & Zefirov, N. S., (2006). *Russian J. of Org. Chem., 42*(8), 1237–1238.
124. Paraskar, A. S., & Sudalai, A., (2006). *Arkivoc, X,* 183–189.
125. Wu, J., Sun, W., Xi, H. G., Sun, X., (2006). *Org. Biomol. Chem., 4,* 1663–1666.
126. Meenen, E. V., Moonen, K., Acke, D., & Stevens, C. V., (2006). *Arkivoc., I,* 31–45.
127. Kaboudin, B., & Moradi, K., (2005). *Tetrahedron Lett., 46,* 2989–2991.
128. Dziemidowicz, J., Witt, D., Sliwka-Kaszynska, M., & Rachon, J., (2005). *Synthesis, 4,* 569–574.
129. Ghosh, R., Maiti, S., Chakraborty, A., & Maiti, D. K., (2004). *Journal of Molecular Catalysis A: Chemical, 210,* 53–57.
130. Xu, F., Luo, Y., Deng, M., & Shen, Q., (2003). *Eur. J. Org. Chem.,* 4728–4730.
131. Matveeva, E. D., Podrugina, T. A., Tishkovskaya, E. V., Tomilova, L. G., & Zefirov, N. S., (2003). *Synlett., 15,* 2321–2324.
132. Song, B. A., Wu, Y. L., Yang, S., Hu, D. Y., He, X. Q., & Jin, L. H., (2003). Synthesis and bioactivity of α-aminophosphonates containing fluorine. *Molecules, 8,* 186–192.
133. Saidi, M. R., & Azizi, N., (2002). *Synlett., 8,* 1347–1349.
134. Chandrasekhar, S., Jaya, P. S., Jagadeshwar, V., & Narsihmulu, C., (2001). *Tetrahedron Lett., 42,* 5561–5563.
135. Green, D., Elgendy, S., Patel, G., Jehan, A. B., Skordalakes, E., Husman, W., Kakkar, V. V., & Deadman, J., (1996). *Phosphorus, Sulfur, and Silicon and the Related Elements, 113*(1), 303–306.
136. Gancarz, R., (1993). *Phosphorus, Sulfur and Silicon, 83,* 59–64.
137. Akiyama, T., Sanada, M., & Fuchibe, K., (2003). *Synlett.,* 1463–1464.
138. Yadav, J. S., Reddy, S. V. B., Sarita, R. K., Bhaskar, R. K., & Prasad, R. A., (2001). *Synthesis, 15,* 2277–2280.
139. Ranu, B. C., Hajra, A., & Jana, U., (1999). *Org Lett., 1,* 1141–1143.
140. Kaboudin, B., & Nazari, R., (2001). *Tetrahedron Lett., 42,* 8211–8213.
141. Yadav, S. J., Reddy, S. V., & Sreedhar, R., (2002). *Green Chem., 4,* 436–438.
142. Bhattacharya, A. K., & Kaur, T., (2007). *Synlett.,* pp. 745–748.
143. Ambica, S., Kumar, S. C., Taneja, M. S., Hundal, K., & Kapoor, K., (2008). *Tetrahedron Lett., 49,* 2208–2212.
144. Chandrasekhar, S., Prakash, S. J., Jagadeshwar, V., & Narsihmulu, C., (2001). *Tetrahedron Lett., 42,* 5561–5563.
145. Mitragotri, S. D., Pore, D. M., Desai, U. V., & Wadgaonkar, P. P., (2008). *Catal. Commun., 9,* 1822–1826.

CHAPTER 6

An Overview of Polyelectrolyte-Based Nanoplex

SWATI G. TALELE, SHWETA S. GEDAM, and AKSHADA A. BAKLIWAL

Department of Pharmaceutics, Sandip Institute of Pharmaceutical Sciences, Mahiravani, Nashik, Maharashtra, India,
E-mail: swatitalele77@gmail.com (S. G. Talele)

ABSTRACT

A recent investigation shows that nanoparticles (NPs) based on nanotechnology widely preferred over conventional drug delivery system to conquer the challenges in it. New inclination in nanotechnology is nanoplex formulation, design, and development. The cationic and anionic charge on drug nanoparticle with oppositely charged polyelectrolytes as a dextran sulfate is an essential component of nanoplex complex formation. The enormous literature reveals that the nanoplex shows better efficiency as compared to other NPs. Nanoplex has noteworthy benefits over other NPs such as ease of preparation, cost-effective and minimal requirement of sophisticated instruments with less number of solvents. One of the additional benefits of nanoplex is enhancing the solubility of BCS Class II and Class IV drugs. Characterization of nanoplex formulation is done by scanning electron microscopy (SEM), differential scanning calorimetry (DSC), x-ray diffraction and dialysis studies. Production yield, complexation efficiency, drug loading, particle size, and zeta potential are some of the evaluation parameters of nanoplex. Nanoplex shows ample of applications including the drug delivery of protein and peptides, Drug targeting in cancer treatment, Enhancement of solubility and dissolution rate, Drug delivery to the brain, nasal, and Nanoplex for gene delivery. The key components of the chapter are preparation, mechanism, features, evaluation parameters, and applications of nanoplex.

6.1 INTRODUCTION

Nanoscience and nanotechnology deals with the extremely small things which can be used in biology, physics, chemistry, material science, and engineering. In particular, nanotechnology accompanied at the nanoscale i.e., about 1 to 100 nanometers and with the process which occurs at the molecular level. Red blood corpuscles (RBC), DNA, viruses, and water molecules are some of the structures with nanodimensions [1]. Nanotechnology has leads to tremendous progress in the multidisciplinary field which comprises both basic sciences and applied disciplines containing bioengineering, biophysics, and molecular biology [2]. Nanotechnology has an influential effect on numerous fields of medicines such as immunology, endocrinology, oncology, ophthalmology, cardiology, and pneumology [3–5]. Also in the gene delivery, brain targeting and oral vaccine formulation it has extremely specific effects.

Nanotechnology modify or develop the materials within the 1–100 nm size and makes the material lighter, stronger, faster, smaller, and more durable. In other terms, nanotechnology is the ability to construct the material using different methods and instruments to give better enactment formulation. Nanotechnology has the substantial impact on almost all areas society and industry [6].

6.2 OUTLETS OF NANOTECHNOLOGY

There are various branches of nanotechnology under which vigorous products are available. Nowadays, nanotechnology has its impact on most of the industrial applications such as several research fields. Here are the some branches which implement the nanotechnology [7]:

1. **Nanoengineering:** It is the branch emphasizes the engineering aspect rather than the applied science having a unit of measurement is nanometer. There are two techniques of nanoengineering i.e., scanning tunneling microscope (STM) and molecular self-assembly. STM deals with the small structure like the single atom while molecular self-assembly is the sequence which is synthesized and create proteins or amino acids [8].
2. **Green Nanotechnology:** It specifically embraces the green nano-technology products to minimize the upcoming human health and environmental issues by replacing the existing products with the environmental friendly nano-products [9, 10]. Green nanotechnology includes water treatment, nano remediation and solar cells [11].

3. **Wet Nanotechnology:** It is also called wet nanotech and it involves working with the large masses from small ones. This technique requires water in which the process occurs. In this, chemists, and biologists put together individual molecules to get larger scales [12].

6.3 NANOPARTICLES (NPS)

Nanoparticles (NPs) are the solid particles ranging from 10–100 nm in size. NPs are of different types and possess highly desired and specific properties.

6.3.1 *TYPES OF NANOPARTICLES (NPS)*

1. **Nanosuspensions:** It is defined as a very finely colloid, biphasic, dispersed, solid drug particle in an aqueous vehicle having a size below 1 μm. there are various routes of administration for nanosuspension such as oral, topical, parenteral, ocular, and pulmonary. In this, surfactant, and polymer are mixed using a suitable method of preparation [13–15].
2. **Nanoemulsions:** Nanoemulsions/submicron emulsions are defined as thermodynamically stable, transparent (translucent) dispersions of oil and water having the droplet size less than 100nm. Nanoemulsions are used for the stabilization of interfacial films of surfactant and co-surfactant molecules [16].
3. **Solid Lipid Nanoparticles:** These are the novel potential colloidal carrier systems which are used as alternative materials to polymers. They are similar to o/w (oil in water) emulsion, only the difference is lipid-lipid in this emulsion is replaced by the solid-lipid. In this, lipophilic drugs show better results. They have low toxicity, good compatibility and are physically stable [17, 18].
4. **Nanofibers:** These are defined as fibers having high surface-to-weight ratio with diameterless than 50–500 nm [19, 20].
5. **Nanocomposites:** These have good mechanical, electrical, thermal, and optical properties with superior strength and fracture toughness [21].
6. **Carbon Nanotubes (CNTs):** These are defined as the nanostructures of allotropes of carbon having a length-to-diameter ratio greater than 1,000,000. These CNTs are targeted drug delivery system which exhibits good carrier properties. These biomolecules

include a miscellaneous assortment of compounds such as drugs, vaccines, small peptides, proteins, nucleic acids, vitamins, and sugars [22–24].

7. **Nanopores:** These are the materials having the pore sizes in nanometers. Nanopores having the exceptional properties regarding the thermal insulation, controllable material separation, release, and also used as fillers for chemistry and catalysis. Due to their nanosize, they are used in broad range for industrial application. The most common example of nanopores is aerosol which is produced using sol-gel chemistry [25].
8. **Liposomes:** These are a minute spherical sac of phospholipid molecules which encloses a water droplet to carry drugs or other substances into the tissues. Liposomes consist of at least one lipid bilayer. It is defined as the small artificial vesicles ranging from the 50–100 nm made up of phospholipids. They are used as a drug carrier because they have the ability to prevent drug degradation, minimizes side effects and acts as a targeted drug delivery system. Liposomes mostly used in the treatment of ocular diseases and transdermal drug delivery [26].
9. **Polymeric Nanoparticles:** The solid colloidal particles having the diameter ranging from 1 nm to 1000 nm are called as polymeric NPs. Polymeric nanoparticulate drug delivery systems established using biodegradable and biocompatible polymers. Due to the biodegradable, and biocompatible polymers used in polymeric NPs they provide drug targeting and controlled drug delivery [27].
10. **Polymeric Micelles:** These are formed in the aqueous solutions which contain the amphiphilic block polymers. The polymeric micelles are self-assembled core-shell nanostructures. When the concentration of the block copolymer increases above a critical micelle concentration or certain concentration, there is the formation of micelles known as polymeric micelles [28].
11. **Dendrimers:** These are defined as the central core unit with a high degree of molecular uniformity, specific characterization of size and shape and a constricted molecular weight distribution with the highly functionalized terminal surface. Dendrimers are the branches of macromolecules. In this, the cellular uptake takes place through the endocytosis process and is bound to dendrimers within cells [29–31].
12. **Nanocapsules:** It consists of a nanoscale shell which is made up of non-toxic polymer. They are the vesicular or reservoir systems.

In this, oil/water is fundamentally restrained to a cavity which is surrounded by a tiny polymeric membrane. In vesicular systems, the polymeric membranes consist of an inner liquid core at the nanoscale. Nanocapsules are the targeted drug delivery system, i.e., they can be targeted to specific cells or locations inside the body. Intravenous or subcutaneous routes of administration are mostly preferred for the nanocapsules [32].

13. **Nanoplex:** It is a drug nanoparticle complex with an oppositely charged polyelectrolyte. Cationic or anionic drugs are made to reach with oppositely charged polyelectrolytes to form Nanoplex [33].
14. **Magnetic Nanoparticles:** It is one of the types of engineering particulate materials having a particle size less than 100nm which is influenced by the external magnetic field. In this, a core of polymer is coated with an inorganic substance. For example, dextran which is a core of polymer is coated with an inorganic substance such as iron oxide [16].
15. **Quantum Dots (QD):** These are defined as the semiconducting materials which comprise of semiconductor core with a shell for improving the optical properties. These are ranging from 10–100 A° in radius with a significant effect on imaging, in-vivo, and in-vitro detection, immunoassay, and DNA hybridization. Their main solicitation is in labeling cells and as therapeutic tools for cancer treatment [34].
16. **Carbon Allotrope Graphenes:** Graphene is made up of pure carbon. The graphenes are similar to that of graphite in case of the atoms which are arranged in the regular hexagonal pattern but in a one-atom thick sheet. Graphenes are the allotrope of carbons in which sp^2 bonded carbon atoms are densely packed in a crystal lattice of honeycomb. Graphenes is also known as the mother of graphite and CNTs [35].

6.4 NANOPLEX: AN INNOVATIVE TREND IN NANOTECHNOLOGY

The nanoplex is defined as the drug nanoparticle comprises of an oppositely charged polyelectrolyte. In this, cationic, and anionic drugs react with an oppositely charged polyelectrolyte. For nanoplex formulation, mostly BCS class II and class IV drugs are used having low aqueous solubility (Table 6.1). The goal of nanoplex formulation is to enhance the solubility and bioavailability of these types of drugs. Producing the crystalline NPs, transformation of API into its salt form or giving the API in its salt form can help to improve the bioavailability of the drugs.

TABLE 6.1 BCS Classification System

Class	Solubility	Permeability
BCS Class I	High	High
BCS Class II	Low	High
BCS Class III	High	Low
BCS Class IV	Low	Low

6.4.1 OBJECTIVES OF NANOPLEX

- Nanoplex formulation enhances the solubility.
- Also to enhance dissolution rate the nanoplex formulation are prepare.

6.4.2 PREPARATION OF NANOPLEX

Drugs dissolved in an acidic or basic aqueous solution depending upon the solubility to form cations and anions respectively. When the amphoteric drugs dissolved in aqueous acetic acid solutions, they form cations while when the acidic drugs dissolved in aqueous potassium hydroxide solution they form anions. Then this ionized drug solution is added into polyelectrolyte solution having opposite charge and soluble drug-polyelectrolyte complex is formed by the electrostatic interaction between the drug and polyelectrolyte (Figure 6.1).

6.4.3 FEATURES OF NANOPLEX

1. Nanoplex preparation method is simple.
2. In this, only mixing of two solutions of drug and polyelectrolyte is required.
3. There is no need to use heavy solvents in this.
4. It is a fast process.
5. The energy required for the formulation of nanoplex is minimum than the other NPs.
6. Also, there is no need to use sophisticated instrument for nanoplex formulation.

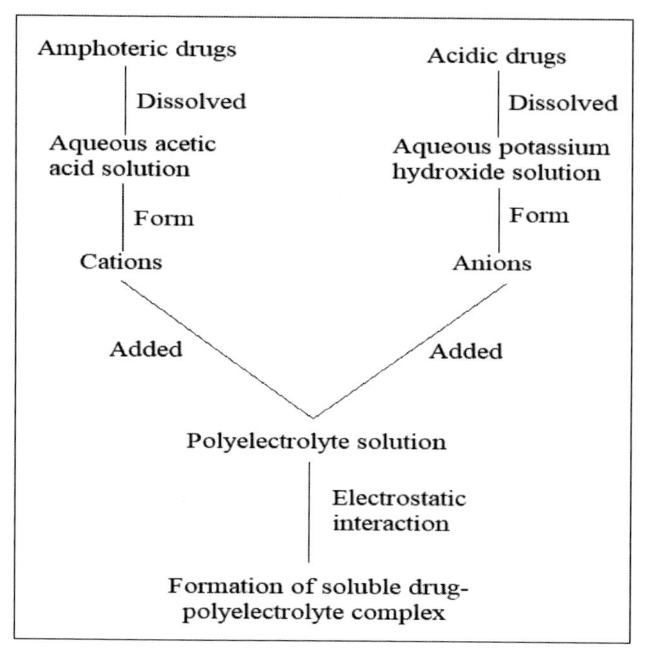

FIGURE 6.1 Preparation of nanoplex.

6.4.4 MECHANISM OF NANOPLEX FORMATION [36]

In this, drug is firstly dissolved in an acidic or alkaline solution to forms anionic or cationic drug solution. Then this ionized drug solution is mixed into an oppositely charged polyelectrolyte solution. Due to the electrostatic interaction, drug-polyelectrolyte complex forms leading to neutralization of the charges. Because of neutralization, drug solution gets to its original sparingly soluble form which leads to loss of solubility causing rapid precipitation and there is formation of nanoscale amorphous drug-polyelectrolyte nanoparticle complex (Figure 6.2).

6.4.5 ROLE OF SALT IN NANOPLEX FORMULATION [37]

In the self-assembly of nanoplex, salt plays an important role due to the electrostatic charge shielding function. Absence of the salt leads to the inhibition of spatial interactions which is due to the intermolecular repulsions between the like-charged polyelectrolyte chains. In regard with the oppositely charges interactions, consequently, there is no nanoplex precipitation in the absence of a salt (Figure 6.3). Due to the soluble amphiphilic drug and polyelectrolytes, the soluble drug-polyelectrolyte complexes may be present. The polyelectrolyte chains are in closer contact with each other due to the shielding charge in presence of the salt. Similarly, there is formation of a core-shell nanoplex owing to the contact between the drug molecules, which then precipitates upon the drug. On the other hand, the charge-shielding effect of the salt must inhibit the electrostatic interactions between the drug molecules and polyelectrolyte which are involved in the complex formation or neutralized the drug-polyelectrolyte complex charge which leads to decomplexation.

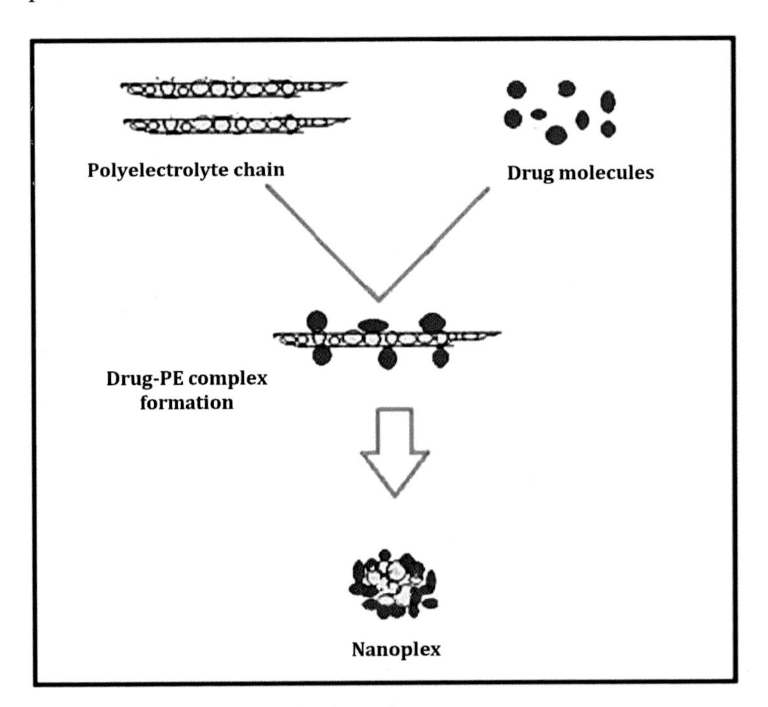

FIGURE 6.2 Mechanism of nanoplex formation.

6.4.6 EVALUATION OF NANOPLEX

1. **Flow Properties of Freeze-dried nanoplex sample [38]**

   i) **The Angle of Repose:** The angle of repose of nanoplex sample was determined by the static funnel method (Table 6.2). The accurately weighed quantity, 5 gm of S-SNEDDS was taken and was allowed to flow through the funnel freely on to the surface. The height of the funnel was adjusted in such a way that the tip of the funnel just touched the apex of the S-SNEDDS. The diameter and height of the powder cone were measured and angle of repose was calculated using the equation:

$$\theta = \tan^{-1} h/r$$

where, h, and r are the height and radius of the powder cone.

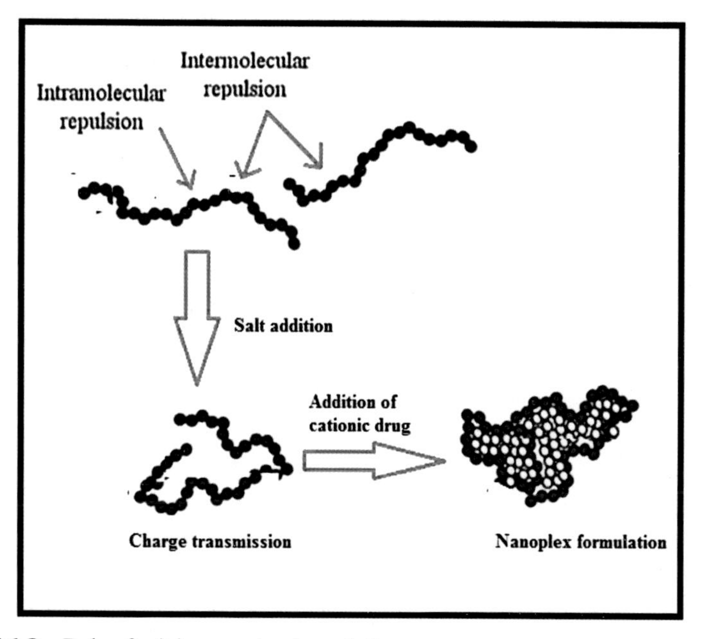

FIGURE 6.3 Role of salt in nanoplex formulation.

TABLE 6.2 Indicates the Type of Flow After Determining the Angle of Repose

Sr. No	Angle of Repose	Type of Flow
1.	< 25	Excellent
2.	25–30	Good
3.	30–40	Passable
4.	> 40	Very poor

ii) Compressibility Index: Compressibility index of the Nanoplex sample was determined by using the following equation. Carr's compressibility index (Table 6.3) is a simple test to evaluate the bulk and tap density of a powder and the rate at which it is packed down.

$$\text{Percent Carr's Index} = \frac{Tapped\ Density - Bulk\ Density}{Tapped\ Density} \times 100$$

TABLE 6.3 Indicates Carr's Index

Sr. No	Angle of Repose	Type of Flow
1.	5–15	Excellent
2.	12–16	Good
3.	18–21	Fair to passable
4.	23–35	Poor
5.	33–38	Very poor
6.	> 40	Extremely poor

iii) **Bulk Density (BD):** It was determined by pouring nanoplex sample into a graduated cylinder and the volume was recorded. It is expressed in g/mL and is given as:

$$\text{Bulk density} = \frac{Mass\ of\ powder\ (M)}{Tapped\ volume\ of\ powder\ (vt)} \times 100$$

iv) **Tapped Density (TD):** It was calculated by tapping the cylinder using a digital BD apparatus. The final volume was then read and used to calculate the TD. The tapped volume was measured after 100 tapping till constant volume is achieved. It is expressed in g/mL and is given by equation:

$$\text{Tapped density} = \frac{Mass\ of\ powder\ (M)}{Tapped\ volume\ of\ powder\ (vt)} \times 100$$

v) **Hausner's Ratio:** It is a number that is correlates to the flow ability of a powder. It was calculated by:

$$\text{Hausner's ratio} = \frac{Tapped\ Density}{Bulk\ density} \times 100$$

Ideally Hausner's ratio should be less than 1.25 indicating good flow.

2. **Angle of Spatula [39]:** The sample was dropped from a height of 2 cm on spatula which consists of steel and the spatula was slowly lifted vertically and the angle of heap formed on the spatula was calculated as the angle of spatula (Figure 6.4) and it denoted as α_S.

α_S less than 40 indicates a good flow.

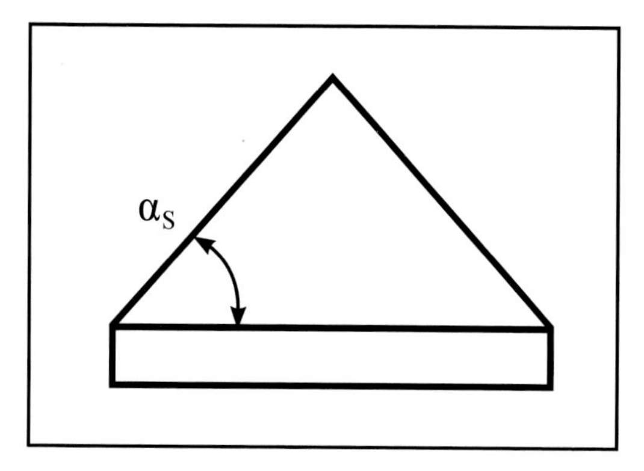

FIGURE 6.4 Representation of angle of spatula.

3. **Complexation Efficiency:** It is represented as the mass of a drug that forms a complex with the poly-electrolyte relative to the initially added drug. It is calculated by measuring the absorbance of the supernatant after the first centrifugation of the nanoplex suspension [42].
4. **Production Yield [38]:** It is obtained by considering the total nanoplex mass-produced by freeze-drying to the total mass of drug and polyelectrolyte as dextran sulfate initially added. It was calculated by using the following formula:

$$Production\ yield = \frac{Practical\ mass\ of\ nanoplex}{Theoretical\ mass\ (drug + polyelectrolyte)} \times 100$$

5. **Drug Loading [38]:** This is the actual amount of drug present in the nanoplex powder. It is calculated by dissolving 5 mg of the nanoplex powder in 20 ml of ethanol or phosphate buffer solution and measuring the absorbance of the solution after centrifugation and filtration [42].

$$Loading\ efficiency = \frac{Actual\ drug\ content}{Theoretical\ drug\ content} \times 100\sqrt{a^2 + b^2}$$

6. **Particle Size Analysis:** The particle size of the nanoplex is analyzed using a Malvern particle size analyzer by using water as dispersion medium. Particle size determination of the naoplex was determined using Dynamic light spectroscopy. The mean droplet size and poly dispersability index were obtained by photon correlation spectroscopy

(PCS) utilizing a Malvern Zetasizer. The light-scattering phenomenon was utilized for detection in which the statistical intensity fluctuations of the scattered light from the particles in the measuring cells are calculated. Prior to the measurements, to produce suitable scattering intensity all formulated samples of nanoplex was diluted with double-distilled water. The polydispersability index and Z-average were obtained at an angle of 90° by utilizing disposable cells, which consist of polystyrene at 25°C, which were equilibrated for 110 s [40, 41].

7. **Polydispersibility Index (PDI):** Polydispersibility which estimates size range of particles in formulation is determined by:

$$\frac{No.of particles\ having\ size\ greater\ than\ 100nm}{No\ of\ particles\ having\ size\ less\ than\ 100nm}$$

It is expressed in terms of polydispersibility index (PDI). An ideal nanoplex should be widely distributed with particles less than 100 nm and so PDI should be less than 0.3 [40].

8. **Zeta Potential:** It was measured by determining the electrophoretic mobility using the Malvern Zetasizer, (Nano ZS 90, Malvern Ltd., and Malvern, UK) reflecting the electric charge on the particle surface and indicating the physical stability of colloidal systems. The nanoplex measurements were performed following dilution in double-distilled water. It was measured using the Dip cell by applying field strength of 20 V/cm. Zeta potential determined by the Zeta-meter was monitored at 25°C at a scattering angle 173° [40].

9. **Infra-Red Spectroscopy:** It is used to estimate the interaction between nanoplex and the drug molecules in the solid state. Nanoplex bands often change only slightly upon complex formation and if the fraction of the guest molecules encapsulated in the complex is less than 25%, bands which could be assigned to the included part of the guest molecules are easily masked by the bands of the spectrum of nanoplex. The technique is not generally suitable to detect the inclusion complexes and is less clarifying than other methods. The application of the Infra-red spectroscopy is limited to the drugs having some characteristic bands, such as carbonyl or sulfonyl groups. Infrared (IR) spectral studies give information regarding the involvement of hydrogen in various functional groups. This generally shifts the absorbance bands to the lower frequency, increases the intensity, and widens the band caused by stretching vibration of the group involved in the formation of the

hydrogen bonds. Hydrogen bond at the hydroxyl group causes the largest shift of the stretching vibration band.

10. **Differential Scanning Calorimetry (DSC):** DSC is performed to determine the interactions between the drug and polyelectrolyte. Melting point variations is an indication of interaction between drug and polyelectrolyte. DSC was performed on Metler Toledo DSC instrument. Thenanoplex heated form 30°C to 400°C for 10°C/min. in an aluminum pan [40].
11. **Powder X-Ray Diffraction:** These patterns of samples are determining using a powder X-ray diffractometer at a scan rate of 10 min^{-1}. With 2Ø range from 10 to 80. The X-Ray diffractogram of drug and nanoplex were recorded by means of a Panalytical X'pert Pro MPD with Cu Kα as an anode which used as source of radiation. Standard run were performed at a voltage of 40 KV and at a current of 30 MA. In 2θ angle range which was given in between 5 to 50° at a rate of 1°/min, the diffractograms were taken. Detector was used as Xcelerator with diffracted beam monochromator. Powder X-ray diffraction of nanoplex gives a suggestion about nature of the sample as crystalline or amorphous [40, 41].
12. **Scanning Electron Microscopy (SEM):** The surface morphology of the sample is determined by scanning electron microscope, where the samples was fixed on a brass stub using double-sided carbon tape and made electrically conductive by coating with a thin layer of gold by sputter coater [40, 41].
13. **Dissolution Study:** Most probably, the dialysis bag method is utilized to determine dissolution testing of the nanoplex and then determine the drug release [42].
14. **Saturation Solubility Study [38, 40]:** The solubility of the drug and that of the formulation are determined using the orbital flask shaker method. The concentration of the drug is determined from the absorbance, through spectrophotometric analysis [41].
15. **Drug and Nanoplex Kinetic Models [38]:** There are various kinetic models that show the release of the drug (see Table 6.4).
16. **Stability Study:** The stability of the nanoplex is determined by placing it in stability chamber and as per ICH guidelines, its drug content is determined. The same process was followed for both drug and nanoplex. The absorbance was taken by UV-Visible spectroscopy.

TABLE 6.4 Dissolution Kinetic Models

Model	Mathematical Equation	Release Mechanism
Zero-order	C=C0-K0t	Diffusion Mechanism
First-order	logC=logCo-Kt/2.303	Fick's first law, diffusion mechanism
Higuchi model	C=[D(2qt-Cs)Cs t]1/2	Diffusion medium based mechanism in Fick's first law
Korsemeyer-peppas model	Ct/C∞=Ktn	Semi-empirical-based mechanism model diffusion
Hixson-Crowell model	C01/3-Ct1/3=KHCt	Erosion release mechanism
Weibull model	C=C0 [1-exp[ ]	Empirical model, life-time distribution function
Baker-Lonsdale model	f1=3/2[1-(1-Ct/C∞)2/3] Ct/C∞=Kt	Release of drug from spherical matrix
Hopfenberg model	Ct/C∞=1-[1-K0t/CL a]n	Erosion mechanism
Gompertz model	Ct=Cmaxexp[-αeβ log t]	Dissolution model

6.5 APPLICATIONS OF NANOPLEX

6.5.1 *IMPROVEMENT IN SOLUBILITY AND DISSOLUTION RATE OF POORLY WATER-SOLUBLE DRUGS*

Nanoplex formation can be applicable for BCS Class II and for IV drugs. This technique shows improvement in solubility and dissolution rate and leads to enhancement in bioavailability.

The enormous literature reveals that ciprofloxacin, cipodoximeproxetil ibuprofen, and curcumin efficiently converted into amorphous form of nanoplex.

6.5.2 *NANOPLEX DELIVERY TO THE BRAIN*

The blood-brain barrier (BBB) is the most important factor limiting the development of a new drug for the central nervous system. The BBB is characterized by relatively impermeable endothelial cells with tight junctions and it regulates the movement of drug. It also prevents transport of water-soluble molecules from the blood circulation into the CNS and with the action of enzymes can decrease the concentration of lipid-soluble molecules in the brain. Subsequently, the BBB only allows selective transport of those molecules that are vital for brain function. From the literature survey, it has been concluded that NPs may be utilized as non-viral gene delivery vectors and widely it is used in various diseases such as HIV-1, AIDS, dementia, and cerebral ischemia. Nanoplex have been evaluated for the specificity and efficiency of QD complexes [with MMP-9-siRNA (nanoplex) in down-regulating the expression of the MMP-9 gene in the brain microvascular endothelial cells (BMVECs) that constitute the BBB. Adela Bonoiu and the group discussed the use of a novel nanoplexs iRNA delivery system in modulating MMP-9 activity in BMVECs and other MMP-9-producing cells. This application suggests that nanoplex keep up the integrity of the BBB with reduction in neuroinflammation [44].

6.5.3 NANOPLEX FOR GENE DELIVERY

Various nonviral vector gene therapies accepted owing to its noteworthy benefits including biodegradability, least toxicity, and biocompatibility with low immunogenicity. At present researchers, focuses on dendrimers, liposomes, and cationic polymers as a non-viral vector for the drug delivery system. Cationic polymer has great potential to neutralize anionic nature of DNA and effectively get complex and enter the cells without any barrier. For barrier-free delivery of DNA, various cationic polymers reported including carbohydrate-based polymers such as chitosan and dextran derivatives, synthetic amino acid polymers such as poly(1-lysine) and polyethyleneimine (PEI), natural DNA binding proteins such as histones [45].

6.5.4 DRUG TARGETING IN CANCER TREATMENT

Many different cationic lipids have been synthesized and a cationic lipid consists of the following parts firstly, a hydrophobic lipid group, which is necessary for the formation of a micellar structure and have the ability to interact with cell membranes with high transfection activity with low toxicity. Second part as a linker group, such as an ester, an amido group, or a carbamate which control the conformational flexibility, degree of stability, biodegradability, and gene transfection efficiency and third part as a positively charged head-group, mainly composed of cationic amines. Cationic NPs composed of OH-Chol (NP-OH) could deliver siRNA with high transfection efficiency in vitro when the nanoparticle/siRNA complex (nanoplex) is formed in a NaCl solution [46].

6.5.5 DRUG DELIVERY OF PROTEINS AND PEPTIDES

Major innovation in biotechnology and biochemistry has led to finding of a large number of bioactive molecules and vaccines based on peptides and proteins. The development of proper carriers remains a challenge because proteins and peptides more prone to degradation in GIT due to digestive enzymes and leads to bioavailability problems. Proteins carry charge and forms complex with polyelectrolyte very effectively therefore, they are considered as an appropriate candidate for complex formulation [47].

6.5.6 NANOPLEX FOR NASAL DRUG DELIVERY

Development of nasal Ropinirole nanoplex which was applicable in Parkinsonism without any hurdle formulated and histological studies revealed that enhancement in permeability and liphophicity and controlled drug release as compared to pure drug was elucidated. To grab pharmaceutical market for this novel nanocarrires, brain pharmacokinetic and neurotoxicity study should be widened [48].

6.6 CONCLUSION

Efficiently nanoplex formulation can be utilized as a novel approach for various chemical entities which are water insoluble. Most of the studies also revealed that drugs showing amphiphilic characteristics and that are enormously soluble in weak acid and base can be complexed with polyelectrolyte. This leads to the formation of amorphous nanoplex and considered as simple and green technology with ease of manufacturing and scale up. Nanoplex shows high encapsulation efficiency with better production yield, spherical shaped NPs and successfully converted into its amorphous form.

Finally, nanoplex formulation is inoculated as capsule dosage forms and therefore, it is essential that nanoplex should be exploited for the development of suitable dosage forms.

KEYWORDS

- **dextran sulfate**
- **drug delivery system**
- **gene delivery**
- **nanoplex**
- **peptides**
- **polyelectrolyte complex**

REFERENCES

1. Ochekpe, N. A., Olorunfemi, P. O., & Ngwuluka, N. C., (2009a). Nanotechnology and drug delivery. Part 1: Nanostructures for drug delivery. *Trop. J. Pharm. Res., 8*, 265–274.
2. Manivannan, R., (2011). Nanotechnology: A review. *J. Appl. Pharm. Sci., 1*, 8–16.
3. Bhattacharyya, D., Singh, S., Satnalika, N., Khandelwal, A., & Jeon, S., (2009). Nanotechnology, big things from a tiny world: A review. *Int. JUE-Ser. Sci. Technol., 2*, 29–38.
4. Malakar, J., Ghosh, A., Basu, A., & Nayak, A., (2012). Nanotechnology: A promising carrier for intracellular drug delivery system. *Int. Res. J. Pharm., 3*, 36–40.
5. Lamba, D., (2006). A brief review on nanotechnology. *Impulse, 2*, 58–63.
6. *Center Responsible for Nanotechnology*. http://www.crnano.org/whatis.html (accessed on 26 February 2020).
7. Ellin, D. M. (2006). *Nanotechnology: A Brief Literature Review.* PhD Food Research Institute, University of Wisconsin–Madison, Madison, WI 53706.
8. Lusk, M. T., & Lincoln, D. C., (2008). "Nanoengineering defect structures on graphene." *Physical Review Letters, 100*(17), 175503.
9. *Environment and Green Nano-Topics–Nanotechnology Project*. Retrieved 11 September 2011.
10. *National Nanotechnology Initiative*. http://www.nano.gov (accessed on 26 February 2020).
11. *Nanotechnology in Water Treatment*. Retrieved 3 November 2013.
12. M. Kaur, G. Singh, & K. Khanna (2005). *Nanotechnology: A Review,* http://faculty.tamu-commerce.edu/dyeager/599/newtechnologyparti_files/v3_slide0205.htm (accessed on 26 February 2020), Contemporary Tech.
13. Nagare, S. K., Ghughure, S. M., Salunke, S. B., Jadhav, S. G., & Dhore, R. J., (2012). A review on: Nanosuspension an innovative acceptable approach in the novel delivery system. *Uni. J. Pharm., 1*, 19–31.
14. Prabhakar, C. H., & Balakrishna, K., (2011). A review on nanosuspension in drug delivery. *Int. J. Pharma. Bio. Sci., 2,* 549–558.

15. Verma, A., & Bindal, M. C., (2012). Nanosuspension: Advantages and disadvantages. *Ind. J. Nov. Drug Delivery, 4*, 179–188.
16. Indira, T. K., & Lakshmi, P. K., (2010). Magnetic nanoparticle: A review. *Int. J. Pharm. Sci. Nanotechnol., 3*, 1035–1042.
17. Ekambaram, P., Sathali, A., & Priyanka, K., (2012). Solid lipid nanoparticle: A review. *Sci. Rev. Chem. Commun., 2*, 80–102.
18. Pragati, S., Kuldeep, S., Ashok, S., & Satheesh, M., (2009). Solid lipidnanoparticle: A promising drug delivery technology. *Int. J. Pharm. Sci. Nanotech., 2*, 509–516.
19. Kattamuri, S. B., Potti, L., Vinukonda, A., Bandi, V., Changantipati, S., & Mogili, R. K., (2012). Nanofibers in pharmaceuticals: A review. *Am. J. Pharmtech. Res., 2*, 187–212.
20. Dineshkumar, B., Krishnakumar, K., Bhatt, A. R., John, A., Paul, D., Cherian, J., et al., (2012). Nanofibers: Potential applications in wound care management. *Adv. Poly. Sci. Tech., 2*, 30–32.
21. Sambarkar, P. P., Patwekar, S. L., & Dudhgaonkar, B. M., (2012). Polymer nanocomposites: An overview. *Int. J. Pharm. Pharma. Sci., 4*, 60–65.
22. Hirlekar, R., Yamagar, M., Garse, H., Mohit, V. I. J., & Kadam, V., (2009). Carbon nanotubes and its application: A review. *Asian J. Pharm. Clin. Res., 2*, 17–27.
23. Mishra, R., & Mishra, A., (2013). Review on potential applications of carbon nanotubes and nanofibers. *Int. J. Pharm. Rev. Res., 3*, 12–17.
24. Gurjar, P. N., Chouksey, S., Patil, G., Naik, N., & Agrawal, S., (2013). Carbon nanotubes: Pharmaceutical applications. *Asian J. Biomed Pharm. Sci., 3*, 8–13.
25. Wanunu, M., (2012). Nanopores: A journey towards DNA sequencing. *Phys, Life Rev., 9*, 125–158.
26. Ochekpe, N. A., Olorunfemi, P. O., & Ngwuluka, N. C., (2009b). Nanotechnology and drug delivery. Part 2: Nanostructures for drug delivery. *Trop. J. Pharm. Res., 8*, 275–287.
27. Jawahar, N., & Meyyanathan, S. N., (2012). Polymeric nanoparticles for drug delivery and targeting: A comprehensive review. *Int. J. Health Allied Sci., 1*, 217–223.
28. Xu, W., Ling, P., & Zhang, T., (2013). Polymeric micelles, a promising drug delivery system to enhance bioavailability of poorly water-soluble drugs. *J. Drug Delivery,* 1–15.
29. Mishra, I. N., (2011). Dendrimer: A novel drug delivery system. *J. Drug Delivery Ther., 1*, 70–74.
30. Shishu, G., & Maheshwari, M., (2009). Dendrimers: The novel pharmaceutical drug carriers. *Int. J. Pharm. Sci. Nanotech., 2*, 493–502.
31. Trivedi, V., Bhimani, B., Patel, U., Daslaniya, D., Patel, G., & Vyas, B., (2012). Dendrimer: Polymer of 21st century. *Int. J. Pharm. Res. Biosci., 1*, 1–21.
32. Dineshkumar, B., Krishnakumar, K., John, A., Paul, D., Cherian, J., & Panayappan, L., (2013). Nanocapsules: A novel nano-drug delivery system. *Int. J. Res. Drug Delivery, 3*, 14–16.
33. Cheow, W. S., & Hadinoto, K., (2012). Self-assembled amorphous drug polyelectrolyte nanoparticle complex with enhanced dissolution rate and saturation solubility. *J. Colloid Interface Sci., 367*, 518–526.
34. Patel, R. P., & Joshi, J. R., (2012). An overview on nanoemulsion: A novel approach. *Int. J. Pharm. Sci. Res., 3*, 4640–4650.
35. Bera, D., Qian, L., Tseng, T., & Holloway, P. H., (2010). Quantum dots and their multimodal applications: A review. *Materials, 3*, 2260–2345.

36. Cheow, W. S., & Hadinoto, K., (2012b). Green preparation of antibiotic nanoparticle complex as potential anti-biofilm therapeutics via self-assembly amphiphile-polyelectrolyte complexation with dextran sulfate. *Colloids Surf. B. Biointerfaces, 92*, 55–63.
37. Hugerth, A., & Sandlot, L. O., (2001). The effect of polyelectrolyte counterion specificity, charge density, and confirmation on polyelectrolyte amphiphile interaction: The carrageenan/ furcellaran amitriptyline system. *Biopolymers, 58*, 186–194.
38. Bakliwal, A. A., Talele, S. G., et al., (2018). Formulation and evaluation of nateglinide nanosponges. *Indian Drugs, 55*(1), 27–35.
39. Vikas, A., Alaadin, A., Akhtar, S., & Sami, N., (2012). Powdered self-emulsified lipid formulations of meloxicam as solid dosage forms for oral administration. *Drug Development and Industrial Pharmacy*, pp. 1–9. Early Online.
40. Swati, T., & Derle, D. V., (2017). Response surface methodology as a tool for optimization of self-nanoemulsified drug delivery system of *Quetiapine fumarate*. *Asian Journal of Pharmaceutics*, *11*(4), 319.
41. Cheow, W. S., & Hadinoto, K., (2012a). Green amorphous nanoplex as a new supersaturating drug delivery system. *Langmuir*, *28*(15), 6265–6275.
42. Cheow, W. S., & Hadinoto, K., (2012b). Green preparation of antibiotic nanoparticle complex as potential anti-biofilm therapeutics via self-assembly amphiphile-polyelectrolyte complexation with dextran sulfate. *Colloids Surf. B. Biointerfaces*, *92*, 55–63.
43. Cheow, W. S., & Hadinoto, K., (2012c). Self-assembled amorphous drug-polyelectrolyte nanoparticle complex with enhanced dissolution rate and saturation solubility. *J. Colloid Interface Sci.,* 367(1), 518–526.
44. Bonoiu, A., Mahajan, S. D., Ling, Y., Ding, H., Ken-Tye, Y., Nair, B., Reynolds, J. L., Sykes, D. E., Imperiale, M. A., Bergey, E. J., Schwartz, S. A., & Prasad, P. N., (2009). MMP-9 gene silencing by a quantum dot-siRNA nanoplex delivery to maintain the integrity of the blood brain barrier. *Brain Res.*, *1282*, 142–155.
45. Thomas, J. J., Rekha, M. R., & Sharma, C. P., (2010b). Dextran-protamine polycation: An efficient nonviral and hemocompatible gene delivery system. *Colloid Surface B. Biointerfaces*, *81*(1), 195–205.
46. Hattori, Y., Hagiwara, A., Ding, W., & Maitani, Y., (2008). NaCl improves siRNA delivery mediated by nanoparticles of hydroxyethylated cationic cholesterol with amido-linker. *Bioorg. Med. Chem. Lett.*, *18*(19), 5228–5232.
47. Ranjan, A., Pothayee, N., Seleem, M., Jain, N., Sriranganathan, N., Riffle, J. S., & Kasimanickam, R., (2010). Drug delivery using novel nanoplex against a salmonella mouse infection model. *J. Nanopart. Res.*, *12*, 905–914.
48. Chandrakantsing, V. P., & Veena, S. B., (2017). Ropinirole-dextran sulfate nanoplex for nasal administration against Parkinson's disease: *In silico* molecular modeling and *in vitro-ex vivo* evaluation. *Artificial cells, Nanomedicine, and Biotechnology, 45*(3), 635–648.

CHAPTER 7

Frontiers of Applications of Nanotechnology in Biological Sciences and Green Chemistry

SUKANCHAN PALIT

43, Judges Bagan, Post-Office-Haridevpur, Kolkata–700082, India, Tel.: 0091-8958728093, E-mails: sukanchan68@gmail.com, sukanchan92@gmail.com, sukanchanp@rediffmail.com

ABSTRACT

The world of science and technology are moving forward towards a new visionary era. Nanotechnology is a marvel of science and engineering today. It has diverse applications in every branches of scientific endeavor today. Global warming, loss of eco-systems and frequent environmental disasters are urging the scientific and engineering domain to gear forward towards new scientific innovations and new scientific instinct. Biological sciences and green revolution are part and parcel of many nations around the world. The applications of nanotechnology in human society and human scientific progress need to be re-envisioned as science and engineering of nanotechnology moves forward. Agricultural sciences, biotechnology, and food engineering are the backbones of economic progress of a nation. Thus the need of application of nanotechnology in the field of biological sciences. The authors in this treatise focus on the vast and varied applications of nanotechnology in biological sciences. The challenges, the vision, and the deep scientific ingenuity of the field of biological sciences and biotechnology are described in detail in this chapter. Green revolution and scientific advancements in agricultural sciences are in the path of newer glory and new scientific profundity. Nanobiotechnology is the scientific wonder of today. Technological and engineering ingenuity in the field of biological sciences and agricultural engineering are the other hallmarks of this paper. The needs of human civilization today are energy and environmental sustainability. This treatise opens up newer thoughts and new future recommendations in the field of biological sciences, agricultural biotechnology, food technology, and nanotechnology as well. A new world of science and engineering will begin if the green revolution and food self-sufficiency will be achieved with immense scientific vision and might.

7.1 INTRODUCTION

In the global scientific scenario, nanotechnology is creating immense scientific wonders and vast scientific profundity. Science, engineering, and technology in developed and developing nations around the world are moving towards a new age of scientific reformation and scientific forbearance. In a similar vision, nanotechnology, and its applications are creating vast scientific wonders. Biological sciences and agricultural engineering are today in the path of new scientific divination. The challenges and the vision behind application of nanotechnology in biological sciences are immense and groundbreaking. Technological and scientific divination and scientific redeeming will all be the true torchbearers towards a new visionary area in science and engineering of nanotechnology today. Today, in the scientific world, there is a huge gap between science in the laboratory and scientific applications in the human society. Arsenic and heavy metal drinking water contamination and its remediation are the unanswered research questions of scientific vision today. Nano-engineering is a relatively new area of scientific research pursuit today. The author in this chapter deeply depicts the success and vision of science in both nanotechnology and biological sciences. Water treatment and wastewater treatment are the burdens of human scientific progress today. This treatise with vast scientific farsightedness and enquiry opens up a new chapter in the field of nanotechnology applications in biological sciences, agricultural sciences, and food technology. The other areas of immense importance are the field of nanobiotechnology which is an integration of nanotechnology and biological sciences. This scientific emancipation in the field of nanotechnology is touched upon in this treatise.

7.2 THE AIM AND OBJECTIVE OF THIS STUDY

The vision and the challenges of nanotechnology and nano-engineering are immense and path-breaking today. The world of science and engineering of biological sciences, biotechnology, agricultural sciences, and environmental remediation are in the process of deep scientific introspection and vast scientific redeeming. In this paper, the author reiterates and vastly pronounces the need of nanotechnology in every branch of science and engineering mainly biological sciences, food technology and environmental protection. The author describes and elucidates on recent advancements in nanotechnology and biological sciences. In industrialized and developing nations around the world, water treatment, and wastewater, treatment is a huge scientific burden. Scientific enquiry, vast scientific enthusiasm, and scientific truthfulness are the backbones of research endeavors in nanotechnology. The author depicts with cogent and lucid insight the interface between nanotechnology and biological sciences. This paper answers difficult research questions in the field of nanotechnology, biological sciences, biotechnology, and agricultural sciences. Heavy metal drinking water and groundwater poisoning are veritably world's largest environmental catastrophe. This paper widens the scientific enquiry and deep profundity in the field of nanotechnology, agricultural sciences, food technology, and biological sciences. Green revolution and biotechnology are today changing the face of scientific progress in developing and disadvantaged nations around the world. This paper pinpoints the success, the scientific divination and the vast scientific needs of research in nanotechnology and biological sciences in developing as well as highly industrialized nations around the world [28–31]. The vision of nanobiotechnology and green sustainability are depicted profoundly in this paper [34, 35].

7.3 ENVIRONMENTAL AND ENERGY SUSTAINABILITY AND THE VISION FOR THE FUTURE

Sustainable development whether it is environmental or energy are the success of science and engineering today. Mankind stands bedeviled and mesmerized with the challenges and the vision of nanotechnology. The amelioration of global warming and industrial pollution are the cornerstones of research pursuit in engineering and science today. Human society is today in the critical juncture of scientific forbearance and vision. The realization of sustainable development goals is the cornerstone of every research endeavor today. Humanity is in the global scenario of immense catastrophe as global warming and environmental disasters shatter the deep scientific firmament. Environmental sustainability and social sustainability are the challenges of human society today. The question of heavy metal poisoning of drinking water in industrialized and developing countries still remains answered. The status and stance of environmental remediation globally is extremely grave and thought-provoking. Environmental and social sustainability will surely solve many research questions in the field of environmental protection and natural resource management. Technological vision and sound motivation, scientific verve and immense scientific divination will surely be the forerunners towards a new era in the field of environmental remediation [28–31, 34, 35].

7.4 SUSTAINABLE DEVELOPMENT GOALS AND NANOTECHNOLOGY

Sustainable development goals as defined by the United Nations Organization are today path-breakers of science and engineering globally today. The world today is not free from poverty. Provision of basic human needs such as water, energy, food, shelter, and education are virtually absent as civilization treads forward towards new goals. The application of sustainable development and nanotechnology will surely open new doors of innovation in scientific progress and research and development initiatives globally.

United Nations Report [1] described and discussed in minute details The United Nations Sustainable Development Report. It depicts profoundly the needs of sustainability in human society and human scientific progress [1]. The 17 sustainable development goals open new directions in the area of development in both developed and developing nations around the world [1]. Environmental, energy, and social sustainability are the utmost necessities of global scientific emancipation today. This United Nations agenda is a plan of action for people, planet, and prosperity. The main aim of this report is to eradicate poverty in all its forms and dimensions, including extreme poverty, and the greatest global challenge and an indispensable requirement for sustainable development [1]. Scientific verve and alacrity and technological vision are the hallmarks of this well-researched treatise. According to this report, all countries acting in a collaborative partnership will effectively implement this plan. The 17 Sustainable Development Goals and 169 targets which were announced during the convention demonstrate the scale and the ambition of the new agenda [1]. Mankind's immense scientific will, girth, and determination will play a major role in the true realization of this report globally [1]. This well-researched report depicts profoundly the success of sustainable development in the global scenario. According to this report, the goals and targets of Sustainable Development Goals will stimulate action over the next 15 years in areas of critical importance for human civilization and the planet [1]. The cornerstone of this report is the absolute eradication of poverty in human civilization. Peace and prosperity are the other areas of immense endeavor in this report [1]. The need of human society today is the successful realization of 17 United Nations Sustainable Development Goals in the coming decades. The full implementation of these goals according to this report is to be done by 2030. This can only be done if technological and scientific verve and motivation are realized with immense vision and might [1].

Nanotechnology and nano-engineering are the veritable marvels of immense vision in human society today. Sustainable development needs to be reframed as mankind moves forward today. Social and economic sustainability are the areas of immense introspection and contemplation today. In this paper, the author deeply comprehends the success and vision of the energy and environmental sustainability as a means towards the furtherance of science and technology [30, 31, 34, 35].

7.4.1 GREEN CHEMISTRY, GREEN SUSTAINABILITY, AND THE VAST VISION FOR THE FUTURE

Civilization and science are today marching forward towards a new scientific age. Green chemistry and green sustainability are in the middle of immense scientific ingenuity and engineering prowess. Sustainable development is the utmost need of developed as well as developing nations around the world. Human society is in the middle of immense peril as arsenic and heavy metal groundwater contamination destroys the scientific fabric. Green sustainability is the coined word of tomorrow. The vast vision of green chemistry and green sustainability needs to be reiterated with the march of science in present-day human civilization. Technological and scientific ingenuity and validation are the utmost needs of human scientific progress today. Green chemistry also called sustainable chemistry is an area of chemistry and chemical engineering which is focused on the designing of products and processes that minimize hazardous and toxic substances release from a chemical process plant. While environmental chemistry is focused on the effects of polluting chemicals on nature, green chemistry is focused on the environmental impact of chemistry including sound and safe chemical processes to prevent pollution and reducing consumption of non-renewable resources. Engineering and scientific might and integrity and the futuristic vision of scientific determination will all open the doors of innovation and scientific instinct in the field of green chemistry today. Green sustainability is the other side of the visionary scientific coin. Sustainable development whether it is energy or environmental is the necessity of human society today. The overall goals of green chemistry-more resource efficient and inherently safer design of molecules, materials, and products which can be pursued in a wide range of contexts. Today green chemistry and green sustainability needs to be merged together towards to the future emancipation of science and engineering. Water and wastewater treatment and heavy metal groundwater remediation are in the middle of immense crisis and an unmitigated disaster. Here comes the importance of green chemistry, green engineering and green sustainability [34, 35].

7.5 NANOTECHNOLOGY AND THE VAST DOMAIN OF BIOLOGICAL SCIENCES

Nanotechnology is today integrated with diverse areas of science and engineering such as biological sciences and biological engineering. The world of science of biological sciences today is in the middle of scientific vision and scientific forbearance. Food technology is in the midst of scientific introspection and also needs to be comprehended as science and civilization progresses. Green revolution in developing countries is changing the face of civilization. So the need of nanotechnology in devising new innovations and new scientific ingenuity in the field of agricultural and biological sciences. Water science and technology also needs to be reorganized and reframed in the similar vein [30, 31]. Nanotechnology research and development forays needs to be envisioned and re-envisaged with the progress of science. Today biological sciences and nanotechnology are in the avenues of new regeneration. This paper will truly open up new doors of ingenuity and insight in the field biological sciences, biotechnology, and nanotechnology [34, 35].

7.6 BIOLOGICAL SCIENCES AND THE MARCH OF SCIENCE AND ENGINEERING

Biological sciences, agricultural engineering, and food technology are the utmost needs of human civilization today. Vast technological fervor, the futuristic vision of nanotechnology, and the vast scientific ingenuity will all open up new dimensions of research pursuit in the field of biological sciences. Today the march of science and engineering is path-breaking in the global scenario. Water treatment principles and concepts of chemical engineering are today integrated with environmental remediation. Thus, scientific adjudication and scientific candor are the necessities of scientific regeneration in nanotechnology today. Scientific prowess and deep technological vision in the field of biological sciences and agricultural engineering will be the forerunners of human civilization today. Science and technology in the global scenario are moving at a rapid pace. In this paper, the author reiterates and pronounces the vision, the scientific divination, and the necessities of modern civilization which include biological sciences, biotechnology, agricultural sciences, water purification, and nanotechnology. Thus, a new dawn will truly usher in the field of science and engineering [28–31, 34, 35].

7.7 RECENT DEVELOPMENTS IN THE FIELD OF APPLICATION OF NANOTECHNOLOGY IN BIOLOGICAL SCIENCES

Nanotechnology applications in agricultural sciences are in the middle of deep scientific introspection and ingenuity. Today, the world of science and engineering are based on the fundamental principles of nanotechnology. Nanotechnology is surely creating wonders of science today. In this section, the author with vast scientific insight and provenance describes recent advancements in nanotechnology applications in agricultural sciences.

Dhewa [2] discussed with immense scientific farsightedness nanotechnology applications in agriculture. The applications of nanotechnology are vastly path-breaking yet in the dormant stages of science and engineering [2]. The rapid and drastic advancements in nanosciences have a significant impact on agricultural practices and food manufacturing industries. Nanotechnology has immense potential to offer smarter, stronger, cost-effective packaging materials, biosensors for the rapid detection of the food pathogens, toxins, and other contaminants [2]. Technological vision and motivation are the forerunners in research pursuit in nanotechnology today. Nanotechnology also plays an important role in developing new generation of pesticides with the safe carriers, preservation, and packaging of food and food additives, strengthening of natural fibers, removal of various contaminants from the soil and water bodies, and improving the shelf-life of vegetables [2]. There are many such applications of nanotechnology in agriculture [2]. The safety and regulatory issues of the application of nanotechnology for human beings, environment, and eco-systems are required to be highly debated, particularly in the developing and poor countries around the world [2]. There are some important points of direct human exposure to nanomaterials along with the food-chains and the immense threat of the possibility of the nano-particles reaching sites which can result in safety and health problems. Keeping this deep scientific vision in mind and the risks concerning safety aspects of nanotechnology, an effective risk management framework needs to be envisioned [2]. Moreover, an extremely stable governance model should be adopted during the continuous interactions of nanomaterials with the surrounding environment. The term "nanotechnology" has defined as a branch of science and engineering that deals with the understanding and control of matter at the dimensions of about 1–100nm as defined by the United States Environmental Protection Agency (EPA) [2]. It includes controlling, building, and restructuring of the devices and other materials of physical, chemical, and biological features at nanoscale level i.e., on the scale of atoms and molecules. In the twenty-first century, nanotechnology has veritably emerged with a great influence on the global industry and economy [2]. In this well-written treatise, the author discusses in detail some of the applications of nanotechnology in agriculture such as the detection of nutrients and pathogens, nanoscale carriers for targeting delivery, wastewater treatment, and disinfection and the vast world of bioremediation [2]. Quality enhancement of agri-products and identification, tracking of agri-foods and the enhancement of shell-life of agricultural products are the other cornerstones of this paper [2]. In spite of various prospective uses of nanotechnology in numerous and path-breaking sectors, such as medical, agriculture, space research technology and engineering, there are certain safety concerns which need to be addressed and envisioned. Some key limitations and risks associated with nanotechnology applications in agriculture are as follows:

- Exposure of nanomaterials to human beings and accumulation in human food chains and agri-food chains [2].
- Interaction of nanoparticles (NPs) with the non-target sites which lead to certain health and environmental issues [2].
- Higher production costs.
- Developments in agricultural sectors are immensely limited due to low investment.
- Public and citizens are not aware of the potential applications of nanotechnology.

- The need of the labeling of the products of nanotechnology further prevents the innovative applications of this technology in agricultural sciences [2].

The author in this paper [2] deeply discusses and comprehends future perspectives in the application of nanotechnology in agriculture. Nanotechnology has today immense applications in agriculture. This scientific ingenuity and technological profundity in the intricacies of nanotechnology applications are brought to the forefront in this paper [2].

Petre [30] discussed and elucidated in details advances in applied biotechnology. The vision and the challenges of biotechnology and biological science are in the visionary avenues of scientific grit and determination. The authors in this paper deeply deliberated biotechnology for agricultural wastes, molecular biotechnology, and genetic engineering and biotechnology applications of tissue engineering. Today biotechnology is a vision of science and engineering globally. The avenues of science touched upon are biotechnology of agricultural wastes, fermentation processes, microbial biotechnology, biopharmaceutical technology, molecular biotechnology, and genetic engineering. Biotechnological applications in tissue engineering are the other cornerstones of this book [30]. In the global scenario, biotechnology, and genetic engineering are creating wonders. They are the marvels of science today. A scientist's deep vision and the technological stance of biotechnology will veritably open up new wonders of science and engineering in decades to come [30]. The first part of this book deals with environmental biotechnology and also deals with biodegradation and bioconversion processes and the second avenue of research endeavor is the biotechnology of biopolymer production. Civilization's immense scientific stance and engineering integrity in the field of bio-engineering and biotechnology. The world of challenges and the difficulties in the application of tissue engineering are depicted profoundly in this book [30].

Hautea et al. [31] deeply discussed with scientific and engineering insight plant biotechnology in Asia. Crop improvement facilitated by modern biotechnology is one of the remarkable developments in plant biotechnology research and development today. The success and the targets of industrial research and development initiatives in biotechnology are in the roads towards new innovation and novel scientific instinct. Key strategies for achieving food security and sustainable agriculture are the hallmarks of civilization and scientific progress today. Harnessing biotechnology applications for the benefit of poor requires considerable attention for science and technology research and development initiatives globally today [31]. Agri-biotechnology is the other area of science and technology research and development. This paper depicts with immense scientific insight the larger vision of bio-nanotechnology applications in agricultural sciences. Biological sciences and biotechnology will thus usher in a new era in developing and less developing nations around the world. The authors deeply discussed the current status of plant biotechnology in Asia [31]. The success and scientific imagination of biotechnology are yet to be unraveled in developing nations in Asia. This paper depicts profoundly the success of biotechnology in revolutionizing science and technology in nations across Asia [31].

Rao et al. [32] reviewed comprehensively on biopolymers. The vast environmental impact of plastic waste is raising global concern and the disposal methods are limited. Biopolymers have vast applications in medicine, food, and petroleum industries [32]. Microorganisms can produce a large number of polysaccharides in simple but costly in product conditions. New applications in agronomy, foods, cosmetics, and therapeutics are revolutionizing the status of biopolymer science today. The authors deeply discussed functions and synthesis of polymers, production of biopolymers, intracellular versus extracellular production of biopolymers, genetic engineering, and biopolymer technology [32]. Streamlining of commercialization of biopolymers stands as a major vision of research initiatives and research direction globally. This paper unfolds the scientific ingenuity in the application areas of biopolymers [32].

Adeosun et al. [33] described with vast scientific and engineering foresight green polymer nanocomposites. There has been lots of attention to the use of bio-reinforced composites in automotive, construction, packaging, and medical applications due to the increasing concerns of environmental sustainability [31]. The technological and engineering still today remains as science trudges forward. Green polymer nanocomposites show unique properties of combining the advantages of natural fillers and organic polymer materials. Nanocomposites and the vision of green technology and green chemistry are challenging the global scientific integrity. This paper discusses thermoplastics starch-based composites, green polymers, polylactic acid-based composites, cellulose-based composites, green fillers, and nanocomposites processing methods [31]. The challenges of application of biopolymers and nanocomposites still today remain unraveled. This paper is a clear eye-opener towards the research directions in the field of green polymers [31].

Abd-Elrahman et al. [3] discussed with insight and scientific rigor applications of nanotechnology in agriculture in an overview. Agriculture provides food for humans from time immemorial and now is a subject of immense scientific introspection and deep contemplation. So it is absolutely necessary to use modern technologies such as nanotechnology in agriculture [3]. Nanotechnology has been rigorously defined as relating to materials, systems, and processes which veritably operates in the scale of 100 nanometers. Today technology is at its helm as regards application of nanotechnology in diverse areas of science and engineering [3]. Nanotechnology has applications in all stages of production, processing, storing, packaging, distribution, and transport of agricultural and food products [3]. Nanotechnology today will revolutionize agriculture and food industry by innovation in science and new techniques such as precision farming, enhancing the ability of plants to absorb nutrients, targeted use of inputs, disease detection and control of diseases, the tools to withstand environmental pressures and effective systems for processing and storage of agricultural products [3]. Research on smart seeds programmed to germinate under favorable conditions with nanopolymer coating are encouraging [3]. Nanoherbicides are being extensively developed to address the problems in weed management and truly enhancing the needs of green revolution in developing and developed nations around the world [3]. Modern technologies such as nanotechnology and environmental biotechnology are the veritable needs of science in modern civilization today. Growth of agricultural sector as a sustainable concept is seen as essential in developing countries around the world. As civilization moves forward, bio, and nanotechnologies can play a stellar role in increasing agricultural production in developing as well as industrialized nations around the globe [3]. Food security is a major issue around the world in the last century as well as in the present century. Food needs are today integrated with huge environmental, health, and economic advantages. Nations, communities, and governments are veritably struggling with this food shortage for a long period of time [3]. Recent decades have witnessed more challenges and a larger vision. Nanotechnology is a powerful technology and it allows global citizens to look and investigate at the molecular level and veritably construct nanometer-scale structures. Here comes also the importance of looking into nanotechnology applications in agriculture [3]. Greenhouse construction with high performance and productivity, prevent extinction of plants and animal species, and overall nanotechnology is re-envisioning the science of agriculture. In recent decades, agricultural land and soil pollution with hazardous elements and compounds present in industrial wastewater are the important factors limiting crop and food production around the world [3]. Nanocatalysts are capable of removing harmful chemicals and hazardous substances in agricultural systems. Thus, the research and development initiatives are in this direction. The authors in this paper stressed on precision farming, applications of nanotechnology in agronomy, application of nanotechnology in pests and plant disease management, applications of nanotechnology in food science and animal science, and nanotechnology solutions in environmental remediation [3]. Nanotechnology solutions in climate change are the other pillars of this well-researched treatise. Climate change is today a burden to human civilization. To bring climate change to a halt, emission of greenhouse gases should be totally ameliorated. Here technology comes into play [3]. All sectors including buildings, industry, energy production, agriculture, forestry, and waste management could contribute to the overall mitigation efforts, through greater energy efficiency. Thus science and engineering need to be re-envisioned and revisited as regards application of nanotechnology [3].

Fawzy et al. [4] deeply discussed with lucid and cogent insight nanotechnology in agriculture and the current and future situation. In the developing countries around the world, the agriculture sector is a very important sector for the economy of the country [4]. A much bigger problem is there which is food scarcity in agriculture [4]. Indeed, by 2050, as the total population increases, there will be a larger scarcity of food that needs to be solved with a sustainable angle. Further, new technologies are developed that show the potential increase in agricultural productivity as well as reduce the environmental cost and the vast resources related to agricultural production [4]. Vast scientific ingenuity and determination and the futuristic vision of scientific research pursuit will surely open up new dimensions in agriculture research in years to come [4]. The absolute need is scientific vision, grit, and determination [4]. Technological abundance and scientific determination are the pillars of science and engineering in modern civilization today. The main objective of nanotechnology in agricultural sciences is to reduce the use of fertilizers and pesticides to protect plants and crops, to reduce nutrient losses, and to increase yields of plant crops through nutrient management [4]. Nanotechnology and NPs are used in nutrient absorption from plants and nanotechnology devices derived from plant breeding and genetic transformation. In this paper, the authors discussed in minute details the application of NPs in various areas of agricultural science [4]. A new dawn in the field of agriculture will usher in with immense scientific might and vision.

Joseph et al. [5] discussed with immense scientific vision and far-sightedness nanotechnology in agriculture and food. The authors discussed in detail nanotechnology in the food market, precision farming, smart delivery systems, packaging, and food safety and the vast world of food processing. The current global population is nearly 6 billion with 50% living in Asia [5]. A large population in Asia resides with immense water and food shortage due to environmental impacts and political instability. Here come the needs of new innovations in science and engineering such as nanotechnology [5]. In the developed countries, the food is in surplus. For developing countries, the drive and scientific motivation is to develop drought and pest-resistant crops, which also veritably increase yield. The vast potential of nanotechnology to revolutionize the health care, textile, material science, information, and communication technology, and energy sectors have been vastly publicized [5]. Nanotechnology has been described as the next global industrial revolution and both developed and developing nations around the world are fervently investing in the application of this technology [5]. At present, the United States of America leads with a 4 year, 3.7 billion dollars USD investment through its National Nanotechnology Initiative. The USA is followed by Japan and the European Union, which have both committed substantial funds (750 million and 1.2 billion, including individual country contributions, respectively per year) [5]. More than 400 companies around the world are veritably and reassuringly involved in nanotechnology research and development and this number is expected to increase to more than 1000 within the next 10 years [5]. The European Union's vision of "knowledge-based economy" and as a part of this, it plans to maximize the potential of biotechnology for the EU economy, society, civilization, and the environment. Technology has gone so far in modern civilization [5]. Nanotechnology and biotechnology are the pillars of scientific endeavor today. This report is a well-researched treatise in the field of agriculture and food technology and widens the vision of its application in human society. Nanotechnology has the immense potential to revolutionize the agricultural and food industry with new treatments for the molecular treatment of diseases, rapid disease detection, effectively enhancing the plants to absorb nutrients, etc. [5]. The authors discussed an agricultural methodology widely used in the USA, Europe, and Japan which efficiently utilizes modern technology for crop management and is called Controlled Environment Agriculture [5]. The technology of agricultural sciences has marched forward at a rapid pace overcoming one scientific boundary over another. The authors in this report also discussed precision farming. Precision farming has been a long-desired goal to maximize output (i.e., crop yields) while minimizing input (i.e., fertilizers, pesticides, herbicides, etc.) [5]. Technological and scientific candor and deep ardor and the futuristic vision of nanotechnology and nano-engineering will open and unravel new doors of innovation in the field of diverse areas of science and engineering which includes agricultural sciences and food technology [5]. The integration of biotechnology and nanotechnology in sensors will surely create new equipment of increased sensitivity allowing an earlier response to environmental changes and a greater emancipation of environmental remediation. Ultimately, precision farming along with the help of smart sensors will allow increased productivity in agriculture by providing accurate information thus helping farmers to take better decision [5]. Technological stance, the utmost needs of agricultural sciences, and the needs of global farmers will surely ensure a new emancipation in the field of science and engineering. Smart delivery systems are the other cornerstones of this report [5].

7.8 WATER PURIFICATION AND THE APPLICATION OF NANOTECHNOLOGY

Water science and technology are today integrated with nanotechnology and nano-engineering. Civilization's scientific hope and scientific determination are at a disaster as environmental crisis looms large over the scientific horizon. The question of environmental remediation and the successful realization of environmental sustainability are of immense importance in the progress of civilization today. The burning issue of water purification and water and wastewater treatment are today linked with the success of environmental sustainability. Nanotechnology and nano-engineering can surely ameliorate the global water crisis. Nanomaterials and engineered nanomaterials are the success and scientific vision of civilization today. The frontiers of nanotechnology applications in water science and agricultural sciences need to be reorganized and revamped. In today's scientific world, basic, and fundamental research questions still remains unanswered yet the civilization and science needs to move forward. Here comes the importance of nanoscience and nanotechnology and its applications in diverse areas of science and engineering. Nanomaterials and engineered nanomaterials need to be applied in the field of environmental remediation and environmental engineering. Today graphenes and carbon nanotubes (CNTs) are used as nano-adsorbents in various industrial wastewater treatments and in the field of water purification. Today health effects and health perspectives are veritably challenging the global nanomaterials scenario. Yet the world of science and engineering stands mesmerized with the effectiveness of nanomaterials and engineered nanomaterials application in human society and scientific progress. The progress of science of nanotechnology is absolutely remarkable as regards its scientific robustness, efficiency in its applications and the vision. Water treatment and wastewater treatment as well as agricultural sciences are in the path of new discovery and scientific ingenuity today. This challenge is deeply handled by the author as this paper opens up new windows of scientific and engineering profundity in the field of agricultural sciences.

Werkneh et al. [6] deeply discussed with lucid and cogent insight applications of nanotechnology and biotechnology for sustainable water and wastewater treatment. Water pollution and drinking water scarcity have become a serious global scientific issue. It is a concern for both public health engineering and the environment [6]. To ameliorate these challenges various treatment strategies have been adopted. Among these technologies, nanotechnology, and biotechnology based tools are usually applied separately for water (domestic purposes) and wastewater reuse treatment. This paper focuses on new and emerging nano- and biotechnologies for the sustainable removal of pollution causing constituents during water and wastewater treatment [6]. In this paper, the authors also discuss the toxicological and safety aspects of different nanotechnologies and their vast scientific perspectives [6]. Civilization's vast scientific prowess, the needs of human society, and the true vision of nanotechnology will veritably open new doors of innovation and scientific instinct in the field of water and wastewater treatment. Increasing concentrations of toxic pollutants including heavy metals, organic, and inorganic pollutants, and other complex compounds are being discharged in the huge volume of industrial wastewater in the present global scenario [6]. Here comes the vast importance of scientific innovation and new tools for water and wastewater treatment [6]. The conventional water treatment technologies used for the remediation of water pollutants are the activated carbon-based adsorption, membrane filtration, ion exchange, coagulation, and flocculation, reverse osmosis, flotation, and extraction, electrochemical treatment, advanced oxidation processes and bioremediation [6]. Today the challenges of the science of industrial pollution control and water treatment are vast and versatile. Scientific and technological validation and the deep scientific verve will all be the true forerunners towards a new age in the field of water purification. Most of the conventional treatment technologies have certain drawbacks in terms of its operational methods, energy requirements, processing efficiency, and economic benefits. From an environmental sustainability point of view, the use of microbes, nanomaterials, and engineered nanomaterials for the removal of pollutants have received tremendous attention in the scientific research community [6]. Nanomaterials are very small in size, i.e., approximately 1–100 nm and deeply show unique characteristics that enable and ensures their applications in innovative applications in wastewater treatment and water purification and others [6]. They exhibit high surface area to volume ratio which is very much significant to produce "highly reactive surface area than the bulk counterparts." Technological and scientific motivation, verve, and validation are today in the path of new vision and ingenuity as civilization trudges forward. Environmental sustainability and nanotechnology are the two opposite sides of the coin today [6]. Nano-oxides (silver, gold, iron, and titanium) are common nanomaterials which have been employed for environmental remediation. The aim of the environmental remediation is to ameliorate pollutants in contaminated water and soil environments. The vast and ever-growing field of environmental biotechnology offers lots of feasible solutions in water and wastewater treatment. This visionary area of science and technology solves complex environmental problems in the ecosystems and the natural environment [6]. The world of environmental biotechnology is thus ushering in a new era in scientific might, forbearance, and vision. According to the International Society of Environmental Biotechnology, environmental biotechnology is defined as an "environment that helps to promote and envision the development, use, and regulation of biological systems for the remediation of contaminated land, air, and water environments that highly works efficiently to sustain an environment-friendly human society" [6]. Nanotechnology offers several advantages because of their unique physico-chemical characteristics such as large specific surface area, higher reactivity, and small size. Technological and scientific profundity is at its vast helm as nanotechnology and environmental biotechnology reaches its might, verve, and vision.

Kunduru et al. [7] discussed and described with cogent insight nanotechnology for water purification and applications of nanotechnology methods in wastewater treatment. Water is the most important asset of human civilization today. It is a basic human necessity. The problem of water scarcity will veritably increase with time [7]. Only 2.5% of the world's oceans and seas supplies fresh water. Only less than 1% of freshwater can be used for drinking. Globally greater than 700 million people do not have proper access to potable water [7]. A grave problem is destroying the vast scientific horizon. This problem is severe in developing countries and sub-Saharan countries [7]. Technological and engineering vision, the vast domain of scientific challenges and the futuristic vision will all be the forerunners towards a newer scientific inquiry and scientific contemplation [7]. Water contaminants may be organic, inorganic, or biological. Some contaminants are highly toxic and carcinogenic and have disastrous effects on humans and eco-systems [7]. Thus, nanotechnology is in the path of new scientific regeneration. Some heavy metals are highly notorious pollutants with high toxicity [7]. Arsenic is one of the deadliest elements well known since medieval times. Other heavy metals water pollutants with high toxicity are cadmium, chromium, mercury, lead, zinc, nickel, copper, and so on; they have serious toxicities [7]. Nitrates, sulfates, phosphates, fluorides, chlorides, selenides, chromates, and oxalates show extremely hazardous effects at high concentrations. Here comes the need of a detailed vision and inquiry in the field of hazardous materials [7]. The authors in this paper deeply discusses the importance of nanotechnology in water purification, major limitations associated with conventional water purification methods, nanoadsorption, carbon-based nanoadsorbents, removal of organic compounds, removal of heavy metal ions, the application areas of metal-based nanoadsorbents, the application of polymeric nanoadsorbents, membranes, and membrane processes [7]. The vast world of nanocomposite membranes, thin-film nanocomposite membranes, and biologically inspired membranes are the other cornerstones of this paper. The other areas of research endeavor are antimicrobial nanomaterials in disinfection and microbial control [7]. Safety, toxicity, and environmental impact of nanomaterials are the other hallmarks of this well-researched treatise. Water purification and wastewater treatment processes by nanotechnology show immense promise in laboratory studies [7]. Some of these technologies are highly marketed and others require significant research before they can be highly scaled up to the industrial scale [7]. The commercialization of industrial processes today lies in the midst of deep vision, provenance, and strong scientific divination. Studies should be conducted under realistic conditions to assess the efficiency of available nanotechnology and nanoengineering to validate nanomaterial-based sensing. This paper is a comprehensive research endeavor on different areas of chemical engineering separation processes and novel separation techniques such as membrane science. This paper will go a long way in the true realization of nanoscience and nanotechnology in human society in the future [7].

Das et al. [8] discussed with vast scientific far-sightedness recent advances in nanomaterials for water protection and monitoring. The effective handling of water and wastewater pollutants is a must since they are depleting freshwater resources and causing environmental catastrophe of unlimited proportions [8]. The vision of this paper is to undertake a detailed investigation of current research trends with a focus on nanomaterials to considerably improve the performance of classical wastewater treatment technologies [8]. This paper targets the investigation of classical water treatment technologies such as adsorption, catalysis, separation, and disinfection. In addition to these endeavors, the authors also dealt with Nanomaterials based sensor technologies since they have been significantly used in the removal of hazardous pollutants from contaminated water [8]. Technological advancements and vast scientific inquiry today lies in the midst of vision, might, and scientific provenance. This paper is an eye-opener towards the scientific needs of nanomaterials and engineered nanomaterials in water science and technology. Sustainability whether it is energy or environmental are the hallmarks of every research pursuit today [8]. Thus the need of nanotechnology and environmental remediation science. The world population has increased at a rate of 80 billion per year, increasing potable water demand by 64 billion cubic meters per annum thus resulting in an ever-growing scientific concern. Thus the need of innovations, scientific profundity, and vision [8]. The United Nations has reported with vast scientific vision that almost 2 billion people did not have access to safe and clean water in 2013 and by 2025, nearly 1.8 billion people will be living underwater scarcity [8]. The alarm and the vision are practically unimaginable. Currently, more than 750 million people in the world do not have proper water facilities and proper sanitation and it is mostly in Asia, Central, and South America and Africa. This water shortage will surely be detrimental to the successful implementation of the United Nations millennium development goals (MDG) [8]. Therefore, a concerted effort is needed in public and government policies. The success of civilization and science is at a deep stake [8]. In order to meet the ever-increasing need of potable water, various wastewater treatment technologies such as screening, centrifugal separation, sedimentation, coagulation, flocculation, aerobic, and anaerobic treatments are the conventional methods of use. Other methods of water treatment are distillation, crystallization, evaporation, solvent extraction, oxidation, precipitation, ion exchange, microfiltration, ultrafiltration, reverse osmosis, forward osmosis, adsorption, electrolysis, and electrodialysis [8]. The disadvantage of these technologies is that they are operationally intensive and depend on large systems. This also results in high capital costs and engineering prowess. In this chapter, the authors deeply elucidate nanomaterials for wastewater treatment, adsorption, the application of carbon nanostructures, the application of nanofibers, the domain of metal and metal oxide NPs, and nanopolymers [8]. The authors discussed in minute detail the chemical kinetics of photocatalysis with a visionary view in the progress of science and engineering of nanomaterials and engineered nanomaterials [8].

Yadav et al. [9] discussed and elucidated in deep detail emerging adsorbents and current demand for defluoridation of water and the bright future in water sustainability. Technological verve, profundity, and vision are today is in new regeneration [9]. The application of nanotechnology in water purification is touching new heights today [9]. Fluoride contamination of groundwater is a serious environmental engineering concern in several countries around the world because the intake of excessive fluoride caused by the drinking of contaminated water [9]. Today arsenic groundwater contamination in industrialized and developing nations around the world is a monstrous environmental engineering problem. Geological and anthropogenic factors are highly the cause of contamination of groundwater with fluoride [9]. This paper deeply reviews the current available methods and the vast emerging approaches for defluoridation of water particularly drinking water [9]. The scientific challenges and the scientific ardor are immense as the developing nations particularly South Asian countries are in the grip of an unending global environmental crisis that is the contamination of groundwater by arsenic and heavy metals. Fluoride is a widely distributed monoatomic anion of fluorine characterized by a small radius. The radius is 0.133 nm [9]. The sources of fluorine in water and soil are mostly geogenic and also are made of many rock forming minerals [9]. Some of these minerals including cryolite, fluorite, and fluoropatite are highly soluble in water and release fluoride ions in the contaminated water stream. Fertilizer, iron, and aluminum manufacturing industries release fluoride as an unwanted by-product. Technology and engineering thus needs to be revamped as civilization and scientific rigor crosses difficult boundaries [9]. In this chapter, the authors discussed with immense lucidity technologies for fluoride removal such as membrane process, reverse osmosis, nanofiltration, electrodialysis, adsorption processes, ion exchange processes, the application of alumina and aluminum-based adsorbents, carbon-based adsorbents, calcium-based adsorbents, the application of nanomaterials, natural materials, building materials, and industrial waste adsorbents, and the factors affecting the process of defluoridation [9]. A worldwide status of fluoride is the other cornerstone of this paper. The contamination of drinking water by fluorine is absolutely disastrous and needs concern at its utmost. This paper reviews the ultimate success and the transforming vision in the field of fluoride removal in industrial water and wastewater [9, 30, 31, 34, 35].

Science and technology of nano-science and nano-engineering are reaching vast scientific zenith as academic rigor in this field treads forward. In this chapter, the author reiterates a proactive vision that is the application of nanotechnology in water and wastewater treatment. The civilization and scientific progress stands in the midst of vision, scientific fortitude, and provenance. The challenges and the ardor of science and engineering of nanotechnology and nano-engineering are immense, vast, and path-breaking. The author validates strongly the needs of nanotechnology in further emancipation of water purification and drinking water treatment. This challenge is the eye-opener of this entire treatise [30, 31, 34, 35].

7.9 NANOTECHNOLOGY APPLICATIONS IN FOOD TECHNOLOGY AND FOOD ENGINEERING

Food technology and food engineering applications are the utmost necessities of mankind today. Nanotechnology today is a marvel of science and engineering. Nanotechnology has vast and varied applications in food sciences and agricultural sciences also. In this paper, the author also elucidates some of the applications of nanotechnology in food technology and food engineering. Mankind's scientific and engineering stance and ingenuity and the vision of agricultural sciences will surely open new windows of innovation and profundity in the diverse domains of engineering and technology. Today is the age of nuclear science and space technology. In the similar manner, nanotechnology is a burgeoning area of scientific domain. Technology and engineering has reshaped the human civilization today. Nanotechnology is a burgeoning area of science and engineering and is revolutionizing the scientific scenario. Food technology and food engineering are truly one of the many stupendous examples of nanotechnology applications. This paper will surely open up new thoughts and new scientific understanding in the field of nanotechnology and nano-engineering. Food engineering and food technology along with biological sciences are boundless as regards application of nanotechnology. The author in this paper adjudicates the success, the challenges, and the targets of application of nanotechnology in food technology and agricultural sciences. Green revolution in developing nations is a forerunner towards a new emancipation of agricultural sciences. These scientific thoughts forms the basis of this research pursuit [10–17, 34, 35].

7.10 FUTURE RECOMMENDATIONS AND FUTURE FLOW OF SCIENTIFIC THOUGHTS

Future recommendations and future flow of thoughts in biological and agricultural sciences should be towards greater scientific realization of science and engineering. Here nanotechnology also comes into play. Futuristic vision in the field of nanotechnology, chemical engineering, and environmental engineering needs to be broadened as civilization trudges forward towards a new age. Sustainable development whether it is energy, environmental, social, or economic should have a new scientific redeeming. Everywhere in the world, citizens are faced with the monstrous issue of provision of clean drinking water. Here comes the need of a concerted urge from scientists, engineers, governments, and the civil society. Thus, future of civilization and scientific progress is extremely grave and thought-provoking. Membrane science and novel separation processes are in the path of new scientific divination and ingenuity. This tool of chemical engineering needs to be reorganized as mankind moves forward [18–31, 34, 35].

7.11 THE MARCH OF MODERN SCIENCE AND THE RESEARCH FORAYS IN NANOTECHNOLOGY

Mankind and scientific advancements are moving from one scientific boundary over another. In the similar manner, nanotechnology is a challenge of science and engineering in the global scenario. The application of nanotechnology in medical science, biological sciences, biotechnology, and other diverse areas of science and engineering needs to be mentioned and envisioned as mankind moves forward. Water treatment and wastewater treatment are the needs of civilization today. Here nanotechnology is creating immense wonders [10–12, 30, 31]. Arsenic and heavy metal groundwater and drinking water poisoning are challenges of science and engineering today. Thus, the march of modern science is highly challenged and needs to be veritably streamlined as civilization moves forward [18, 19, 27–31, 34, 35].

7.12 CONCLUSION AND FUTURE SCIENTIFIC PERSPECTIVES

Technological advances in nanotechnology applications in biological sciences, agricultural sciences, and agricultural engineering today are in the midst of deep scientific introspection and contemplation. Water science, water purification, and environmental remediation are the marvels and needs of human society today. Today is the world of technological divination and engineering innovations. The futuristic vision of the application of nanotechnology in biological sciences, agricultural sciences, and food engineering are vast and versatile. Thus the need of a concerted effort from scientists and engineers towards a greater scientific emancipation of nanotechnology and nano-engineering. The status of the global environment is exceedingly grave and thought provoking. Loss of ecological biodiversity and the grave concerns of climate change are opening new doors of scientific vision, ingenuity, and profundity in the field of nanotechnology and diverse areas of science and technology. The future perspectives of the application of nanotechnology in agricultural sciences needs to be re-envisioned and reframed today as human scientific understanding and discernment moves forward. Nano-vision is really a challenge of human civilization today. In this paper, the author deeply delineates the scientific success, the scientific redeeming, and the vast scientific stance in the applications of nanotechnology in not only biological sciences and agricultural sciences but also water purification science. Heavy metal and arsenic drinking water and groundwater poisoning are really challenging the vast scientific fabric in industrialized and developing countries around the world. This paper opens a new chapter in the field of biological sciences as well as environmental protection. The challenges of arsenic groundwater contamination still remain unresolved. So the immediate necessity of a concerted effort of government, scientists, engineers, and the civil society in confronting the global water issue. Biological sciences and environmental protection are allied branches of science and engineering. So the need of nanotechnology and nano-engineering in greater scientific understanding of application and vision. Thus, human scientific progress will usher in a new era in nanotechnology or nano-engineering. A new dawn in the field of biological sciences and biotechnology will usher in with an immense vision, might, integrity, and scientific forbearance in years to come.

KEYWORDS

- **biological sciences**
- **biotechnology**
- **green chemistry**
- **green sustainability**
- **nano-engineering**
- **nanomaterials**
- **nanotechnology**

REFERENCES

1. United Nations Organization Report, (2015). *Transforming our World: The 2030 Agenda for Sustainable Development*, (A/RES/70/1).
2. Dhewa, T., (2015). Nanotechnology applications in agriculture: An update. *Octa Journal of Environmental Research, 3*(2), 204–211.
3. Abd-Elrahman, S. H., & Mostafa, M. A. M., (2015). Applications of nanotechnology in agriculture: An overview. *Egyptian Journal of Soil Science, 55*(2), 1–18.
4. Fawzy, Z. F., Yunsheng, L., Shedeed, S. I., & El-Bassiony, A. M., (2018). Nanotechnology in agriculture-current and future situation. *Research and Reviews: Journal of Agriculture and Allied Sciences, 7*(2), 73–76.
5. Joseph, T., & Morrison, M., (2006). *Nanotechnology in Agriculture and Food.* A nanoforum Report. (www.nanoforum.org).
6. Werkneh, A. A., & Rene, E. R., (2019). Applications of nanotechnology and biotechnology for sustainable water and wastewater treatment. In: Bui, X. T., et al., (eds.), *Chapter 19, Water, and WASTEWATER TREATMENT TECHNOLOGIES: Energy, Environment and Sustainability* (pp. 405–430). Springer Nature Singapore Pte. Ltd., Singapore.
7. Kunduru, K. R., Nazarkovsky, M., Farah, S., Pawar, R. P., Basu, A., & Domb, A. J., (2017). Nanotechnology for water purification: Applications of nanotechnology methods in wastewater treatment. In: Grumezescu, A., (ed.), *Chapter 2, Water Purification* (pp. 33–74). Academic Press, London, United Kingdom.
8. Das, R., Vecitis, C. D., Schulze, A., Cao, B., Ismail, A. F., Lu, X., Chen, J., & Ramakrishna, S., (2017). Recent advances in nanomaterials for water protection and monitoring: Chemical society reviews. *The Royal Society of Chemistry, 46*, 6946–7020.
9. Yadav, K. K., Gupta, N., Kumar, V., Khan, S. A., & Kumar, A., (2018). A review of emerging adsorbents and current demand for defluoridation of water: Bright future in water sustainability. *Environment International, 111*, 80–108.

10. Palit, S., (2016). Filtration: Frontiers of the engineering and science of nanofiltration-a far-reaching review. In: Ubaldo, O. M., Kharissova, O. V., & Kharisov, B. I., (eds.), *CRC Concise Encyclopedia of Nanotechnology* (pp. 205–214) (Taylor and Francis).
11. Palit, S., (2015). Advanced oxidation processes, nanofiltration, and application of bubble column reactor, In: Boris. I. K., Oxana. V. K., & Rasika, D. H. V., (eds.), *Nanomaterials for Environmental Protection* (pp. 207–215) (Wiley, USA).
12. Palit, S., (2016). Nanofiltration and ultrafiltration: The next generation environmental engineering tool and a vision for the future. *International Journal of Chem. Tech Research, 9*(5), 848–856.
13. Palit, S., (2015). Frontiers of nano-electrochemistry and application of nanotechnology: A vision for the future. *Handbook of Nanoelectrochemistry*. Springer International Publishing, Switzerland.
14. Palit, S., (2011). Dependence of order of reaction on pH and oxidation-reduction potential in the ozone-oxidation of textile dyes in a bubble column reactor. *International Journal of Environmental Pollution Control and Management, 3*(4), 69–78.
15. Palit, S., (2015). Microfiltration, groundwater remediation, and environmental engineering science: A scientific perspective and a far-reaching review. *Nature, Environment and Pollution Technology, 14*(4), 817–825.
16. Palit, S., & Hussain, C. M., (2018). Biopolymers, nanocomposites, and environmental protection: A far-reaching review. In: Shakeel, A., (ed.), *Bio-Based Materials for Food Packaging* (pp. 217–236). Springer Nature Singapore. Pvt Ltd.
17. Palit, S., & Hussain, C. M., (2018). Nanocomposites in packaging: A groundbreaking review and a vision for the future. In: Shakeel, A., (ed.), *Bio-Based Materials for Food Packaging* (pp. 287–303). Springer Nature Singapore. Pvt Ltd.
18. Palit, S., (2017). Advanced environmental engineering separation processes, environmental analysis and application of nanotechnology: A far-reaching review. In: Hussain, C. M., & Kharisov, B., (eds.), *Chapter-14, Advanced Environmental Analysis: Application of Nanomaterials* (Vol. 1, pp. 377–416). The Royal Society of Chemistry, Cambridge, United Kingdom.
19. Hussain, C. M., & Kharisov, B., (2017). *Advanced Environmental Analysis: Application of Nanomaterials* (Vol. 1). The Royal Society of Chemistry, Cambridge, United Kingdom.
20. Hussain, C. M., (2017). Magnetic nanomaterials for environmental analysis. In: Hussain, C. M., & Kharisov, B., (ed.), *Chapter-19, Advanced Environmental Analysis: Application of Nanomaterials* (Vol. 1, pp. 3–13). The Royal Society of Chemistry, Cambridge, United Kingdom.
21. Hussain, C. M., (2018). *Handbook of Nanomaterials for Industrial Applications*. Elsevier, Amsterdam, Netherlands.
22. Palit, S., & Hussain, C. M., (2018). Environmental management and sustainable development: A vision for the future. In: Chaudhery, M. H., (ed.), *Handbook of Environmental Materials Management* (pp. 1–17). Springer Nature Switzerland A.G.
23. Palit, S., & Hussain, C. M., (2018). Nanomembranes for environment. In: Chaudhery, M. H., (ed.), *Handbook of Environmental Materials Management* (pp. 1–24). Springer Nature Switzerland A.G.
24. Palit, S., & Hussain, C. M., (2018). Remediation of industrial and automobile exhausts for environmental management. In: Chaudhery, M. H., (ed.), *Handbook of Environmental Materials Management* (pp. 1–17). Springer Nature Switzerland A.G.

25. Palit, S., & Hussain, C. M., (2018). Sustainable biomedical waste management. In: Chaudhery, M. H., (ed.), *Handbook of Environmental Materials Management* (pp. 1–23). Springer Nature Switzerland A.G.
26. Palit, S., (2018). Industrial vs. Food Enzymes: application and future prospects. In: Mohammed, K., (ed.), *Enzymes in Food Technology: Improvements and Innovations* (pp. 319–345). Springer Nature Singapore Pvt. Ltd., Singapore.
27. Palit, S., & Hussain, C. M., (2018). Green sustainability, nanotechnology and advanced materials: A critical overview and a vision for the future. In: Shakeel, A., & Chaudhery, M. H., (eds.), *Chapter-1, Green and Sustainable Advanced Materials: Applications* (Vol. 2, pp. 1–18). Wiley Scrivener Publishing, Beverly, Massachusetts, USA.
28. Palit, S., (2018). Recent advances in corrosion science: A critical overview and a deep comprehension. In: Kharisov, B. I., (ed.), *Direct Synthesis of Metal Complexes* (pp. 379–410). Elsevier, Amsterdam, Netherlands.
29. Palit, S., (2017). Nanomaterials for industrial wastewater treatment and water purification. In: *Handbook of Ecomaterials* (pp. 1–41). Springer International Publishing, AG, Switzerland.
30. Petre, M., (2011). *Advances in Applied Biotechnology*. In Tech Publishers, Rijeka, Croatia (Book).
31. Hautea, R. A., & Escaler, M., (2004). Plant biotechnology in Asia. *Ag. Bio Forum, 7*(1/2), 2–8.
32. Rao, M. G., Bharathi, P., & Akila, R. M., (2014). A comprehensive review on biopolymers. *Scientific Reviews and Chemical Communications, 4*(2), 61–68.
33. Adeosun, S. O., Lawal, G. I., Balogun, S. A., & Akpan, E. I., (2012). Review of green polymer nanocomposites. *Journal of Minerals and Materials Characterization and Engineering, 11*(4), 385–416.
34. www.wikipedia.com (accessed on 27 February 2020).
35. www.google.com (accessed on 27 February 2020).

IMPORTANT WEBSITES FOR REFERENCES

www.azonano.com/article.aspx?ArticleID=3141 (accessed on 4 March 2020).
www.frontiersin.org/articles/10.3389/fenvs.2016 (accessed on 4 March 2020).
www.insituarsenic.org/gallery.html (accessed on 4 March 2020).
www.ipcbee.com/vol94/rp0025_ICWT2016-W0011.pdf (accessed on 4 March 2020).
www.iwapublishing.com/books/9781780406343/situ (accessed on 4 March 2020).
www.nanoafe.ca (accessed on 4 March 2020).
www.nanotechproject.org/./new_report_on_nanotechnology_in (accessed on 4 March 2020).
www.nanowerk.com/spotlight/spotid=37064.php (accessed on 4 March 2020).
www.ncbi.nlm.nih.gov (accessed on 4 March 2020).
www.sciencedirect.com/science/article/pii/S (accessed on 4 March 2020).
www.teoma.co.uk/Whatisananotechnology/Seenow (accessed on 4 March 2020).

CHAPTER 8

Ecofriendly Polymers: A Need of the Day

JAYESH BHATT,[1] MONIKA JANGID,[1] RAKSHIT AMETA,[2] and SURESH C. AMETA[1]

[1]*Department of Chemistry, PAHER University, Udaipur–313003, Rajasthan, India, E-mail: ameta_sc@yahoo.com (S. C. Ameta)*

[2]*Department of Chemistry, J. R. N. Rajasthan Vidyapeeth (Deemed to be University), Udaipur–313001, Rajasthan, India*

ABSTRACT

Biodegradable polymers cover a large area of high molecular weight compounds. These are used in packaging, agriculture, medicine, and other areas. In recent years, there has been an increase in interest in biodegradable polymers. Two classes of biodegradable polymers can be distinguished: synthetic or natural polymers. There are polymers produced from feedstocks derived either from petroleum resources (non-renewable resources) or from biological resources (renewable resources). In general, natural polymers offer fewer advantages than synthetic polymers. The following review presents an overview of the different biodegradable polymers that are currently being used and their properties, as well as new developments in their synthesis and applications.

Presently consumers have attached little or no added value to biodegradability; thus, forcing the industry to face the challenge to lower down the cost as compared to familiar products. In addition, no suitable infrastructure for the disposal of non-biodegradable materials is available to the desired extent. As a result, biodegradable materials are the need of the day.

8.1 INTRODUCTION

The name "polymer" is derived from two Greek words polus, which means "some, much" and "meros," meaning "parts." A polymer is an extensive molecule (called macromolecules), which has been made out of numerous rehashed subunits, known as monomers. These monomers can be connected together in different ways to give direct, fanned, and cross-connected polymers and so on.

In the last few decades, every material has been replaced by some or other types of polymer, because of its tunable properties like flexibility, mechanical strength, molding ability, long term stability, etc. It may be wood, cloth, fiber, furniture, capsule, tyre, thread, body part of automobile, toy, bottle, packaging material, disposable, carry bag, etc. It is therefore nothing wrong, if this era has been named as "Polymer Era."

Although various kinds of polymers have made our life comfortable from many angles, but it has also created a lot of problems regarding disposal of such polymeric materials as they are long lasting. Some of the countries are using their nearby oceans as dumping stations, but that is not the true solution as their presence in water affects life of aquatic animals and plants adversely. Dumping waste or used polymers anywhere or burning will not provide a solution to the polymer pollution in air, or water, but this is just shifting of a pollution from one phase to the another phase. Therefore, there is an urgent need to develop ecofriendly polymers, which can be degraded in nature biochemically to less harmful smaller fragments than these macromolecules.

Here biodegradable polymers enter the scene. Biodegradable polymers are those polymers, which degrades after their use and breakdown into some simple products such as water, biomass, inorganic salts, etc. They can be natural as well as synthetic polymer. On the other hand, non-biodegradable polymers do not breakdown after their use. They persist in the environment for a long period and are responsible for water and air pollution.

These polymers are found in both forms: naturally and synthetically prepared. They largely consist of ester, amide, and ether functional groups. Their properties and breakdown mechanism are determined by their exact structures. These polymers are often synthesized by condensation reactions, ring opening polymerization, metal catalysts, etc. There are lot more examples of biodegradable polymers and their applications.

Plastics are mainly produced from petrochemical products, but there is an increasing demand for eco-friendly plastics, which are produced from renewable resources. These plastics are degraded in the environment, and it helps us in providing solution of global environmental and waste management problems.

Block copolymers have excellent thermal and mechanical properties, much better than their corresponding random copolymers and homopolymers, but, it is difficult to synthesize such block copolymers, which comprises of different polyester segments via copolycondensation due to the serious transesterification reaction.

The disposal of large amounts of waste, (used polymers) is a major problem in the current era. Effective utilization of bio-renewable materials derived from natural sources can be considered to be a potential solution to this problem.

8.2 HISTORY

Biodegradable polymers have a long history as many of them are natural products. Hence, the precise timeline of their discovery and use cannot be easily traced out accurately. One of the first medicinal uses of a biodegradable polymer was the catgut suture, which dates back to at least 100 AD [1]. The first catgut sutures were made from the intestines of sheep, but modern catgut sutures are made from purified collagen extracted from the small intestines of cattle, sheep, or even goats [2].

The specialty in any biodegradable polymer is that it degrades rapidly as compared to non-biodegradable polymer. Sidewise their ultimate products of breakdown are eco-friendly in nature. These are biocompatible like carbon dioxide, water, methane, and inorganic compounds or biomass, which can be easily scavenged by any suitable microorganism. Biodegradable polymers can be classified into two large groups based on their structure and synthesis. One of these groups is group of agro-polymers, or those derived from biomass. The other group consists of biopolyesters, which are derived from microorganisms or these are synthetically prepared from either naturally or synthetic monomers.

8.3 BIODEGRADABLE POLYMERS

Foury et al. beautifully classified biodegradable polymer in four major categories. These are:

Biomass products (agropolymers);

- From microorganisms (obtained by extraction);
- From biotechnology (conventional synthesis from bio-derived monomers); and
- From oil-products (conventional synthesis from synthetic monomers).

1. **Agropolymers/Natural Biodegradable Polymers:** Agropolymers include polysaccharides, like starches found in potatoes or wood, and proteins, such as whey (animal-based) or derived (plant) gluten. Polysaccharides consist of glycosidic bonds, which utilize a hemiacetal of a saccharide to bind with alcohol loosing water. Proteins are made from amino acids, containing different functional groups. These amino acids undergo condensation reactions to form peptide bonds, consisting amide of functional groups.

   Natural polymers are interesting class of biodegradable polymers as these are associated with certain benefits: These are:

   - Derived from regular sources;
   - Easily accessible; and
   - Relatively shabby.

   Albumin, gelatin, collagen, starch, dextran, pectin, etc. are good examples of biodegradable polymers.

2. **Collagen:** It is the most generally discovered protein in well-evolved creatures, which are the real suppliers of quality to tissue. A lot of biomedical applications are there, where collagen has been used, like different sorts of surgery, beauty care products, bioprosthetic embeds and tissue designing of numerous organs, etc. There are some disadvantages also such as poor dimensional solidness, variability in medication discharge energy and poor mechanical quality. It is majorly utilized as a part of visual medication conveyance framework. It is utilized as sutures, dressings, and so on.

3. **Albumin:** It is a plasma protein part, which represents more than 55% of aggregate protein in human plasma and utilized to plan particulate medication conveyance frameworks. Albumin small scale circles are utilized to convey drugs like insulin, sulphadiazene, 5-fluorouracil, prednisolone, and others. It is used as a part of chemotherapy to accomplish high nearby medication fixation for moderately more time.

4. **Dextran:** It is a complex extended polysaccharide, which is made of numerous glucose atoms joined into chains of variable lengths. It comprises of α-D-1,6-glucose-connected glucan with side-affixes connected to the foundation of polymer. Its molecular weight ranges between 1000–2,00,000 Daltons. It is used for colonic conveyance of medication as gels.

5. **Gelatin:** It is a blend of peptides and proteins, which are created by fractional hydrolysis of collagen, extricated from the bubbled bones, connective tissues, organs, and a few digestion systems of some creatures. It is an irreversible hydrolyzed type of collagen, whose physicochemical

properties relies on the wellspring of collagen, extraction strategy, and warm corruption. It is used as covering material and its micropellets are used for oral controlled conveyance of medications.

As starch is a utilized in obtaining and storing energy in plants, starch-rich plants were used from time inmemorial as sources of food, Starch is commonly extracted for use in industry. It is stored in granules, which contain linear amylose and branched amylopectin. If starch and water are passed through an extruder, it produces thermoplastic starch (TPS). TPS is not stable and due to this, its retrodegradation is an issue. It tries to revert to its natural starch form. Here, main process is the gelatinization of the starch granules causing the swelling of the amorphous parts of the granules. Therefore, glycol and sugars are added to stabilize the TPS plasticizers [4].

Biodegradable polymers can be classified into two types depending upon the method of preparation. These are:

- Polymers synthesized from renewable sources, and
- Mineral oil derived polymers which are biodegradable.

The most important biodegradable polymers are (biobased) aliphatic polyesters. There are a number of aliphatic biodegradable polyesters but a limited number are commercially available. Some biobased polyesters are:

- Polylactic acid (PLA),
- Polyglycolic acid (PGA),
- Poly-ε-caprolactone (PCL),
- Polyhydroxybutyrate (PHB), and
- Poly(3-hydroxy valerate).

Out of these polymers, PHB, and PLA are biodegradable thermoplastic polyesters, which were extensively investigated. Both of them are truly biodegradable and biocompatible with a relatively high melting point (160 to 180°C).

Synthetic biodegradable polyesters include poly(ethylene succinate) (PESu), poly(propylene succinate) (PPSu) and poly(butylene succinate) (PBSu). These polymers are produced from the reaction of a diacid or acid anhydride with a diol with the elimination of water. Most of the aliphatic polyesters are frequently produced from fossil fuels, but some of these polyesters can also be produced using monomers from renewable resources.

6. **Mechanism of Breakdown:** Biodegradable polymers break down to form simple gases, salts, and biomass. When complete biodegradation

of such polymers is over, there are no oligomers or monomers left at the end. Such breakdown of these biodegradable polymers depend on a number of factors such as polymer and also the environment in which polymer is present. These factors include pH, temperature, micoorganisms present, water, etc.

There are two basic mechanisms and these are:

- Physical decomposition through reactions like hydrolysis and photodegradation, leading to partial or complete degradation; and
- Biological processes, which break it down further into aerobic and anaerobic processes. The general equations are:

$$C_{polymer} + O_2\ C_{residue} + C_{biomass} + CO_2 + H_2O \quad (1)$$

$$C_{polymer}\ C_{residue} + C_{biomass} + CO_2 + CH_4 + H_2O \quad (2)$$

8.3.1 POLY(AKLYLENE SUCCINATE)

Oishi et al. [4] prepared PBSu copolymers from succinic acid (SA), diglycollic acid, and 1,4-butanediol, in the presence of titanium tetraisopropoxide and magnesium hydrogen phosphate trihydrate. It was reported that all the polymers exhibited a higher number-average molecular weight than 65,000. It was revealed that composition of the copolymers was found to be almost the same as the feed composition. The copolymers exhibited a higher break strain as compared to corresponding homopolymer. The biodegradability of copolymers showed a maximum on adding 2% diglycollic acid, but the addition of more amount of diglycollic acid decreased their biodegradability. PESu copolymers, including diglycollate moiety, were also prepared by SA, diglycollic acid, and ethylene glycol, using the same reaction conditions. Unlike the PBS copolymers, the molecular weights of the PES copolymers decreased on increasing amount of diglycollate, but the biodegradability was enhanced on increasing ratio of diglycollate in the copolymers, in the range 0–20%.

A series of PESu/poly(ε-caprolactone) block copolymers were prepared by Seretoudi et al. [5] using a two-step process. These steps are:

- PESu was synthesized by the melt polycondensation process; and
- Block copolymers were prepared by ring-opening polymerization of ε-caprolactone (ε-CL) in the presence of stannous octoate.

The copolymers had final compositions very similar to the initial feed compositions as evident from 1H NMR spectroscopy. It was reported that increase in the ε-caprolactone content also led to an increase in the molecular weights of the copolymer, but polydispersity was narrowed. It was observed that there were two melting points for copolymers containing 75 and 50 mol% PESu, which confirmed that block copolymers were prepared. PESu is a brittle material as it has low molecular weight. On the contrary, copolymer containing 75 mol% ε-CL possessed sufficient tensile strength (TS) and elongation at break. Enzymatic degradation was also performed using Rhizopus delemar lipase in a buffer solution at 37°C. It was revealed that biodegradation rates were mainly affected by the crystallinity of copolymers, and not by their molecular weight. It was interesting to note that PESu and copolymer films were degraded into small fragments on six months burial in soil, but polycaprolactone remained almost unaffected.

Kint et al. [6] investigated the reactive blending of a series of mixtures of poly(ethylene terephthalate) (PET) and poly(1,4-butylene succinate) (PBS) at 290°C, which led to the formation of block PET/PBS copolyesters. It was revealed that block lengths of the resulting copolymers decreased with the severity of the treatment. Copolyesters with different PET/PBS molar compositions (90/10, 80/20, 70/30, and 50/50) were prepared. It was observed that glass transition temperature (Tg0, melting point temperature ™, and crystallinity of the copolymers decreased on increasing the content in PBS, but degree of randomness increased. It was found that elastic modulus (EM) and TS of the copolymers also decreased with the content of PBS, while the elongation at break increased. A pronounced hydrolytic degradability was exhibited by PET/PBS copolymers, when the content in 1,4-butylene succinic units was increased.

A noncovalent adduct of the antineoplastic drug cis-diamminedichloroplatinum (cDDP) and a biocompatible graft copolymer of poly(l-lysine) and methylpoly(ethylene glycol) succinate has been prepared by Bogdanov et al. [7]. Highly soluble and long circulating adducts were formed upon incubation of cDDP with [O-methylpoly(ethylene glycol)-O'-succinyl]-N-ε-poly(l-lysine)n–N-ε-succinate, n = 250–270, containing 4.3% of platinum by weight. About 60% of the polymer-associated drug released was achieved during dialysis against saline or serum albumin containing saline, with a half-time of release of 63 h. These adducts showed a high antineoplastic effect in BT-20 human adenocarcinoma cell cultures. It was observed that the concentration of half-inhibition of [3H]thymidine uptake was 0.9 ± 0.2 μM for the drug–copolymer adduct as compared to 0.3 ± 0.1 μM for free cDDP. This adduct showed a long blood half-life (about 14 h in rats) and it was found to be accumulated in mammary adenocarcinomas at

2.5–3.5% injected dose per of tissue, while a control adduct of cDDP with the backbone portion of the copolymer, poly(l-lysine)–N-ε-succinate, had a very short half-life in the bloodstream (about 30 min) and low accumulation (0.5% injected dose) in tumor. A dual effect of methylpoly(ethylene glycol)succinylpoly(l-lysine)-succinate has been suggested as a carrier of cDDP. These roles are:

- A carrier for systemic release of the active drug from the macromolecule, when it is circulating in the bloodstream;
- As a carrier for on-site delivery resulting from the release of the drug in the tumor as a consequence of accumulation of the copolymer in the tumor.

Multiblock copolyester (PBS-b-PES) containing PBSu and PESu was synthesized by Zhu et al. [8] via chain-extension of dihydroxyl terminated PBS (HO-PBS-OH) and PESu (HO-PESu-OH). They used 1,6-hexmethylene diisocyanate (HDI) as a chain extender. They obtained high molecular weight copolyesters with Mw more than 2.0×10^5 g mol^{-1} through chain extension. It was reported that copolyesters showed a single glass transition temperature, which increases with increasing PES content. On the other hand, melting point temperature, and relative degree of crystallinity (Xc) of the copolyesters first decreased followed by an increase with increasing PES content. Such copolyesters had excellent mechanical properties. It was revealed that PBS5-b-PES5 had a fracture stress of 61.8 MPa and fracture strain of 1173%.

Biodegradable polyurethanes may be used for preparing plastic films and these can replace commonly used films prepared from nondegradable polymers like polyethylene and polypropylene. Prepolymers were synthesized by Lee et al. [9] from PBSu polyol (PBS, Mn 1650), poly(ethylene glycol) (PEG, Mn 1000), and 4,4′-dicyclohexylmethane diisocyanate. Then these were reacted with 1,4-butanediol chain extender to obtain polyurethanes. It was reported that their number average molecular weights and melting temperatures were in the range of 30,000–38,000 and 99–101°C, respectively. The TS (2.0–2.4 kg mm^{-2}) of the polymers were also slightly greater than the reported value (1.6 kg mm^{-2}) of PBS with a molecular weight of 20,000 but their elongations at break (230–330%) were quite higher than that of PBS (42%). It was revealed that they were hydrolytically degraded very rapidly in 3% NaOH solutions even at 37°C so that their weight loss was approximately 30–60% within 2 days; however, it depends on the content of PEG and hard segment. On the contrary, polyurethanes containing only

PBS as soft segments showed a significant slower degradation rate under the identical conditions.

Kondratowicz and Ukielski [10] observed conditions of synthesis of statistical poly(ethylene succinate-co-terephthalate) copolymers (2GTS) and high molecular weight PESu with good hydrolytic and optical parameters, designed for the production of biodegradable products and resins. They prepared copolymers by melt polycondensation of bis-(β-hydroxyethyleneterephthalate) (BHET) and SA in excess of ethylene glycol in the presence of a novel titanium/silicate catalyst and catalytic grade of germanium dioxide as cocatalyst. They carried out hydrolytic degradation in a water solution with range of pH, in garden soil and in compost also. It was observed that highest hydrolytic degradation rate and better hydrolytic degradation values in compost medium was observed at pH 4, when copolyester was prepared in the presence of GeO_2 as polycondensation cocatalyst.

Chrissafis et al. [11] prepared two aliphatic polyesters consisting of SA, ethylene glycol and butylene glycol, PESu and PBSu, using a melt polycondensation process. Their number average molecular weight was found almost similar in both polyesters, i.e., approx. to 7000 g mol^{-1}. It was also reported that the decomposition step appears at a temperature 399 and 413°C for PBSu and PESu, from TG and differential TG thermograms (DTG) respectively, which indicates that PESu is more stable as compared to PBSu and their chemical structures play an important role in this process. In both these polyesters, degradation takes place in two stages:

- First corresponding to a very small mass loss; and
- Second at elevated temperatures being the main degradation stage.

These two stages are attributed to different decomposition mechanisms. It was reported that first mechanism, which is taking place at low temperatures, involves auto-catalysis with activation energy 128 and 182 kJ mol^{-1} for PBSu and PESu, respectively. The second mechanism involves a nth-order reaction with activation energy 189 and 256 kJ mol^{-1} for PBSu and PESu, respectively.

Biodegradable PESu and its three novel copolyesters poly(ethylene succinate-co-ethylene adipate) (P(ESu-co-EA)) with the ethylene adipate (EA) comonomer composition ranging from 5.1 to 15.3 mol% were prepared by Wu and Qiu [12]. It was reported that P(ESu-co-EA) copolyesters have the same crystal structures as PESu, but the crystallinity was slightly smaller in the copolyesters as compared to PESu. Both PESu and P(ESu-co-EA) copolyesters have high thermal stability and moreover, it was improved significantly in case of P(ESu-co-EA) as compared with

PESu in presence of EA composition. It was observed that glass transition temperature, melting point temperature, and equilibrium melting point temperature of P(ES-co-EA) was reduced on increasing the EA composition. The overall crystallization rates of P(ES-co-EA) was also found to decrease with increasing the EA composition and crystallization temperature, but the crystallization mechanism remains unchanged for both. Both of them showed a bell shape for the growth rates in a wide crystallization temperature range, which is not affected by EA composition.

A series of aliphatic polyesters, having different molecular weights, were synthesized by Chrissafis et al. [13] from SA and ethylene glycol, following a melt polycondensation process. It was observed that the molecular weight of polyesters achieved during polycondensation was strongly related to thermal stabilities of initial oligomers as evident from thermo gravimetric analysis (TGA). It was reported the number average molecular weight of oligomers must not be lower than 2300–3000 g mol^{-1} in order to synthesize high molecular weight polyesters, as thermal decomposition begins at temperatures lower than 200°C. However, polycondensation temperatures must not be below 230–240°C. Samples with different molecular weights could be divided into two groups based on different thermal stabilities. In the first group, samples with intrinsic viscosity (IV) = 0.08 dL g^{-1} are there while in the second group all the other samples with IV > 15 dL g^{-1} are included. It was found that degradation of all polyesters takes place in three stages. It was revealed that degradation of samples with low molecular weight is more complex as compared to polyesters having high molecular weights.

Mochizuki et al. [14] also obtained high molecular weight poly(butylene succinate-co-ethylene succinate)s, P(BSu-co-ESu)s, from SA, 1,4-butanediol, and/or ethylene glycol via a polycondensation process. They also carried enzymatic hydrolysis of hot-pressed copolyester films using lipases derived from various microorganisms. The formation of water-soluble total organic carbon (TOC) was also monitored to determine the effect of structure upon enzymatic degradability. The P(BSu-co-ESu) copolyester with about 53 mol% ES exhibited a minimum in systemic crystallinity but a maximum in enzymatic degradation. It was reported that the degree of crystallinity may have some dominant influence upon the rate of degradation.

PESu and its copolyesters containing 7, 10, or 48 mol% butylene succinate (BSu) were synthesized by Chen et al. [15] via a direct polycondensation reaction with titanium tetraisopropoxide as the catalyst. IV (1.08–1.27 dL g^{-1}) indicated that polyesters with high molecular weights are produced. However, distribution of ethylene succinate and BSu units was at random. Thermal stabilities of these polyesters were almost similar. All the synthesized copolymers exhibited a single glass transition temperature. It was indicated that the BSu units were incorporated into PESu significantly, which inhibits the crystallization behavior of PESu.

Salhi et al. [16] observed that cyclodi(ethylene succinate) (C2) easily reacts with PET in the melt, and it leads to the formation of high molar mass PET-poly(ethylene succinate) copolymers (PET-PESu). Copolyesters were synthesized with a PET/C2 starting mass ratio of 90/10, 80/20, 70/30 and 50/50. It was revealed that 50/50 copolyester was almost random, while copolyesters with higher ethylene terephthalate contents exhibited some block copolymer character. It was reported that melting temperatures and crystallinity of 90/10, 80/20 and 70/30 copolyesters were significantly higher as compared to PET-aliphatic polyester copolymers.

Huang et al. [17] synthesized thermally-induced shape memory multi-block poly(ether-ester)s (PBSEGs) comprising crystallizable PBSu hard segments and PEG soft segments, through polycondensation from SA, 1,4-butanediol and PEG diol. It was revealed that all prepared PBSEGs were double-crystalline copolymers, which ensure the formation of the separated crystalline domain, determining the permanent shape and temporary shape. It was also found that Tm of PEG segment (Tm, PEG) of the PBSEGs ranged from 27.54 to 51.04°C, acting as the transition temperature (Ttrans), which can be controlled by the variation of chain length of the soft segment. The mechanical properties of the copolymer films were also assessed. It was found that the copolymers were ductile enabling remarkable reversible deformation. Most of copolymers possessed excellent shape memory effect. It was also demonstrated that the copolymers were more hydrophilic on introducing PEG segment, which suggests that this biodegradable PBSEG multiblock copolymer has excellent shape-memory properties with great potential for application in biomaterials.

Papageorgiou and Bikiaris [18] synthesized series of poly(butylene-co-propylene succinate) (PBPSu) random copolyesters. It was observed that tensile properties decreased with increasing propylene succinate (PSu) content. It was reported that minimum melting point corresponded to 75 mol% PSu. Copolymers with up to 60 mol% PSu units formed PBSu crystals as evident from Wide-angle x-ray diffraction (WAXD) patterns. It was found that copolymer with 11.5 mol% PPSu units formed PPSu crystals. It was reported that defect-free energy decreased for copolymers with high PPSu content. Enzymatic hydrolysis study was carried out using *Rhizopus delemar* and *Pseudomonas cepacia lipases*, which showed that degradation was faster for copolymers with high PSu content, compared to even fast-degrading PPSu.

Synthesis of biodegradable aliphatic polyester PPSu was carried out by Chrissafis et al. [19] using 1,3-propanediol and SA via a two-stage melt polycondensation. It was reported that it has a number average molecular weight 6880 g mol^{-1}, peak temperature of melting (44°C) for heating rate 20°C min^{-1} and a glass transition temperature at −36°C. PPSu showed a very high thermal stability as its major decomposition rate is at 404°C (heating rate 10°C min^{-1}). This is very high as compared to that of aliphatic polyesters and can be compared to the decomposition temperature of aromatic polyesters. It was revealed that PPSu degradation takes place in two stages; the first at low temperatures corresponding to a very small mass loss of about 7%, and the second at elevated temperatures mainly degradation stage.

Some biodegradable polyesters were synthesized such as PBSu, PPSu, and poly(butylene succinate-co-propylene succinate)s (PBSuPSus), by Xu et al. [20] using 1,4-SA with 1,4-butanediol and 1,3-propanediol through a two-step process of esterification and polycondensation. It was revealed that all, the melting temperature, crystallization temperature (Tc), crystallinity (X), and thermal decomposition temperature (Td) of these polyesters decreased gradually on increasing the content of PSu unit. PBSuPSu copolyesters was found to show the same crystal structure as that of the PBSu homopolyester.

A series of poly(ester-ether)s based on PBSu and poly(propylene glycol) (PPG), with various mass fractions and molecular weights of PPG, was prepared by Huang et al. [21] through melt polycondensation. In this biodegradable thermoplastic elastomer, the composition of the copolyesters agreed very well with the feed ratio. It was reported that the glass transition temperature decreased gradually on increasing content of the soft PPG segment while the melting temperature, crystallization temperature and the relative degree of crystallinity decreased. The toughness of PBSu was found to improve significantly. The elongation at break of the copolyesters was about 2–5 times than that of the original PBSu. Most of the poly(ester-ether) specimens were flexible enough. The enzymatic degradation rate of PBSu was also enhanced. It was revealed that the difference in molecular weight of PPG led to properties being changed to some extent among the copolyesters.

The miscibility behavior and biodegradability of poly(ε-caprolactone)/poly(propylene succinate) (PCL/PPSu) blends were investigated by Bikiaris et al. [22]. Different blends of the polymers were prepared by solution-casting with compositions 90/10, 80/20, 70/30 and 60/40 w/w. It was observed that the particle size distribution of PPSu dispersed phase increased on increasing

PPSu content. Enzymatic hydrolysis prepared blends was also performed using *Rhizopus delemar lipase* for several days at pH 7.2 and 30°C. It was revealed that hydrolysis affected mainly the PPSu polymer as well as the amorphous phase of PCL as evident from SEM images. An increase of the melting temperatures and the heat of fusions was recorded for all polymer blends after the hydrolysis. The biodegradation rates were faster for the blends with higher PPSu content.

Multiblock copolymers comprising of two different polyester segments, like crystallizable PBSu and amorphous poly(1,2-propylene succinate) (PPSu), were synthesized by Zheng et al. [23] through chain-extension with hexamethylene diisocyanate (HDI). Amorphous PPSu segment was incorporated just to improve the impact strength of PBSu. It was reported that synthesized multiblock copolymers were having regular sequential structure. It was also revealed that block copolymers possess excellent thermal and mechanical properties with satisfactory TS and extraordinary impact strength achieving about 1900% to that of pure PBSu. The influence of PPSu ratio and chain length of both the segments was investigated on the thermal and mechanical properties. It was reported that incorporation of an amorphous soft segment PPSu imparted high-impact resistance to the copolymers without decreasing the melting point. It was found that mechanical and thermal properties of the copolymers also depend on their regular sequential structure.

Pang et al. [24] prepared biodegradable blends of poly(propylene carbonate) (PPC) and PBSu via melt-preparation and followed by compression-molding. It was reported that the yield strength and the strength at break were found to increase significantly up to 30.7 and 46.3 MPa, respectively, by incorporating PBSu. The immiscibility of the two components was verified by the two independent glass-transition temperatures. It was indicated that the thermal decomposition temperatures ($T_{-5\%}$ and $T_{-10\%}$) of the PPC/PBS blends increased dramatically by 30–60°C as compared with PPC matrix.

Three copolymers were synthesized e.g., poly(propylene isophthalate) (PPI), PPSu, and poly(propylene isophthalate/succinate) (PPI-PPSu) by Soccio et al. [25]. Their thermal behaviors were examined by thermo gravimetric analysis (TGA) and Differential scanning calorimetry (DSC) studies. All these polymers were found to show a good thermal stability. They appeared as semicrystalline materials at room temperature (RT), except 20PPI-PPSu and 30PPI-PPSu. The main effect of copolymerization was in lowering the amount of crystallinity as compared to homopolymers. An increase in T_g was observed with increase in PI units. This behavior may be due to the presence of stiff phenylene groups.

A fast eco-friendly synthesis of PBSu was reported by Velmathi et al. [26] using microwaves (MWs) and 1,3-dichloro-1,1,3,3-tetrabutyldistannoxane as catalyst. They also studied the effect of catalyst concentration, bulk vs. solution polymerization, reaction time, temperature, and stoichiometry of the monomers just to obtain optimum conditions. It was found that PBSu with a weight-average molecular weight of 2.35×10^4 was obtained within a short time of 20 min under optimum conditions.

8.3.2 POLY(LACTIC ACID)

PLA is a biodegradable polymer, which can be obtained from polymerization of sustainable sources as chips sugarcane, starch, corn, etc. Ring-opening polymerization (ROP) of lactide (LA) monomer in presence of catalysts such as Al, Sn or Zn an efficient method for synthesis of PLA. PLA polymerized using this type of catalysts may contain some traces of these elements, which may have carcinogenic nature and therefore, the traces of such catalysts should be removed from the synthesis process. An alternative way of supplying energy is through UV or MW sources, which can prove to be a potential route.

Alternative development of non-metal catalysts is better for the processing of PLA through ROP. PLA layer-based composite materials are finding use in the field of food, medical, etc. as eco-friendly materials. Dubey et al. [29] reviewed on the implementation of alternative energy sources for PLA processing with the possibility of application in the field of nanocomposite materials.

Paclitaxel is one of the most effective antineoplastic drugs. but its current clinical administration is formulated in Cremophor EL, which causes serious side effects. Of course, nanoparticle (NP) technology may provide a solution for such a problem and it can promote a sustained chemotherapy, where biodegradable polymers may play a key role. Zhang and Feng [27] successfully synthesized poly(lactide)-tocopheryl polyethylene glycol succinate (TPGS) (PLA-TPGSu) copolymers of desired hydrophobic-hydrophilic balance for nanoparticle formulation of anticancer drugs. The effects of the PLA:TPGSu composition ratio was observed on drug encapsulation efficiency, *in vitro* drug release, *in vitro* cellular uptake and viability of the PLA-TPGS NP formulation of paclitaxel. They prepared paclitaxel-loaded PLA-TPGSu NPs by a modified solvent extraction/evaporation method. Cancer cell viability of the drug-loaded PLA-TPGSu was measured by

3-(4,5-dimethylthiazol-2-yl)-2,5-diphenyl tetrazolium bromide (MTT) assay. It was found that the PLA:TPGSu composition ratio showed little effects on the particle size and size distribution, but PLA-TPGSu NPs (89:11 PLA:TPGSu ratio) achieved the best results on the drug encapsulation efficiency, the cellular uptake as well as cancer cell mortality of the drug-loaded PLA-TPGS NPs.

Zeng et al. [28] successfully prepared an aliphatic polyester-based poly(ester urethane) (PEU) consisting of poly(l-lactic acid) and PESu via chain-extension reaction of poly(l-lactic acid)-diol (PLLA-OH) and poly(ethylene succinate)-diol (PESu-OH). They used 1,6-hexamethylene diisocyanate (HDI) as a chain extender. PLLA-OH was prepared by direct polycondensation of l-lactic acid in the presence of 1,4-butanediol while PES-OH was synthesized by condensation polymerization of SA with an excess of ethylene glycol. High molecular weight polymers (> 200 000 g•mol^{-1}) were easily synthesized through chain-extension reaction as evident from gel permeation chromatography (GPC) analysis. Such materials with high molecular weight and excellent tensile properties may find some useful applications in the field of biomaterials and environmental friendly materials.

Green nanocomposites composed of cellulose nanofiber (CNF) and PLA were prepared by Khani et al. [30] using a solvent casting method. First, surface modification of CNF was carried out by esterification and then CNF was incorporated into the polymer matrix to improve the dispersion of CNF and its interfacial adhesion with the biopolymer. It was revealed that there was a uniform distribution of NPs in the polymer matrix at low contents (1 and 3 wt%), but at higher content (5 wt%), CNF was also easily agglomerated. However, this affected mechanical properties of the nanocomposites adversely. It was observed that the use of acetylated nanofibers had no significant effect on the permeability of films. Properties like TS and EM of nanocomposites with 1 wt% CNF did not show significant changes, but elongation percentage (E) was increased by more than 60%. TS, EM, and E all the three changed significantly for nanocomposites with 3 and 5 wt% CNF. The effect of reinforcing CNF composition with PLA was a slight increase in glass transition and melting temperatures. It was reported that nanocomposite films showed an almost similar pattern of thermal behavior to that of neat PLA film.

Al-Mulla et al. [31] used fatty nitrogen compounds (FNCs) like fatty amides (FAs), fatty hydroxamic acids (FHAs), and N,N′-carbonyl difatty amides (CDFAs), as organic compounds to modify a natural clay (sodium montmorillonite). This modification was carried out by stirring the clay particles in an aqueous solution of FAs, FHAs, or CDFAs, where it was observed that clay layer distance increased from 1.23 to 2.69, 2.89 and 3.21 nm, respectively. Modified clay was then used for preparing epoxidized palm oil (EPO) plasticized PLA nanocomposites. The presence of these FNCs was also estimated in the clay. The nanocomposites were synthesized by solution casting of the modified clay and a PLA/EPO blend at the weight ratio of 80/20. It was found to have the highest elongation at break. Such PLA/EPO modified clay nanocomposites showed higher thermal stability as well as improvement of mechanical properties as compared to PLA/EPO blend.

Quitadamo et al. [32] prepared blends based on high-density polyethylene (HDPE) and poly(lactic) acid (PLA) with different ratios of both polymers. A blend with of HDPE and PLA (equal amounts), proved to be a better compromise with 50 wt.% each. It allowed a high amount of bioderived charge without much loss in mechanical properties; thus, making outdoor application based on biodegradation behavior. Hence, wood flour (WF) was added as a natural filler in different proportions (20, 30, and 40 wt.%), considering as 100 the weight of the polymer blend matrix. Two compatibilizers were used to modify both HDPE-PLA blend and wood-flour/polymer interfaces. These are: polyethylene-grafted maleic anhydride and a random copolymer of ethylene and glycidyl methacrylate. The most suitable percentage of compatibilizer for HDPE-PLA blends was found to be 3 wt.%, which was used with WF.

PLA/PEO/PLA triblock copolymers having short poly(l-lactic acid) blocks were synthesized by Rushkov et al. [33], via ring-opening polymerization of l-lactide initiated by PEG in the presence of CaH_2. The number average degree of polymerization (DP) of each PLA block was 2, 4, 8, and 12. The length of PEO blocks was varied using parent PEG of different number average degrees of polymerization (14, 26, and 49). It was reported that triblock copolymers did not contain any detectable amount of PLA homopolymer as side product. It was revealed that last two constitutive units located at both ends were of PLA block. PEO and PLA blocks were phase-separated even for copolymers with very short PLA blocks. It was found that an increase in the length of PLA blocks led to a decrease in the crystallinity of PEO blocks, even up to disappearance.

Tingaut et al. [34] prepared novel bionanocomposite materials with tunable properties using a poly(lactic acid) matrix and acetylated microfibrillated cellulose (MFC) as reinforcing agent. It was observed that grafting of acetyl moieties on the cellulose surface not only prevented MFC hornification upon drying but it also significantly improved redispersibility of the powdered nanofibers in chloroform (a PLA solvent of low polarity). It has been demonstrated that the properties of the resulting PLA nanocomposites could be tailored by proper selection of both; acetyl content (Ac%) and the amount of MFC. As-prepared nanomaterials showed improved filler dispersion, higher thermal stability, and reduced hygroscopicity as compared to those prepared with unmodified MFC. The reinforcing potential of both; unmodified and acetylated MFC was observed on the viscoelastic properties of the neat PLA. Much more interesting is the fact an increase in the PLA glass transition temperature was detected on using 8.5% acetylated MFC at 17 wt%, which indicates an improved compatibility at the fiber matrix interface. It was suggested that the final properties of nanocomposite materials can be controlled by adjusting the %Ac of MFC.

Cellulose diacetate-graft-poly(lactic acid)s (CDA-g-PLAs) were synthesized by Teramoto and Nishio [35] over a wide range of composition combination of different methods of graft polymerization. Copolycondensation of lactic acid, a ring-opening copolymerization of L-lactide in dimethylsulfoxide (DMSO), and a copolymerization similar to the second, but in bulk, each initiated at residual hydroxyl positions on CDA. It was revealed that all the copolymer products gave a single glass transition temperature, as evident from DSC, which decreased sharply from 202°C of the original CDA (~60°C), close to Tg of PLA homopolymer. It was observed in tensile measurements for film sheets conducted at 80–100°C of melt-quenched CDA-g-PLAs, that their drawability increased drastically with increasing PLA content. The elongation at rupture reached a maximum of ca. 2000%.

Binary and ternary blends of TPS, poly(lactic acid) and poly(butylene adipate-co-terephthalate) (PBAT) were prepared by Ren et al. [36] using a one-step extrusion process. The concentration of TPS in both the blends (binary and ternary) was fixed at 50 wt%, with the rest being PLA and PBAT. They used a compatibilizer with anhydride functional groups, which improved the interfacial affinity between TPS and the synthetic polyesters. It was observed that addition of a small amount of compatibilizer greatly increased the mechanical properties of the blends. It was also revealed that mechanical properties of the blends exhibited a dramatic improvement in elongation at break on increasing PBAT content. The morphology analysis of the blends showed that most of the TPS particles were melting and these were well dispersed in the polyester matrix for the compatibilized blends as compared to the non-compatibilized blends.

PLA and polycaprolactone (PCL) nanocomposites were prepared by Fukushima et al. [37] adding 5 wt% of a sepiolite (SEPS9). Then these were degraded in compost, leading to effective degradation for all samples. PLA and PLA/SEPS9 were mainly degraded by a bulk mechanism; however, the presence of SEPS9 particles partially delays the degradation, which may be due to a preventing effect of these particles on polymer chain mobility and/or PLA/enzymes miscibility. It was reported that PCL and PCL/SEPS9 showed a surface mechanism of degradation; and in contrast to PLA, sepiolite does not show a considerable barrier effect on the degradation of PCL.

Chuensangjun et al. [38] applied lipase-catalyzed polymerization with lecitase ultra and lipozyme TL IM to synthesize PLA to decrease the utilization of catalysts. It was indicated that low molecular weight PLA could be successfully produced from commercial lactic acid by using the commercial lipases. However, when lipozyme TL IM was used as biocatalyst, Mn, and Mw of PLA were obtained as 7,933 and 194 Da, respectively, while lecitase ultra gave Mn and Mw of PLA as 8,330 and 216 Da, respectively. PLA products as PLA films were prepared blended with commercial PLA beads with different blending ratios by casting on glass plate. Their degradation was studied under controlled soil burial laboratory conditions. The characteristics of PLA blend films were analyzed by visual observations, and measuring weight loss. It was revealed that the different blends ratios of PLA films showed more flexibility than pure PLA film. Different blends of PLA films were disintegrated in soil within the short burial time.

8.3.3 POLY(HYDROXYALKANOATES)

Madison et al. [39] reviewed poly (3-hydroxyalkanoates) (PHAs) members of a family of polyesters. They exist as homopolymers such as poly(3-hydroxybutyrate) (P3HB) or copolymer poly(3-hydroxybutyrate-co-3-hydroxyvalerate) (P(3HB-co-3HV) in nature. PHAs exist as granules of pure polymer in bacteria, which act as medium for energy storage like fat for animals and starch for plants. These are commercially produced using energy-rich feedstock, which is then transformed into fatty acids providing feed to bacteria. For the industrial production of PHAs, cells are isolated and lysed after a few "feast-famine" cycles. Then polymer is extracted from the remains of the cells, it is purified, and processed into pellets or powder.

Witholt and Kessler [40] reported that medium-chain length (MCL) poly(hydroxyalkanoic acids) (PHAs) are polyesters, which are accumulated by fluorescent *Pseudomonads* and other bacteria. It was revealed that genetics of MCL-PHA formation led to polymer synthesis in recombinant bacteria and plants. As a result, several high and medium cost applications are now appearing. It is predicted that with optimized synthesis of bacterial MCL-PHA on low cost agro-substrates and the development of such plant-based MCL-PHAs in the next decade, or so, the economics of production of these bioplastics will be sustainable for bulk applications.

Short-chain-length/medium-chain-length (SCL/MCL) polyhydroxyalkanoate (PHA) was produced by Matsumoto et al. [41] in the plastids of *Arabidopsis thaliana*. They introduced Phe87Thr (F87T) mutated 3-keto-acyl-acyl carrier protein (ACP) synthase III (FabH) from *Escherichia coli*, and Ser325Thr/Gln481Lys (ST/QK) mutated PHA synthase (PhaC1) from *Pseudomonas sp.* 61-3, along with the beta-ketothiolase (PhaA) and acetoacetyl-CoA reductase (PhaB) from *Ralstonia eutropha* (Cupriavidus necator) genes into *Arabidopsis*. Such transgenic *Arabidopsis* produced PHA copolymers, which are composed of monomers consisting of 4–14 carbons. It was reported that the introduction of the engineered PHA synthase resulted in a ten-fold increase in PHA content as compared to plants expressing the wild-type PHA synthase. It was also observed that the expression of the engineered fabH gene in the plastid led to an increase in the amount of the SCL monomer, 3-hydroxybutyrate, incorporated into PHA, and it contributed to supply of MCL monomers for PHA production.

8.3.4 CELLULOSE

George et al. [42] opined that water-soluble polymers are becoming very important in industries, as they are easy to process, low cost, easily available, and more eco-friendly as compared to other polymers. These are widely used as stabilizers, flocculants, thickeners, protective colloids, drug delivery materials, dispersants, materials for oil recovery, etc. Water-soluble polymers, derived from naturally occurring proteins, polysaccharides, etc., as well as synthesized polymers, are not lacking sufficient properties to replace the presently existing non-degradable plastic materials for the majority of applications. However, incorporation of nanomaterials into such polymer matrices is likely to enhance the mechanical properties such as TS, modulus, stiffness, and impact strength. Apart from it, barrier, optical, thermal resistance, nonflammability, etc., can also be improved by the incorporation of nanomaterials. Nanotechnology can play an important role in solving the problem of adverse impact on the environment, if water-soluble nanocomposite materials are developed. Several nanomaterials were studied for reinforcing water-soluble polymers, out of which rod-shaped cellulose nanocrystals (CNs) with high aspect ratios were found to be a potential candidate for such applications.

The biocomposites developed by the use of renewable polymers and naturally available fibers like furfuryl alcohol, poly(lactic acid), gluten, starch, soy flour, etc., are gaining considerable attention these days because of their environment-friendly nature. Wood is a biologically derived biodegradable raw material and it requires minimum processing energy. Hazarika et al. [43] opined that wood polymer composites (WPC) have enormous advantages and it can improve the mechanical, physical, chemical as well as some other properties of the composite, which are suitable for different applications (outdoor and indoor). Such properties of the WPC can be improved to the desired extant by application of nanotechnology, cross-linking agents, flame retardants, grafting, etc. For example, flame retardants, which are obtained from gum of the plant *Moringa oleifera* can efficiently improve the flame retardancy and also some other properties of the composites.

Dobos et al. [44] reviewed excellently the area of cellulose acetate nanocomposites. They stressed on the importance of antimicrobial activity, based on the fact that their interaction mechanisms create inhibitory effects against microbial growth in a solid medium; thus, deciding areas of their applicability. Cellulosic materials can be designed and fine-tuned to have certain properties, which are required in different biomedical areas.

Nanocomposite is a reinforced composite material, which consists of nanoscale reinforcing fillers and matrix polymer. It is known that if fillers are dispersed within nanoscale, that requires lesser amount as compared to conventional reinforcing fillers and also properties of composites are greatly improved. Insignificant deterioration of properties may be there in case of recycling and; therefore, it can also be used as an eco-friendly composite material. Cellulose nanocomposites represent eco-friendly materials than that of nanocomposites reinforced with inorganic nanoscale fillers like montmorillonite nanoclay, mica, and silica. Natural filler such as nanofiber cellulose from palm empty fruit bunch promotes its eco-friendly character. Muhamad et al. [47] prepared CNF through a pretreatment so to remove noncellulosic content. It was followed by acid hydrolysis process. Starch-based nanocomposite film was formed by incorporation of 2–10% CNF per weight of starch into the film matrix. As-prepared film appeared translucent and easy to handle, but the film becomes more opaque, if the percentage of CNF incorporation was increased. It was reported that films with the addition of up to 2% CNF showed higher TS and thermal stability, as well as better barrier properties to water vapor than control films. They also studied the effect of CNF on starch/chitosan composite packaging film to know the influence of CNF toward antimicrobial properties of the composite film. The effects of CNF contents were also investigated on the tensile, dynamic mechanical, thermal properties and barrier properties of the starch/chitosan nanocomposite. There is potential of CNF as filler for antimicrobial packaging, because it enhances antimicrobial efficacy toward food shelf life.

The biodegradable PPC/glycerol-plasticized thermoplastic dried starch (GPTPS) composites were prepared by Ma et al. [42]. Here, dried starch was used, because the moisture in native starch may easily induce the degradation of PPC during the processing. The effects of succinic anhydride (SA) were evaluated on the morphology, thermal properties, dynamic mechanical thermal analysis, and mechanical properties of PPC/GPTPS composites. The use of SA leads to a significant improvement in thermal stability for PPC/GPTPS composites as evident from TGA. SA can also increase mechanical properties of PPC/GPTPS composites. It was reported that TS and elongation at break of PPC/GPTPS/SA(75/25/1) composite can reach up to 19.4 MPa and 88.5%, respectively.

A carboxymethyl cellulose/grapheme oxide nanocomposite (CMC/GO) high-performance film was prepared by Yadav et al. [47] via a simple solution mixing-evaporation method. It was revealed that CMC and graphene oxide were able to form a homogeneous mixture. The TS and Young's modulus of the graphene-based materials improved significantly upon incorporation of 1 wt% graphene oxide by 67 ± 6% and 148 ± 5%, respectively as compared with pure CMC.

Green nanocomposites have been successfully prepared by Park et al. [48] from cellulose acetate (CA) powder, eco-friendly triethyl citrate (TEC) plasticizer and organically modified clay. The effect of the amount of plasticizer was observed by varying it from 15 to 40 wt% on the performance of the nanocomposites. It was reported that cellulosic plastic-based nanocomposites with 20 wt% TEC plasticizer and 5 wt% organoclay had better intercalation and an exfoliated structure than its counterpart with 30/40 wt% plasticizers. It was found that TS, modulus, and thermal stability of this cellulosic plastic reinforced with organoclay had a decreasing trend as plasticizer content was increased from 20 to 40 wt%. Water vapor permeability of cellulosic plastic were reduced by 2 times on nano-reinforcement at the lower volume fractions ($\varphi \leq 0.02$).

8.3.5 OTHER ECOFRIENDLY POLYMERS

Murugan et al. [49] polymerized glycolic acid under vacuum in the presence as well as absence of nano-sized clay. The added clay was found to catalyze the condensation polymerization. An increase in relative intensity of C=O/CH was observed on increasing the amount of clay. It was revealed that the percentage weight residue remained above 750°C for polymer-nano composite system was about 21%, which proved the flame retardancy (char forming) nature.

An efficient and environmentally friendly process has been developed by Kiasat and Mehrjardi [50] for the synthesis of β-hydroxy thiocyanates from epoxides using NH_4SCN and PEG-bound sulfonic acid as catalyst in water without using any organic solvent. They successfully obtained several β-hydroxy thiocyanates, useful synthetic intermediates towards the synthesis of many biologically active compounds, in high yields under mild conditions.

Maleki et al. [51] reported design, preparation, and performance of magnetic cellulose/Ag nanobiocomposite, which can be used as a recyclable and highly efficient heterogeneous nanocatalyst. They also investigated its activity in the synthesis of 2-amino-6-(2-oxo-2*H*-chromen-3-yl)-4-phenyl-nicotinonitrile derivatives, where high yields were achieved in short reaction times. The magnetic property of the nanobiocomposite catalyst makes its separation easy from the reaction mixture by using an external magnet. It was revealed that there was no considerable loss in its catalytic activity.

Jie et al. [52] synthesized a macroporous cross-linked antimicrobial polymeric resin containing *N*-halamine and quaternary ammonium salt moieties in an eco-friendly and cost effective way. A macroporous cross-linked chloromethylated polystyrene (CMPS) resin was treated with the salt of 5,5-dimethylhydantoin (DMH) and trimethylamine (TMA) in water affording a polymeric resin containing hydantoinyl and quaternary ammonium salt moieties, poly(*p*-methylstyrene)-3-(5,5-dimethylhydantoin)-*co*-trimethyl ammonium chloride (PSHTMA). Then hydantoinyl groups in PSHTMA were converted to an antimicrobial *N*-halamine structure by a simple chlorination reaction in dilute NaOCl solution. They also showed that the as-synthesized antimicrobial polymeric resin was capable of 7-log inactivation of *S. aureus* and 8-log inactivation of *E. coli* within 1 min contact time. *N*-halamine moieties in Cl-PSHTMA also exhibited excellent regenerability.

A three-component condensation of aldehydes, β-ketoesters and urea was carried out by Sabitha et al. [53] in water using ceria NPs supported on poly vinylpyridine crosslinked divinylbenzene (4vp-co-dvb) as a catalyst for the preparation of 3,4-dihydropyrimidin-2(1*H*)-ones in good yields. It was reported that this catalyst was recovered easily and can be reused again without loss of its activity.

Azarudeen et al. [54] synthesized anthranilic acid-thiourea-formaldehyde terpolymer resin in an eco-friendly manner using dimethylformamide as a reaction medium. A transition state between crystalline and amorphous nature was established based on SEM images. It was observed that electrical property of this terpolymer resin showed an appreciable change in its conductivity at various concentrations and temperatures. One of the important applications of such polymers is their capability to act as chelating ion-exchangers and therefore, chelation ion-exchange property of the terpolymer were studied which showed a powerful adsorption towards specific metal ions such as Zn^{2+}, Mn^{2+}, Cu^{2+}, Ba^{2+}, and Mg^{2+}.

Carbon nanotubes (CNTs) are nanomaterial produced from carbon in the form of powder, liquid, or gel by Julkapli et al. [55] via acid or chemical hydrolysis. As it has unique properties such as renewability, biodegradability, mechanical, physicochemical properties, and abundance, if it is incorporated in a small quantity of CNTs to polymeric matrices, it may result in enhancement of the mechanical and thermal resistance. Stability of composite of NCC-derived carbon materials does not put any serious threat to the environment and can provide green and renewable biomaterial for lightweight and degradable composites. Surface functionalization of these CNTs may assist in tailoring its properties for dispersion in hydrophilic and hydrophobic media.

Saw [56] provided a simple method for the preparation of nanocomposites with different weight ratios of blends of epoxy novolac (ENR) and diglycidyl ether of bisphenol A (DGEBA) resin, natural coir fiber, and organically modified montmorillonite (OMMT) nanoclay. It was reported that the storage modulus at RT was enhanced by about 100% or more in the case of 50 and 65% ENR-containing matrices; on blending ENR with DGEBA, while the enhancement is only 50% in the case of 20 and 35% ENR-containing matrices than that of the pure matrix. It was also observed by them that the tan δ peak heights of the composite containing 50 and 65% ENR were closer to that of 35% ENR-containing composite. Dynamic mechanical analysis (DMA) showed that the modification of the clay influences the stiffness and glass transition temperature of the nanocomposites strongly. As a result, it is possible to manufacture coir composites with increased stiffness without sacrificing their ductility.

Vegetable triglycerides (TG) are initial renewable resources, which were used primarily in coating applications. It is because of the fact that their unsaturated varieties polymerize as thin films, whenever it is exposed to atmospheric oxygen. They have certain disadvantages like poor mechanical properties. To overcome this problem, and to complete with high price of synthetic biodegradable polymers, various blends and composites have been attempted in last decade. Three kinds of polymers from renewable resources may be used for preparing blends. These are:

- Natural polymers, such as starch, protein, and cellulose;
- Synthetic polymers from natural monomers, such as PLA;
- Polymers from microbial fermentation [57].

Nanofibers have some remarkable increasing applicability in various fields because of their large surface area and controlled morphology. Such nanofibers can be prepared using three main techniques:

- Electrospinning;
- Phase separation; and
- Self-assembly.

Out of these, electrospinning techniques are commonly used. Sherbiny and Ali [58] discussed electrospinning of eco-friendly polymers and polymeric nanocomposites for possible biomedical applications. They find their applications in wound healing and tissue regeneration. Characteristics of the developed electrospun nanocomposites may be controlled via tailoring the collectors used as well as carefully changing their surface chemistry.

Poly(ester amide)s are an emerging group of biodegradable polymers these days. These polymers have both the group's ester and amide on their chemical structure, which are of a degradable nature and provide good thermal and mechanical properties. Here, strong hydrogen-bonding interactions between amide groups may counter some typical weaknesses of aliphatic polyesters like for example poly(ε-caprolactone). Rodriguez-Galan et al. [59] reported that poly(ester amide)s can be prepared from different monomers with different synthetic methodologies leading to polymers with random, blocky, and ordered microstructures. Their properties like hydrophilic/hydrophobic ratio and biodegradability can also be tuned easily. They also reviewed applications like controlled drug delivery systems, hydrogels, tissue engineering and smart materials.

8.4 CONCLUSION

Synthetic polymers are regularly replaced by biodegradable materials, particularly, those derived from renewable or natural resources. Chemical structure of the biopolymer determines its biodegradability and the use of such polymers will open up newer dimensions of packaging films for food products, transport packaging, carry bags, cups, plates, and cutlery, biowaste bags, etc. Bilayer and multi-component films like synthetic packaging materials has to be developed, which have an excellent barrier and mechanical properties. They can be cross-linking, (chemically or enzymatically). Biodegradation of polymers provides an attractive route for environmental waste management. Some bio-based polymers are already in use in the field of medicine, where cost is much less important than function, but these are also to be prepared, which will be cost-effective so that to make them commercially viable.

KEYWORDS

- **biodegradable polymers**
- **biopolymers**
- **cellulose**
- **eco-friendly polymers**
- **poly(lactic acid)**
- **transesterification**

REFERENCES

1. Nutton, V., (2012). *Ancient Medicine* (2nd edn.). Routledge, London.
2. Isabelle, V., & Lan, T., (2009). Biodegradable polymers. *Mater., 2*(2), 307–344.
3. Foury, G. K., Raquez, J. M., Hassouna, F., Odent, J., Toniazzo, V., Ruch, D., & Dubois, P., (2013). Recent advances in high performance poly(lactide): From "green" plasticization to super-tough materials via (reactive) compounding. *Front. Chem., 1*, doi: 10.3389/fchem.2013.00032.
4. Oishi, A., Zhang, M., Nakayama, K., Masuda, T., & Taguchi, Y., (2006). Synthesis of poly(butylene succinate) and poly(ethylene succinate) including diglycollate moiety. *Polym. J., 38*, 710–715.

5. Seretoudi, G., Bikiaris, D., & Panayiotou, C., (2002). Synthesis, characterization and biodegradability of poly(ethylene succinate)/poly(ε-caprolactone) block copolymers. *Polym.*, *43*(20), 5405–5415.
6. Kint, D. P. R., Alla, A., Deloret, E., Campos, J. L., & Guerra, S. M., (2003). Synthesis, characterization, and properties of poly(ethylene terephthalate)/poly(1,4-butylene succinate) block copolymers. *Polym.*, *44*(5), 1321–1330.
7. Bogdanov, A. A., Martin, C., Bogdanov, A. V., Brady, T. J., & Weissleder, R., (1996). An Adduct of cis-diamminedichloroplatinum(II) and poly(ethylene glycol)poly(L-lysine)-succinate: Synthesis and cytotoxic properties. *Bioconjugate Chem.*, *7*(1), 144–149.
8. Zhu, Q. Y., He, Y. S., Zeng, J. B., Huang, Q., & Wang, Y. Z., (2011). Synthesis and characterization of a novel multiblock copolyester containing poly(ethylene succinate) and poly(butylene succinate). *Mater. Chem. Phys.*, *130*(3), 943–949.
9. Lee, S. I., Yu, S. C., & Lee, Y. S., (2001). Degradable polyurethanes containing poly(butylene succinate) and poly(ethylene glycol). *Poly. Degrad Stab.*, *72*(1), 81–87.
10. Kondratowicz, F. L., & Ukielski, R., (2009). Synthesis and hydrolytic degradation of poly(ethylene succinate) and poly(ethylene terephthalate) copolymers. *Polym. Degrad. Stab.*, *94*(3), 375–382.
11. Chrissafis, K., Paraskevopoulos, K. M., & Bikiaris, D. N., (2005). Thermal degradation mechanism of poly(ethylene succinate) and poly(butylene succinate): Comparative study. *Thermochim. Acta.*, *435*(2), 142–150.
12. Wu, H., & Qiu, Z., (2012). Synthesis, crystallization kinetics and morphology of novelpoly(ethylene succinate-co-ethylene adipate)copolymers. *Cryst. Eng. Comm.*, *14*(10), 3586–3595.
13. Chrissafis, K., Paraskevopoulos, K. M., & Bikiaris, D. N., (2006). Effect of molecular weight on thermal degradation mechanism of the biodegradable polyester poly(ethylene succinate). *Thermochimi. Acta.*, *440*(2), 166–175.
14. Mochizuki, M., Mukai, K., Yamada, K., Ichise, N., Murase, S., & Iwaya, Y., (1997). Structural effects upon enzymatic hydrolysis of poly(butylene succinate-co-ethylene succinate)s. *Macromolecules*, *30*(24), 7403–7407.
15. Chen, C. H., Lu, H. Y., Chen, M., Peng, J. S., Tsai, C. J., & Yang, C. S., (2009). Synthesis and characterization of poly(ethylene succinate) and its copolyesters containing minor amounts of butylene succinate. *Appl. Polym.*, *111*(3), 1433–1439.
16. Salhi, S., Tessier, M., Blais, J. C., Gharbi, R. E., & Fradet, A., (2004). Synthesis of aliphatic-aromatic copolyesters by a high temperature bulk reaction between poly (ethylene terephthalate) and cyclodi(ethylene succinate). *Macromol. Chem. Phys.*, *205* (18), 2391–2397.
17. Huang, C. L., Jiao, L., Zhang, J. J., Zeng, J. B., Yang, K. K., & Wang, Y. Z., (2012). Poly (butylene succinate)-poly(ethylene glycol) multiblock copolymer: Synthesis, structure, properties and shape memory performance. *Polym. Chem.*, *3*(3), 800–808.
18. Papageorgiou, G. Z., & Bikiaris, D. N., (2007). Synthesis, cocrystallization, and enzymatic degradation of novel poly(butylene-co-propylene succinate) copolymers. *Biomacromolecules*, *8*(8), 2437–2449.
19. Chrissafis, K., Kevopoulos, K. M. P., & Bikiaris, D. N., (2006). Thermal degradation kinetics of the biodegradable aliphatic polyester, poly(propylene succinate). *Polym. Degrad. Stab.*, *91*(1), 60–68.

20. Xu, Y., Xu, J., Liu, D., Guo, B., & Xie, X., (2008). Synthesis and characterization of biodegradable poly(butylene succinate-co-propylene succinate)s. *Appl. Polym. Sci.*, *109*(3), 1881–1889.
21. Huang, X., Li, C., Zheng, L., Zhang, D., Guan, G., & Xiao, Y., (2009). Synthesis, characterization and properties of biodegradable poly(butylene succinate)-block-poly(propylene glycol) segmented copolyesters. *Polym. Int.*, *58*(8), 893–899.
22. Bikiaris, D. N., Papageorgiou, G. Z., Achilias, D. S., Pavlidou, E., & Stergiou, A., (2007). Miscibility and enzymatic degradation studies of poly(ε-caprolactone)/poly(propylene succinate) blends. *Europ. Polym. J.*, *43*(6), 2491–2503.
23. Zheng, L., Li, C., Huang, W., Huang, X., Zhang, D., Guan, G., Xiao, Y., & Wang, D., (2011). Synthesis of high-impact biodegradable multiblock copolymers comprising of poly(butylene succinate) and poly(1,2-propylene succinate) with hexamethylene diisocyanate as chain extender. *Polym. Adv. Technol.*, *22*(2), 279–285.
24. Pang, M. Z., Qiao, J. J., Jiao, J., Wang, S. J., Xiao, M., & Meng, Y. Z., (2008). Miscibility and properties of completely biodegradable blends of poly(propylene carbonate) and poly(butylene succinate). *Adv. Polym. Sci.*, *107*(5), 2854–2860.
25. Soccio, M., Finelli, L., Lotti, N., Gazzano, M., & Munari, A., (2007). Poly(propylene isophthalate), poly(propylene succinate), and their random copolymers: Synthesis and thermal properties. *Polym. Phys.*, *45*(3), 310–321.
26. Velmathi, S., Nagahata, R., Sugiyama, J., & Takeuchi, K., (2005). A rapid eco-friendly synthesis of poly(butylene succinate) by a direct polyesterification under microwave irradiation. *Macromol. Rapid Commun.*, *26*(14), 1163–1167.
27. Zhang, Z., & Feng, S. S., (2006). The drug encapsulation efficiency, *in vitro* drug release, cellular uptake and cytotoxicity of paclitaxel-loaded poly(lactide)-tocopheryl polyethylene glycol succinate nanoparticles, *Biomater.*, *27*(21), 4025–4033.
28. Zeng, J. B., Li, Y. D., Li, W. D., Yang, K. K., Wang, X. L., & Wang, Y. Z., (2009). Synthesis and properties of poly(ester urethane)s consisting of poly(L-lactic acid) and poly(ethylene succinate) segments. *Ind. Eng. Chem. Res.*, *48*(4), 1706–1711.
29. Dubey, S. P., Thakur, V. K., Krishnaswamy, S., Abhyankar, H. A., Marchante, V., & Brighton, J. L., (2017). Progress in environmental-friendly polymer nanocomposite material from PLA: Synthesis, processing and applications. *Vacuum, 146*, 655–663.
30. Khani, A. A., Hosseinzadeh, J., Ashori, A., Dadashi, S., & Takzare, Z., (2014). Preparation and characterization of modified cellulose nanofibers reinforced polylactic acid nanocomposite. *Polym. Testing*, *35*, 73–79.
31. Al-Mulla, E. A. J., Yunus, W. Md. Z. W., Ibrahim, N. A. B., & Rahman, M. Z. A., (2010). Epoxidized palm oil plasticized polylactic acid/fatty nitrogen compound modified clay nanocomposites: Preparation and characterization. *Polym. Polym. Compos.*, *18*(8), 451–460.
32. Quitadamo, A., Massardier, V., & Valente, M., (2019). Eco-friendly approach and potential biodegradable polymer matrix for WPC composite materials in outdoor application. *Int. J. Polym. Sci.* https://doi.org/10.1155/2019/3894370 (accessed on 4 March 2020).
33. Rashkov, I., Manolova, N., Li, S. M., Espartero, J. L., & Vert, M., (1996). Synthesis, characterization, and hydrolytic degradation of PLA/PEO/PLA triblock copolymers with short poly(L-lactic acid) chains. *Macromolecules*, *29*(1), 50–56.
34. Tingaut, P., Zimmermann, T., & Suevos, F. L., (2010). Synthesis and characterization of bionanocomposites with tunable properties from poly(lactic acid) and acetylated micro fibrillated cellulose. *Biomacromolecules*, *11*(2), 454–464.

35. Teramoto, Y., & Nishio, Y., (2003). Cellulose diacetate-graft-poly(lactic acid)s: Synthesis of wide-ranging compositions and their thermal and mechanical properties. *Polymer*, *44*(9), 2701–2709.
36. Ren, J., Fu, H., Ren, T., & Yuan, W., (2009). Preparation, characterization and properties of binary and ternary blends with thermoplastic starch, poly(lactic acid) and poly(butylene adipate-co-terephthalate). *Carbohyd. Polym.*, *77*(3), 576–582.
37. Fukushima, K., Tabuani, D., Abbate, C., Arena, M., & Ferreri, L., (2010). Effect of sepiolite on the biodegradation of poly(lactic acid) and polycaprolactone. *Polym. Degrad. Stab.*, *95*(10), 2049–2056.
38. Chuensangjun, C., Pechyen, C., & Sirisansaneeyakul, S., (2013). Degradation behaviors of different blends of polylactic acid buried in soil. *Energy Procedia.*, *34*, 73–82.
39. Madison, L. L., & Huisman, G. W., (1999). Metabolic engineering of poly(3-hydroxyalkanoates): From DNA to plastic. *Microbiol. Mol. Biol. Rev.*, *63*, 21–53.
40. Witholt, B., & Kessler, B., (1999). Perspectives of medium chain length poly (hydroxy-alkanoates), a versatile set of bacterial bioplastics. *Curr. Opin. Biotechnol.*, *10*, 279–285.
41. Matsumoto, K., Murata, T., Nagao, R., Nomura, C. T., Arai, S., Arai, Y., Takase, K., Nakashita, H., Taguchi, S., & Shimada, H., (2009). Production of short-chain-length/medium-chain-length polyhydroxyalkanoate (PHA) copolymer in the plastid of *Arabidopsis thaliana* using an engineered 3-ketoacyl-acyl carrier protein synthase III. *Biomacromolecules*, *10*, 686–690.
42. George, J., & Sabapathi, S. N., (2015). Water soluble polymer-based nanocomposites containing cellulose nanocrystals/ *Eco-friendly Polym. Nanocomp.*, *75*, 259–293.
43. Hazarika, A., Baishya, P., & Maji, T. K., (2015). Bio-based wood polymer nanocomposites: A sustainable high-performance material for future. *Eco-Friendly Polym. Nanocomp.*, *75*, 233–257.
44. Dobos, A. M., Onofrei, M. D., & Ioan, S., (2015). Cellulose acetate nanocomposites with antimicrobial properties. *Eco-Friendly Polym. Nanocomp.*, *75*, 367–398.
45. Muhamad, I. I., Salehudin, M. H., & Salleh, E., (2015). Cellulose nanofiber for eco-friendly polymer nanocomposites. *Eco-Friendly Polym. Nanocomp.*, *75*, 323–365.
46. Park, H. M., Misra, M., Drzal, L. T., & Mohanty, A. K., (2004). "Green" nanocomposites from cellulose acetate bioplastic and clay: Effect of eco-friendly triethyl citrate plasticizer. *Biomacromolecules*, *5*(6), 2281–2288.
47. Ma, X., Chang, P. R., Yu, J., & Wang, N., (2008). Preparation and properties of biodegradable poly(propylene carbonate)/thermoplastic dried starch composites. *Carbohyd. Polym.*, *71*(2), 229–234.
48. Yadav, M., Rhee, K. Y., Jung, I. H., & Park, S. J., (2013). Eco-friendly synthesis, characterization, and properties of a sodium carboxymethyl cellulose/graphene oxide nanocomposite film. *Cellulose*, *20*(2), 687–698.
49. Murugan, K. D., Radhika, S., Baskaran, I., & Anbarasan, R., (2008). Clay catalyzed synthesis of bio-degradable poly(glycolic acid). *Chinese J. Polym. Sci.*, *26*(04), 393–398.
50. Kiasat, A. R., & Mehrjardi, M. F., (2008). PEG-SO_3H as eco-friendly polymeric catalyst for regioselective ring opening of epoxides using thiocyanate anion in water: An efficient route to synthesis of β-hydroxy thiocyanate. *Catal. Comm.*, *9*(6), 1497–1500.
51. Maleki, A., Movahed, H., & Ravaghi, P., (2017). Magnetic cellulose/Ag as a novel eco-friendly nanobiocomposite to catalyze synthesis of chromene-linked nicotinonitriles. *Carbohydrate Polym.*, *156*, 259–267.

52. Jie, Z., Yan, X., Zhao, L., Worley, S. D., & Liang, J., (2014). Eco-friendly synthesis of regenerable antimicrobial polymeric resin with *N*-halamine and quaternary ammonium salt groups. *RSC Adv.*, *4*(12), 6048–6054.
53. Sabitha, G., Reddy, K. B., Yadav, J. S., Shailaja, D., & Sivudu, K. S., (2005). Ceria/vinylpyridine polymer nanocomposite: an ecofriendly catalyst for the synthesis of 3,4-dihydropyrimidin-2(1*H*)-ones. *Tetrahedron Lett.*, *46*(47), 8221–8224.
54. Azarudeen, R. S., Ahamed, M. A. R., Jeyakumar, D., & Burkanudeen, A. R., (2009). An eco-friendly synthesis of a terpolymer resin: Characterization and chelation ion-exchange property. *Iran. Polym. J.*, *18*(10), 821–832.
55. Julkapli, N. M., Bagheri, S., & Sapuan, S. M., (2015). Multifunctionalized carbon nano-tubes polymer composites: Properties and applications. *Eco-Friendly Polym. Nanocomp.*, *75*, 155–214.
56. Saw, S. K., (2015). Static and dynamic mechanical analysis of coir fiber/montmorillonite nanoclay-filled novolac/epoxy hybrid nanocomposites. *Eco-Friendly Polym. Nanocomp.*, *75*, 137–154.
57. Roopan, S. M., & Madhumitha, G., (2015). Green synthesis of polymer composites/nanocomposites using vegetable oil. *Eco-Friendly Polym. Nanocomp.*, *75*, 495–511.
58. Sherbiny, I. M. E., & Ali, I. H., (2016). Eco-friendly electrospun polymeric nanofibers-based nanocomposites for wound healing and tissue engineering. *Eco-Friendly Polym. Nanocomp.*, *75*, 399–431.
59. Rodriguez-Galan, A., Franco, L., & Puiggali, J., (2011). Degradable poly(ester amide)s for biomedical applications. *Polym.*, *3*, 65–99.

CHAPTER 9

An Environmentally Benign Synthesis of 2,4,5-Triaryl-1*H*-Imidazoles via Multi-Component Reactions and Its Medicinal Importance

NANA V. SHITOLE

Department of Chemistry, Shri Shivaji College, Basmat Road, Parbhani–431401, Maharashtra, India, E-mail: nvshitole@gmail.com

ABSTRACT

Synthesis of 2,4,5-Triaryl-1*H*-Imidazole and their derivatives is a fertile source of biologically important molecules. Compounds containing imidazole moiety have many pharmacological properties and play important roles in biochemical processes. Multi-component reactions (MCRs) have drawn great interest in modern organic synthesis and medicinal chemistry because they are one-pot processes bringing together three or more components and show high atom economy and high selectivity. With increasing environmental concerns, more, and more chemists are devoting in the area of *"Green Chemistry."* Utilization of non-toxic chemicals, environmentally benign solvents, and renewable materials are some of the key issues that merit important consideration in "green" synthetic strategies. Avoiding organic solvents during the reactions in organic synthesis leads to a clean, efficient, and economical technology (Green Chemistry); safety is largely increased, work-up is considerably simplified, cost is reduced, increased amount of reactants can be used in the same equipment and the reactivates, and sometimes selectivities are enhanced without dilution. Due to all these advantages, therc is an increasing interest in the use of environmentally benign reagents and procedures or, in other words, the absence of solvents coupled with the high yields and short reaction times often associated with reactions of this type make these procedures very attractive for synthesis. In the present thesis, we have described new routes for the synthesis of bioactive 2,4,5-triaryl -1*H*-imidazole Heterocycles using greener approaches such as multi-component one-pot synthesis, solid-state reactions coupled with the application of grinding technique, microwave (MW), and ultrasound irradiation. The present chapter expresses various synthetic methods and biological importance 2,4,5-triaryl -1*H*-imidazoles and their derivatives.

9.1 INTRODUCTION

2,4,5-Triaryl-1*H*-imidazole compounds have gained the remarkable importance due to their widespread biological activities and their use in synthetic chemistry. The substituted imidazoles are well known as inhibitors of P38MAP kinase [1] and therapeutic agents [2]. Imidazole chemistry, because of its use in ionic liquids (ILs) [3] and in N-heterocyclic carbenes (NHCs) [4], gave a new dimension in the area of organometallics and "Green Chemistry." Moreover, the imidazole ring system is one of the most important substructures found in a large number of natural products and pharmacologically active compounds such as the hypnotic agent etomidate [5] and the proton pump inhibitor omeprazole A [6], trifenagrel [7] B is a 2,4,5-triaryl-1*H*-imidazole that reduces platelet aggregation in several animal species and humans (Scheme 9.1).

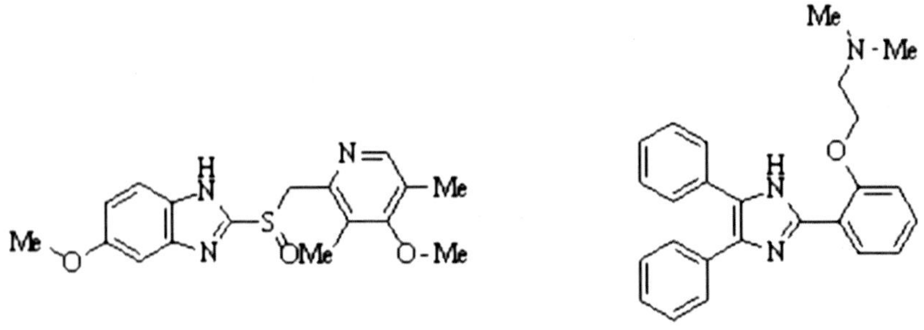

SCHEME 9.1 (A) Omeprazole [6], (B) trifenagrel [7].

Due to their great importance, many synthetic strategies have been developed. In 1882, Radziszewski, and Japp reported the first synthesis of the imidazole from 1, 2-dicarbonyl compound, various aldehydes, and ammonia, to obtain the 2,4,5-triphenyl imidazoles. Also, Grimmett et al.

[8] was proposed the synthesis of the imidazole using nitriles and esters. Another method is the four-component one-pot condensation of a glyoxals, aldehydes, amines, and ammonium acetate in refluxing acetic acid is the most desirable convenient method [9].

The art of performing efficient chemical transformations coupling three or more components in a single operation by a catalytic process avoiding stoichiometric toxic reagents large amounts of solvents and expensive purification techniques represent a fundamental target of the modern organic synthesis [10].

9.2 ORGANIC REACTIONS IN GREEN SOLVENT

Over the past decade, there has been an increased emphasis on the topic of "Green Chemistry and Green Chemical Processes" [11–13]. The growing awareness of the pressing need for greener, more sustainable technologies has focused attention on the use of atom efficient catalytic methodologies for the manufacture of fine chemicals and pharmaceuticals. These efforts aim at the total elimination or at least the minimization of waste generation and the implementation of sustainable processes as discussed by Anastas and Warner [14]. Any attempt at meeting these goals must comprehensively address some of these principles in the design of a synthetic route, chemical analysis, or chemical process [15]. Utilization of non-toxic chemicals, environmentally benign solvents, and renewable materials are some of the key issues that merit important consideration in "green" synthetic strategies.

The toxic and volatile natures of many organic solvents, particularly chlorinated hydrocarbons (HCs) that are widely used in organic synthesis have posed a serious threat to the environment. To address some of these issues, attempts have been made to develop solvent-free processes, which to some extent have succeeded in achieving some chemical transformations [16]. Despite such advances, in performing the majority of organic reactions, solvents play a critical role in making the process "liquids" and allowing molecular interactions to be more efficient. In such situations, the search for alternative reaction media to replace these organic solvents has attracted much attention from academia and industry [17].

One of the recently developed methods is to use water as solvent for organic reactions [18] but its use is limited due to the hydrophobic nature of organic compounds and the sensitivity of many catalysts to aqueous conditions. Fluorous phases [19] have achieved adaptation and enjoy increased utility due to the advantage of being highly hydrophobic. Supercritical fluid media [20] are also attractive solvent alternatives for a variety of chemical and industrial processes but, at the same time, they need high pressure. ILs [21] have attracted interest due to their advantageous properties such as tuneable polarity, high thermal stability, negligible vapor pressure, and recyclability. However, ILs require tedious preparation and their environmental safety is still debatable.

Recently, liquid polymers or low melting polymers have emerged as alternative green reaction media with unique properties such as thermal stability, commercial availability, nonvolatility, immiscibility with a number of organic solvents and recyclability. PEGs are preferred over other polymers because they are inexpensive, completely nonhalogenated, easily degradable and of low toxicity. Many organic reactions have been carried out using PEGs as solvent or co-solvent [22] the use of PEG as a recyclable solvent system for the metal-mediated radical polymerization of methyl methacrylate and styrene has also been reported [23].

9.3 MULTI-COMPONENT REACTIONS (MCRS) IN ORGANIC SYNTHESIS

MCRs have become important tools in modern preparative synthetic chemistry because these reactions increase the efficiency by combining several operational steps without isolation of intermediates or change of the conditions [24–29]. MCRs processes, in which three or more reactants are combined in a single chemical step to produce products that incorporate substantial portions of all the components, naturally comply with many of these stringent requirements for ideal organic syntheses.

MCRs, though fashionable these days, have in fact a long history. Indeed, many important reactions such as the Biginelli dihydropyrimidine synthesis [30], 2,4,5-triarylimidazole synthesis [31], Hantzsch dihydropyridine synthesis [32], Mannich reaction [33] and Ugi four-component reactions (Ugi-4CRs), [34] among others, are all multi-component in nature. In spite of the significant contribution of MCRs to the state of the art of modern organic chemistry and their potential use in complex organic syntheses, little attention was paid to the development of novel MCRs in the second half of the twentieth century. However, with the introduction of molecular biology and high-throughput biological screening, the demand on the number and the quality of compounds for drug discovery has increased enormously. By virtue of their inherent convergence and high productivity, together with their exploratory and complexity-generating power, MCRs have naturally become a rapidly evolving field of research and have attracted the attention of both academic and industrial scientists.

9.4 METHODS FOR THE SYNTHESIS OF 2,4,5-TRIARYLIMIDAZOLE DERIVATIVES

Recently, the literature survey reveals several methods for synthesis of 2,4,5-triaryl imidazoles using reaction condensation like catalyst, solvents, temperature, and ILs such as following:

Balalaie and co-workers [35] has described the microwave (MW)-assisted synthesis of 2,4,5-triarylimidazole from benzil, aldehydes, and ammonium acetate in the presence of Zeolite HY give 54–94% yields, which achieved reductions in times and cleaner reactions (Scheme 9.2).

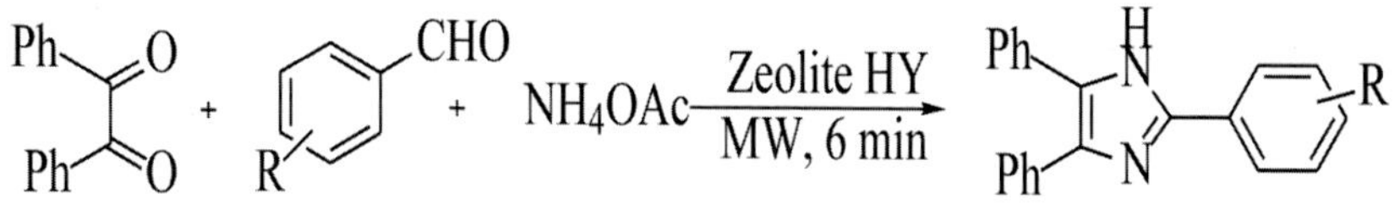

Zeolite HY
MW, 6 min

SCHEME 9.2 The microwave-assisted synthesis of 2,4,5-triarylimidazoles in the presence of zeolite HY.

Srinivasan et al. [36] were reported the synthesis of 2,4,5-triarylimidazole in 85–98% yields by using 1-hexyl-3-methylimidazolium tetrafluroborate [(Hmim)BF_4] IL at 100°C (Scheme 9.3).

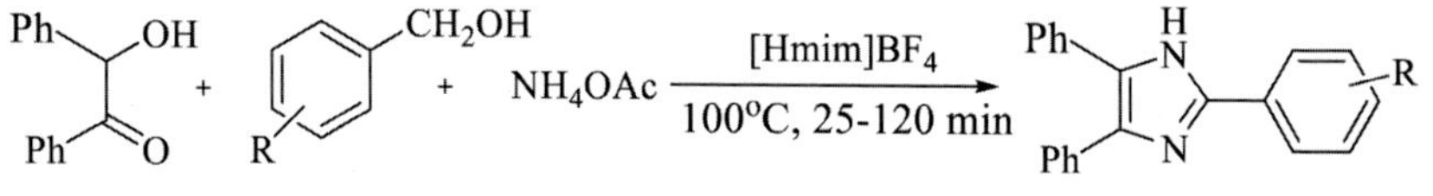

[Hmim]BF_4
100°C, 25-120 min

SCHEME 9.3 The synthesis of 2,4,5-triarylimidazoles by using [(Hmim)BF_4] ionic liquid.

Sharma et al. [37] proposed the synthesis of 2,4,5-triarylimidazole from benzil, aldehydes, and ammonium acetate in the presence of zirconium tetrachloride ($ZrCl_4$) at room temp. give 84–96% yields (Scheme 9.4).

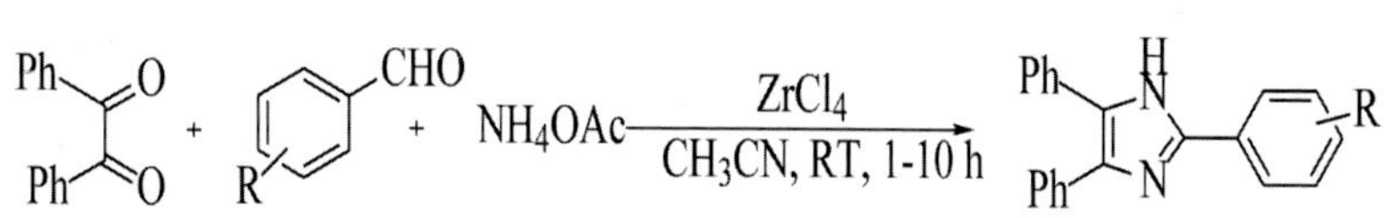

$ZrCl_4$
CH_3CN, RT, 1-10 h

SCHEME 9.4 The synthesis of 2,4,5-triarylimidazoles by using zirconium tetrachloride ($ZrCl_4$).

Khodaei and co-workers [38] have synthesized 2,4,5-triarylimidazole from benzil/benzoin, aldehydes, and ammonium heptamolybdate tetrahydrate in the presence of p-toluenesulfonic acid (p-TSA) heated at 140°C gives 75–95% yields (Scheme 9.5).

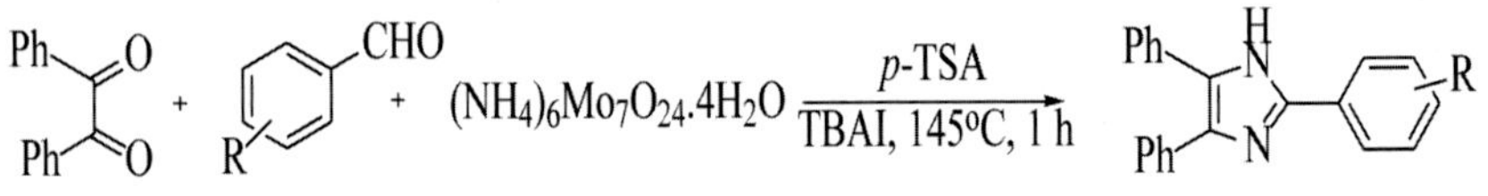

SCHEME 9.5 The synthesis of 2,4,5-triarylimidazoles by using p-toluenesulfonic acid (p-TSA).

Su et al. [39] described for the synthesis of 2,4,5-triarylimidazole using europium triflate [$Eu(OTf)_3$] in ethanol 60°C give 64–93% yield (Scheme 9.6).

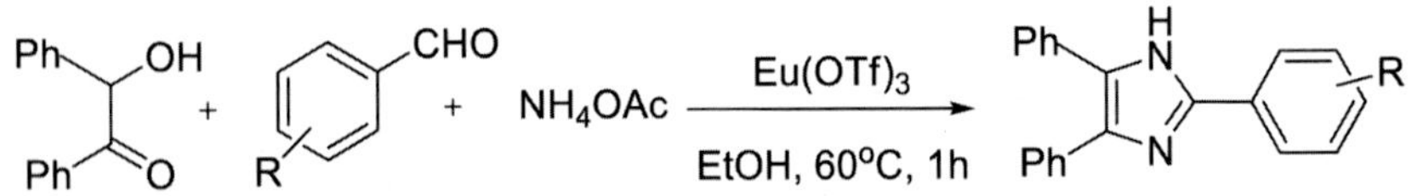

SCHEME 9.6 Synthesis of 2,4,5-triarylimidazole using europium triflate [Eu(OTf)3] as catalyst.

Shinde et al. [40] explored sulphanilic acid ($C_6H_4SO_3HNH_2$) as catalyst in EtOH:H_2O was heated at 80°C for the synthesis of 2,4,5-triarylimidazole in 77–97% yields (Scheme 9.7a).

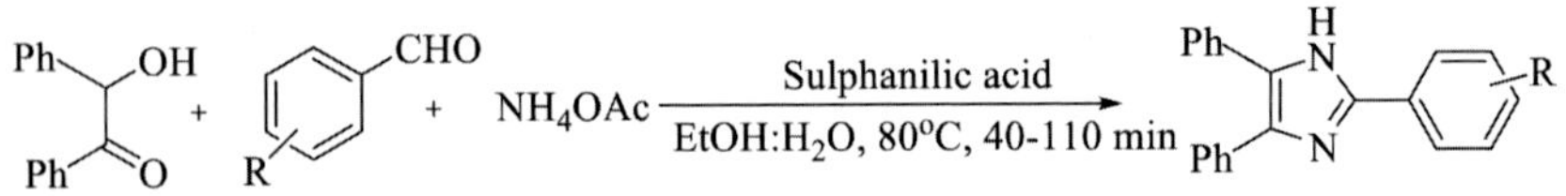

SCHEME 9.7a Synthesis of 2,4,5-triarylimidazoles by using sulphanilic acid catalyst.

Shinde et al. [41] have synthesized 2,4,5-triarylimidazole from benzil/benzoin, aldehydes, and ammonium acetate in the presence of sodium bisulfite ($NaHSO_3$) in EtOH:H_2O medium give 85–98% yields (Scheme 9.7b).

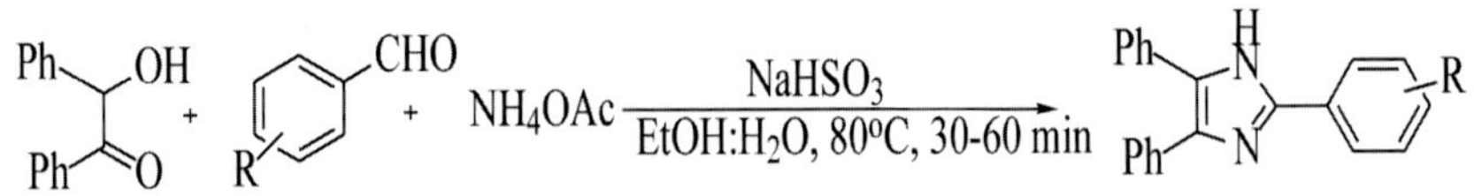

SCHEME 9.7b Synthesis of 2,4,5-triarylimidazoles by using catalyst sodium bisulfite ($NaHSO_3$).

Khosropour [42] have expanded the synthetic scope of the MCR treating benzil, aldehydes, and ammonium acetate producing 2,4,5-triarylimidazole in 1-methylimidazolium hydrogen sulfate chloride [(Hmim)HSO_4] gives 80–97% yields (Scheme 9.8).

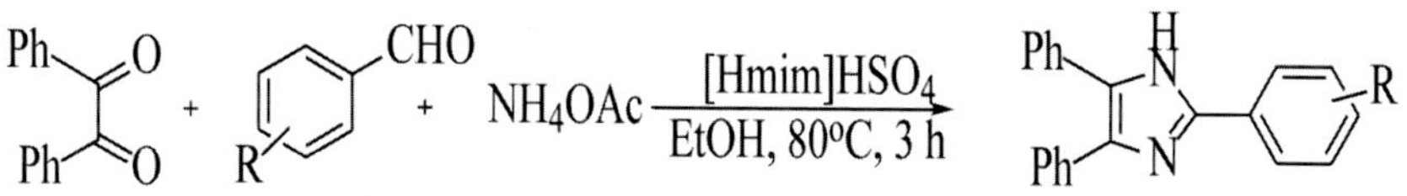

SCHEME 9.8 Synthesis of 2,4,5-triarylimidazoles in [(Hmim)HSO_4] ionic liquid.

Mohammadi and co-workers [43] have introduced a procedure for 2,4,5-triarylimidazole by use of potassium aluminum sulfate (Alum) as a catalyst in ethanol gives 93% yields (Scheme 9.9).

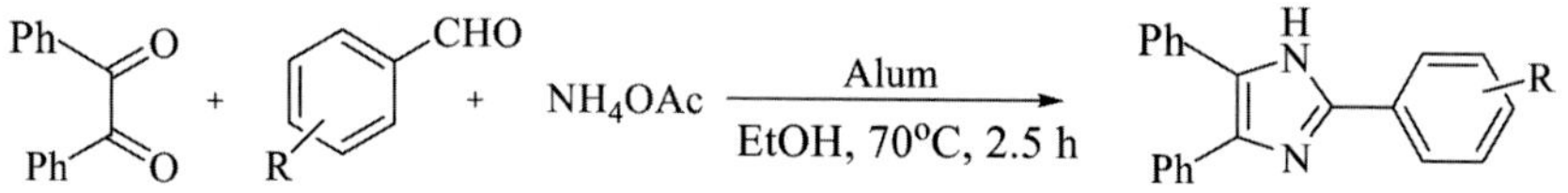

SCHEME 9.9 Synthesis of 2,4,5-triarylimidazole by use of potassium aluminum sulfate (alum) as a catalyst.

Shelke et al. [44] have synthesized 2,4,5-triarylimidazole from benzil/benzoin, aldehydes, and ammonium acetate in the presence of glyoxylic acid under solvent-free condition in MW give 90–98% yields (Scheme 9.10).

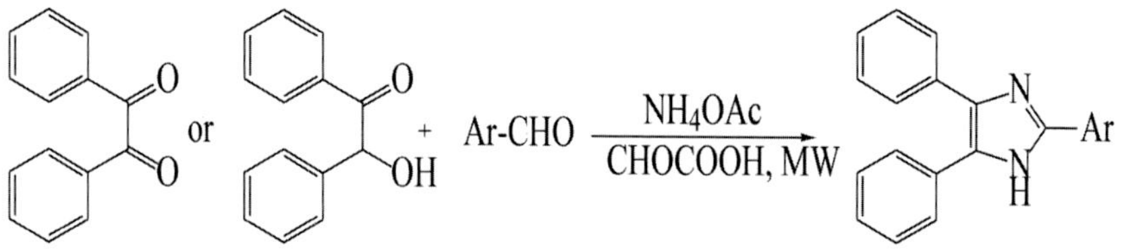

SCHEME 9.10 Synthesis of 2,4,5-triarylimidazole by use of glyoxylic acid as a catalyst.

Shitole and co-worker [45] has been found to be L-Proline an efficient organocatalyst for one-pot synthesis of 2,4,5-triaryl substituted imidazole by the reaction of an aldehyde, a benzil and an ammonium acetate in ethanol gives 75–94% yields (Scheme 9.11).

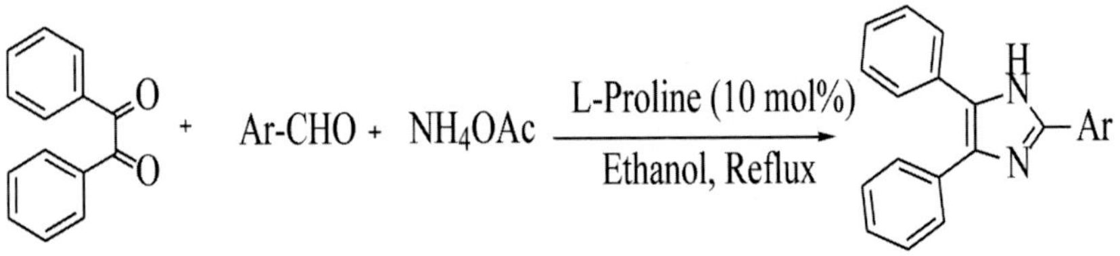

SCHEME 9.11 Synthesis of 2,4,5-triaryl-1*H*-imidazoles catalyzed L-proline.

Shelke et al. [46] have synthesized 2,4,5-triarylimidazole from benzil/benzoin, aldehydes, and ammonium acetate in the presence of Boric acid (BO_3H_3) in EtOH:H_2O(1:1) under ultrasound at room temperature (RT) give 85–94% yields (Scheme 9.12).

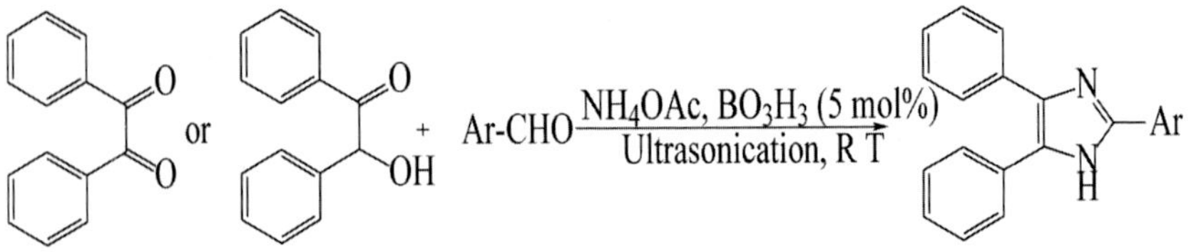

SCHEME 9.12 Synthesis of 2,4,5-triaryl-1*H*-imidazoles by using boric acid (BO_3H_3) as catalyst.

Shelke and co-workers [47] have synthesized 2,4,5-triarylimidazole from benzil/benzoin, aldehydes, and ammonium heptamolybdate tetrahydrate in the presence of ceric (IV) ammonium nitrate (CAN) under ultrasound at RT give 85–94% yields (Scheme 9.13).

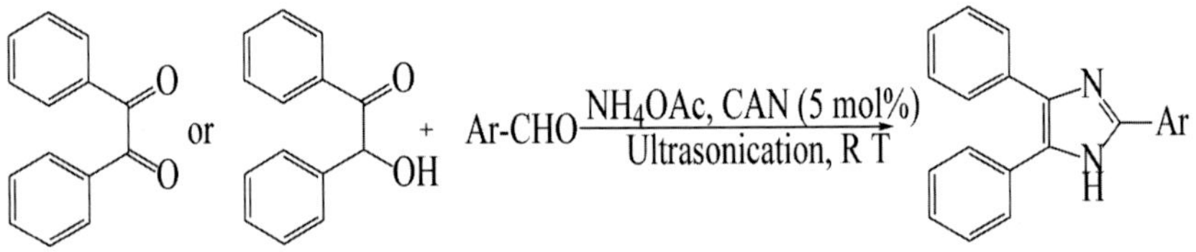

SCHEME 9.13 Synthesis of 2,4,5-triaryl-1*H*-imidazoles by using ceric (IV) ammonium nitrate (CAN) as catalyst.

Shelke et al. [48] proposed the synthesis of 2,4,5-triarylimidazole from benzil, aldehydes, and ammonium acetate in the presence of Cellulose sulfuric acid as a bio-supported and recyclable solid acid catalyst in a MW oven at the power of 180 W at give 91–95% yields (Scheme 9.14).

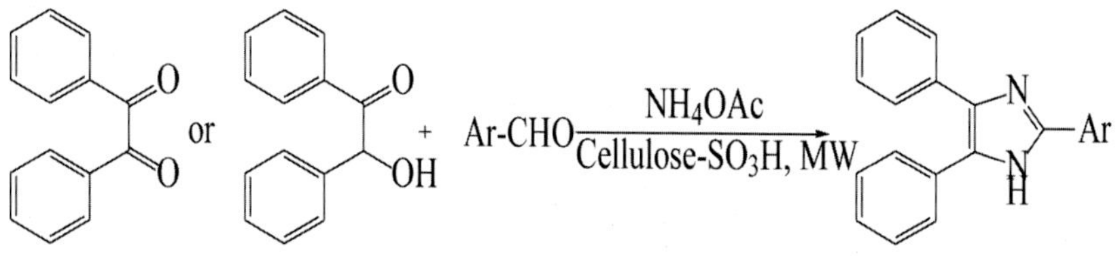

SCHEME 9.14 Synthesis of 2,4,5-triaryl-1*H*-imidazoles by using cellulose sulfuric acid as catalyst.

Nalage and co-workers [49] have been developed an efficient and green procedure for the synthesis of 2, 4, 6-triaryl-1*H*-imidazole in polyethylene glycol under MW irradiation in excellent yield (Scheme 9.15).

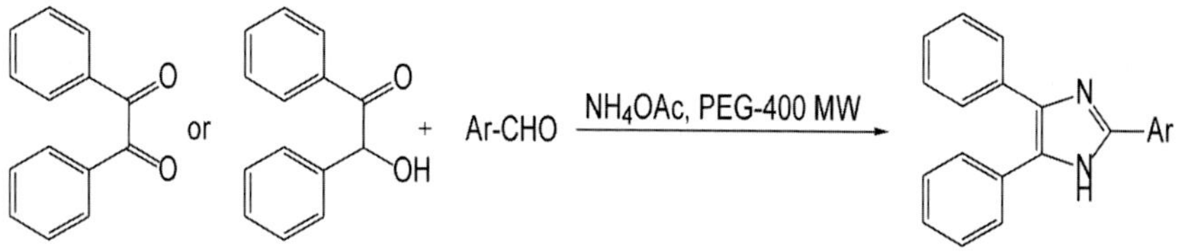

SCHEME 9.15 Synthesis of 2,4,5-triaryl-1*H*-imidazoles by using as polyethylene glycol catalyst.

Bahador et al. [50] proposed the synthesis of 2,4,5-triarylimidazole from aldehydes, benzil, and ammonium acetate in the presence of Fe_3O_4 NPs as catalyst under solvent-free condition gives 70–91% yield (Scheme 9.16).

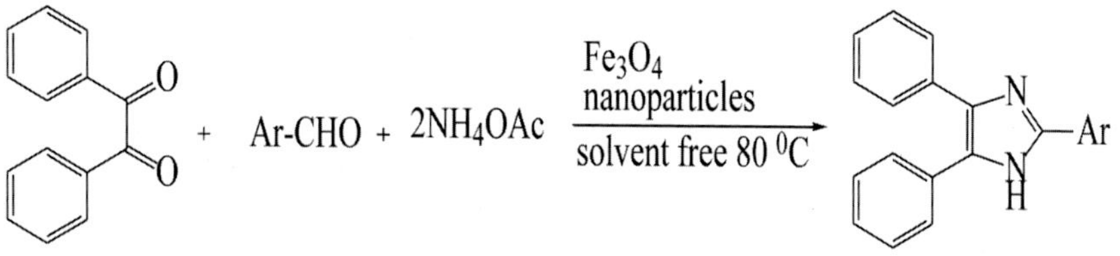

SCHEME 9.16 Synthesis of 2,4,5-triaryl-1*H*-imidazoles by using Fe_3O_4 nanoparticles as catalyst.

Mohammadi and co-workers [51] have introduced a procedure for 2,4,5-triarylimidazole by use of [poly(AMPS-co-AA)] as catalyst under solvent-free conditions (Scheme 9.17).

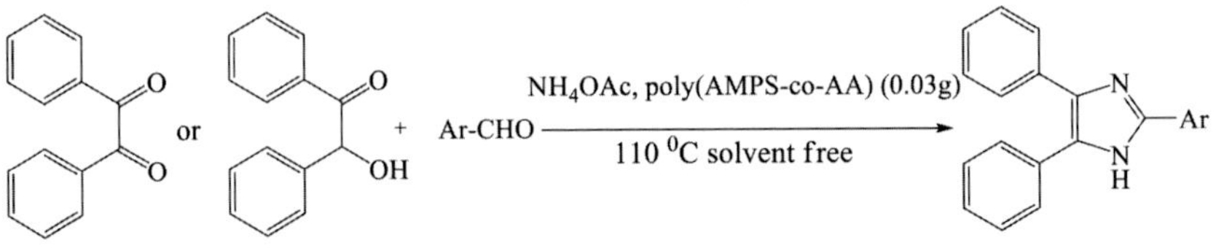

SCHEME 9.17 Synthesis of 2,4,5-triaryl-1*H*-imidazoles by using [poly(AMPS-co-AA)] as catalyst.

Behrooz et al. [52] proposed 2,4,5-Triaryl-1H-Imidazoles by the coupling of Benzil/Benzoin, aldehyde, and ammonium acetate under solvent free condition using the $H_2SO_4 \cdot SiO_2$ as a catalyst gives 85–92% of yields (Scheme 9.18).

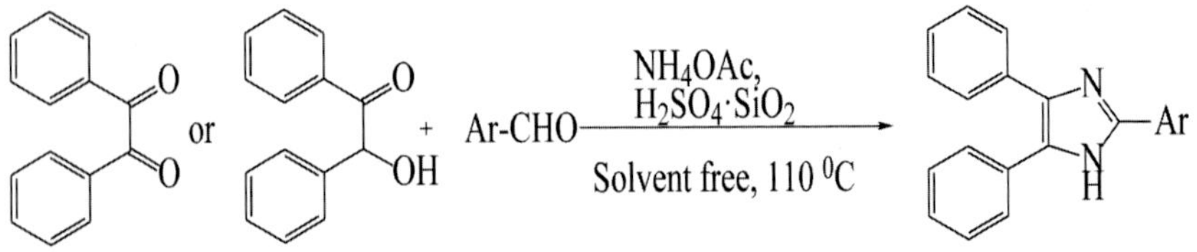

SCHEME 9.18 Synthesis of 2,4,5-triaryl-1*H*-imidazoles by using $H_2SO_4 \cdot SiO_2$ as catalyst.

Dake and co-worker [53] have introduced a procedure for 2,4,5-triarylimidazole by using benzil/benzoin, substituted aromatic aldehydes, and ammonium acetate in an ethanol-water (1:1, v/v) solvent system using sulfated tin oxide catalyst under reflux condition give good yield (Scheme 9.19).

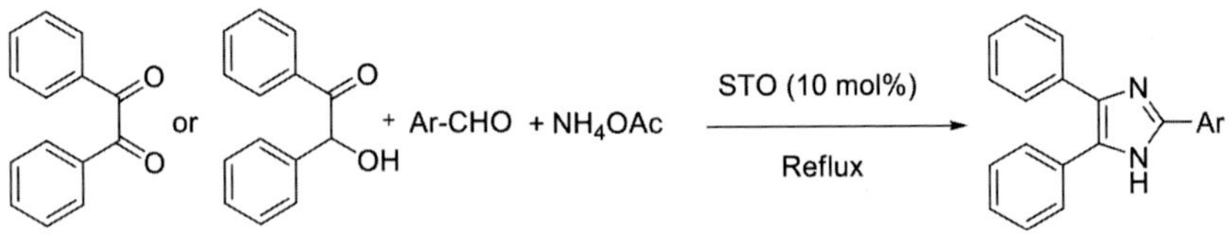

SCHEME 9.19 Synthesis of 2,4,5-triaryl-1*H*-imidazoles by using sulfated tin oxide as catalyst.

Shitole and co-workers [54] have introduced a procedure for 2,4,5-triarylimidazole by use of Tannic acid as a catalyst in ethanol gives 85–93% yields (Scheme 9.20).

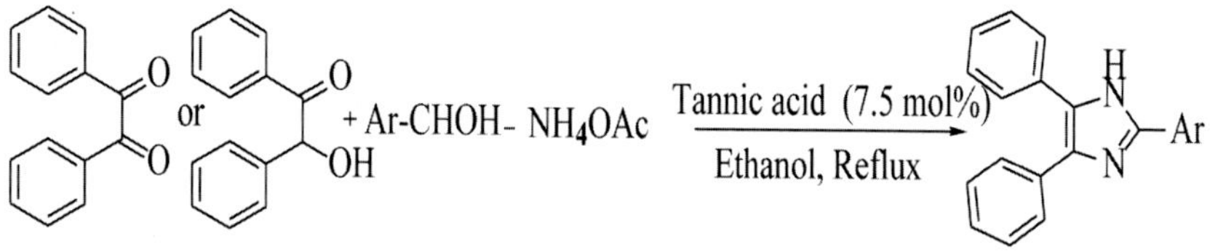

SCHEME 9.20 Synthesis of 2,4,5-triaryl-1*H*-imidazoles by using tannic acid as catalyst.

Adel, and co-workers [55] have synthesized 2,4,5-triarylimidazole from benzil, aldehydes, and ammonium acetate in the presence of diethyl ammonium hydrogen phosphate as catalyst under ultrasound at RT give 85–94% yields (Scheme 9.21).

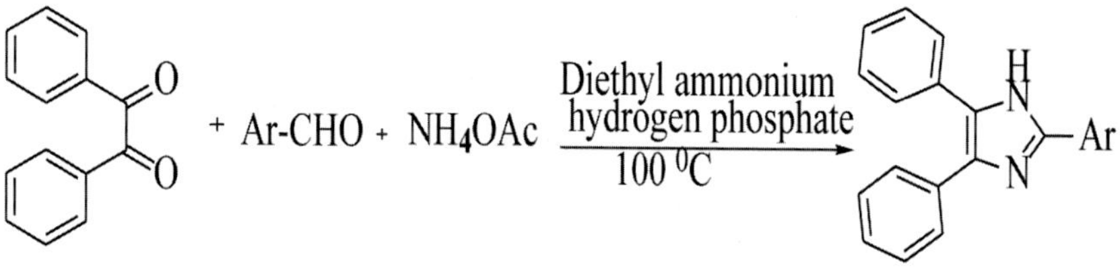

SCHEME 9.21 Synthesis of imidazole derivatives by using diethyl ammonium hydrogen phosphate as catalyst.

Farhad et al. [56] were explored SiO_2-$NaHSO_4$ as catalyst under solvent-free condition was heated at 120°C for the synthesis of 2,4,5-triarylimidazole in 85–89% yields (Scheme 9.22).

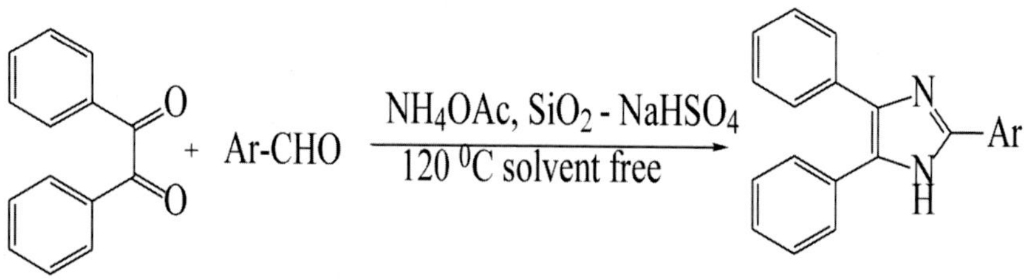

SCHEME 9.22 Synthesis of imidazole derivatives by using SiO_2-$NaHSO_4$ as catalyst.

Hangirgekar et al. [57] proposed the synthesis of 2,4,5-triarylimidazole from benzil, aldehydes, and ammonium acetate in the presence of cupric chloride as a catalyst in a MW oven to give 85–91% yields (Scheme 9.23).

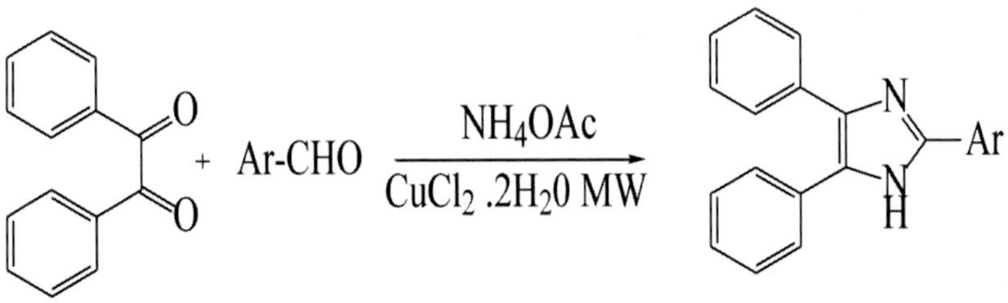

SCHEME 9.23 Synthesis of imidazole derivatives by using cupric chloride as catalyst.

Safari et al. [58] explored cyclo-condensation of benzil, aldehydes, and ammonium were used to synthesize 2,4,5-trisubstituted-1H-imidazole derivatives under MW irradiation with silica-supported $SbCl_3$ ($SbCl_3/SiO_2$) as a heterogeneous catalyst to gives 77–97% of yields (Scheme 9.24).

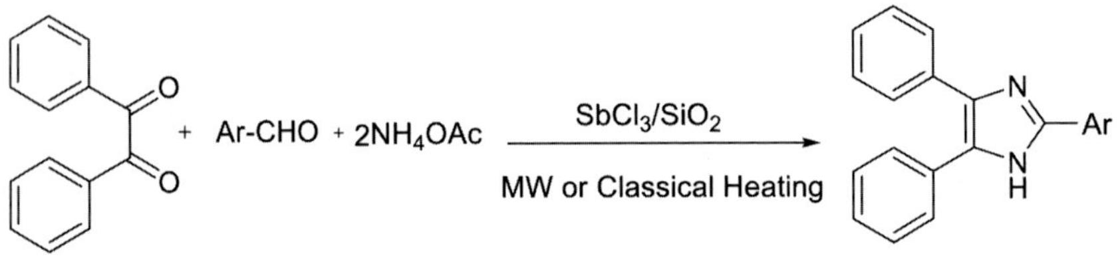

SCHEME 9.24 Synthesis of imidazole derivatives by using silica-supported $SbCl_3$ as catalyst.

Shaikh et al. [59] proposed the synthesis of 2,4,5-triarylimidazole from benzoin, aldehydes, and ammonium acetate in the presence of SO_4^{2-}/CeO_2–ZrO_2 as a catalyst in ethnoal to give 70–85% yields (Scheme 9.25).

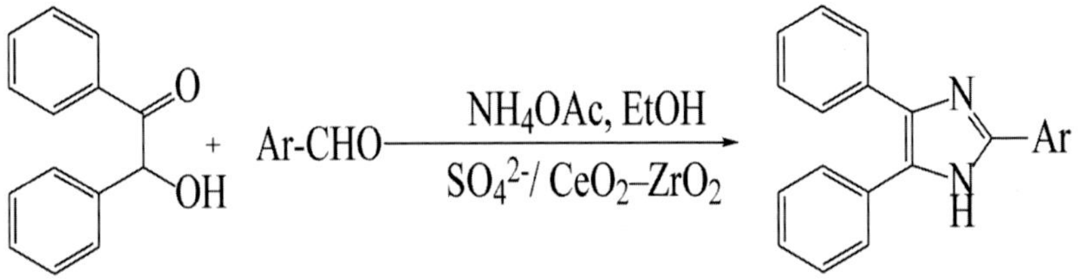

SCHEME 9.25 Synthesis of imidazole derivatives by using SO_4^{2-}/CeO_2–ZrO_2 as catalyst.

Niralwad et al. [60] proposed the Silica sulfuric acid was found to be an efficient catalyst for the green synthesis of 2,4,5-Triaryl-1H-Imidazoles by the coupling of Benzil/Benzoin, aldehyde, and ammonium acetate under

MW-irradiation at ambient temperature for appropriate time to furnish to excellent yield (Scheme 9.26).

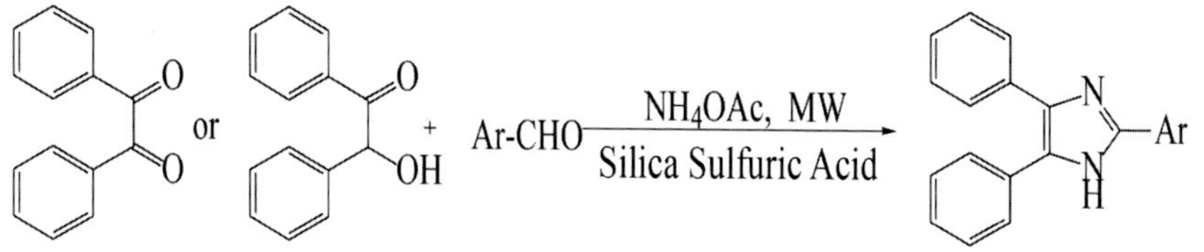

SCHEME 9.26 Synthesis of 2,4,5-triaryl-1*H*-imidazoles by using as catalyst.

Shitole et al. [61] have expanded the synthetic scope of the MCR treating benzil, aldehydes, ammonium acetate and Rochelle salt (10 mol%) as catalyst producing 2,4,5-triarylimidazole in MW oven at the power of 450W to gives 90–97% yields (Scheme 9.27).

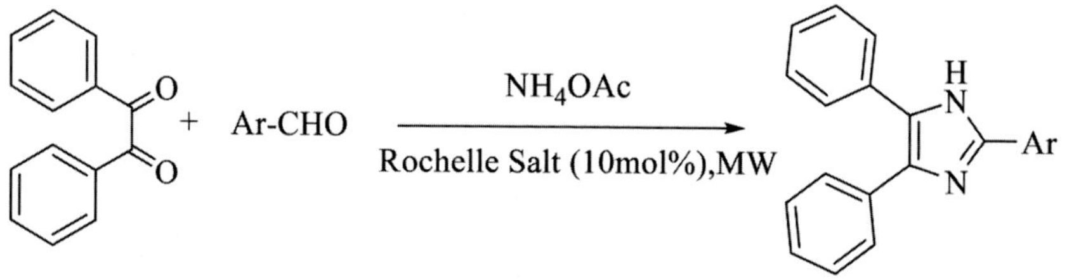

SCHEME 9.27 Synthesis of imidazole derivatives by using Rochelle salt as catalyst.

Chen et al. [62] have expanded the synthetic scope of the MCR of 2,2-dibromo-1,2-diarylethanone, ammonium, aryl aldehyde, and Pri_2NEt in EtOH was carried out at refluxing temperature under argon atmosphere to give good yield (Scheme 9.28).

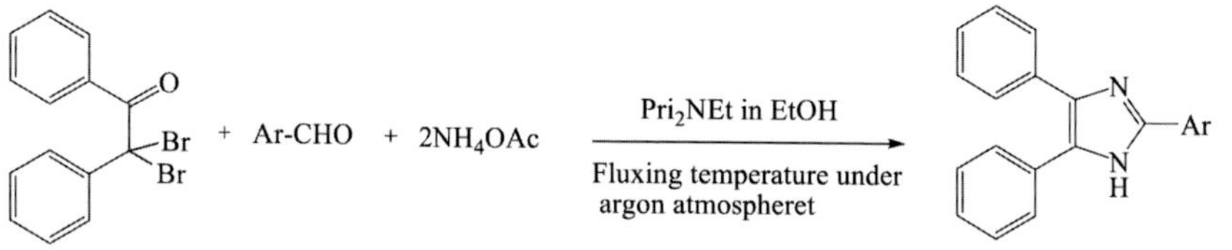

SCHEME 9.28 Synthesis of imidazole derivatives by using as catalyst Pri_2NEt in EtOH.

Ghodsi et al. [63] was proposed the Sulfonic acid functionalized SBA-15 nanoporous material (SBA-Pr-SO_3H) with a pore size of 6 nm was found to be a green and effective solid acid catalyst in the one-pot synthesis of 2,4,5-trisubstituted at 140°C to give better yield (Scheme 9.29).

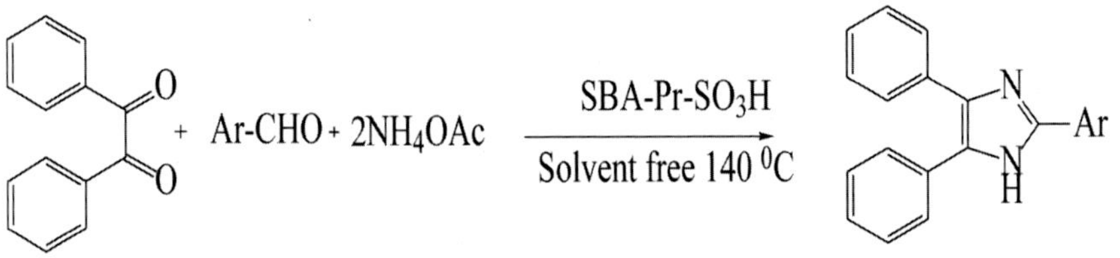

SBA-Pr-SO_3H
+ Ar-CHO + $2NH_4OAc$
Solvent free 140 °C

SCHEME 9.29 Synthesis of 2,4,5-triaryl-1*H*-imidazoles by using SBA-Pr-SO_3H as catalyst.

Narkhede [64] has expanded 2, 4, 5-triaryl-1H-imidazoles was synthesized by multi-component one-pot method from benzoin, ammonium acetate, and aromatic aldehyde by using sulphamic acid as the catalyst to give 92–98% yield (Scheme 9.30).

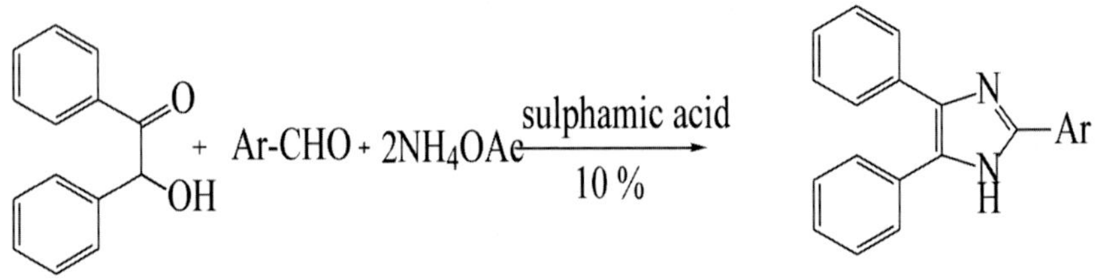

+ Ar-CHO + $2NH_4OAc$
sulphamic acid
10 %

SCHEME 9.30 Synthesis of 2,4,5-triaryl-1*H*-imidazoles by using sulphamic acid catalyst.

Nora et al. [65] was proposed 2,4,5-Triaryl-1*H*-Imidazoles by the coupling of Benzil, aldehyde, and ammonium acetate in ethanol as solvent using the citric acid as a catalyst gives 65–92% of yields (Scheme 9.31).

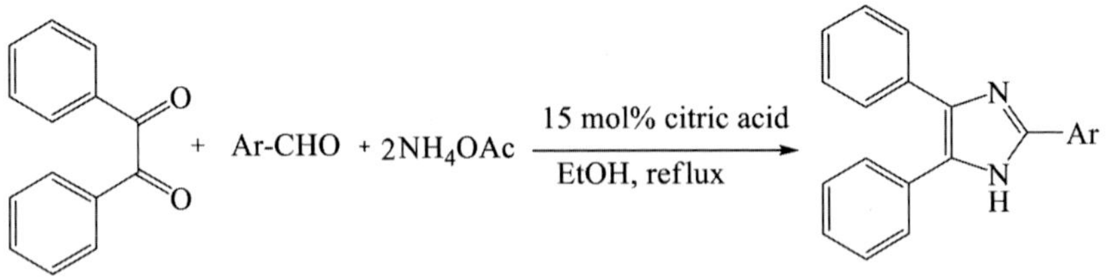

+ Ar-CHO + $2NH_4OAc$
15 mol% citric acid
EtOH, reflux

SCHEME 9.31 Synthesis of 2,4,5-triaryl-1*H*-imidazoles by using the citric acid as a catalyst.

Narkhede et al. [66] has described for the synthesis of 2,4,5-trisubstituted imidazoles through a three component one pot reaction of benzyl, benzaldehyde, and NH4OAc, in the presence of catalytic amount of potassium dihydrogen phosphate (10 mol%) under solvent-free condition at RT gives good yield (Scheme 9.32).

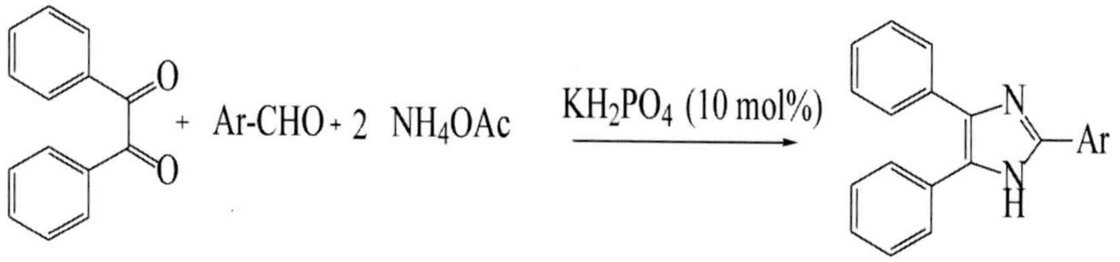

SCHEME 9.32 Synthesis of 2,4,5-trisubstituted imidazoles catalyzed by potassium dihydrogen phosphate.

Patil et al. [67] has been proposed 2,4,5-Triaryl-1*H*-imidazole derivatives is achieved by pathway of MCR involving cyclocondensation of 1,2-dicarbonyl compound, aromatic aldehyde and ammonium acetate as a source of ammonia in presence of anhydrous $PbCl_2$ as a catalyst in ethyl alcohol to gives 88–93% of yields (Scheme 9.33).

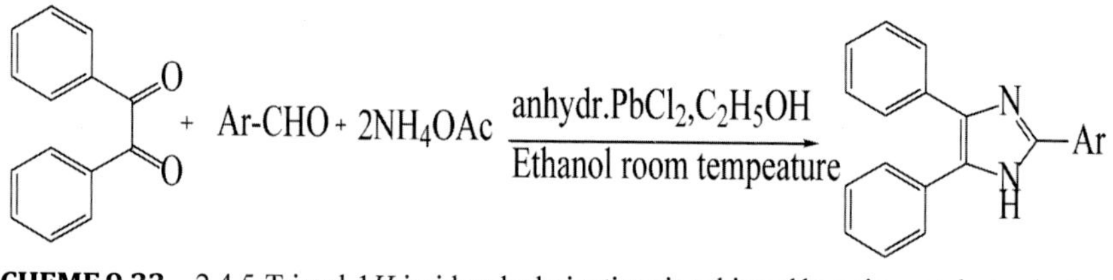

SCHEME 9.33 2,4,5-Triaryl-1*H*-imidazole derivatives is achieved by using catalyst anhydrous $PbCl_2$.

Mohammad et al. [68] was proposed 2,4,5-Triaryl-1H-Imidazoles by the coupling of Benzil/Benzoin, aldehyde, and ammonium acetate under solvent-free condition using the benzyltriphenylphosphonium chloride (BT PPC), as a catalyst gives 81–92% of yields (Scheme 9.34).

SCHEME 9.34 Synthesis of 2,4,5-triaryl-1H-imidazoles by using (BT PPC) as a catalyst.

Korupolu [69] has been expanded 2,4,5-triaryl-1*H*-imidazoles was synthesized by multi-component one-pot method from benzoin, ammonium acetate, and aromatic aldehyde by using in ethanol using magnetically reusable nano nickel-cobalt ferrite ($NiCoFe_2O_4$). as the catalyst to give 89–94% yield (Scheme 9.35).

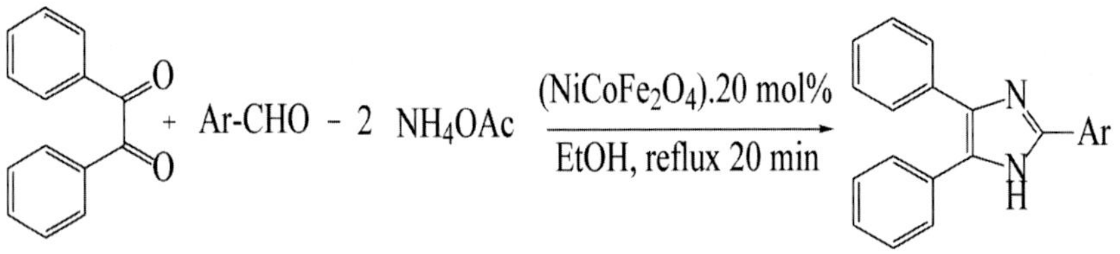
O
O
+ Ar-CHO - 2 NH4OAc
(NiCoFe2O4).20 mol%
EtOH, reflux 20 min
N
Ar
N
H

SCHEME 9.35 Nano nickel-cobalt ferrite ($NiCoFe_2O_4$) catalyzed the synthesis of 2,4,5-triaryl-1H-imidazoles.

Agarwal et al. [70] was proposed 2,4,5-Triaryl-1H-Imidazoles by the coupling of Benzil/Benzoin, aldehyde, and ammonium acetate in water: ethanol (4:1) mixture containing zinc-proline hybrid material (ZnPHM) as catalyst to afford good yield (Scheme 9.36).

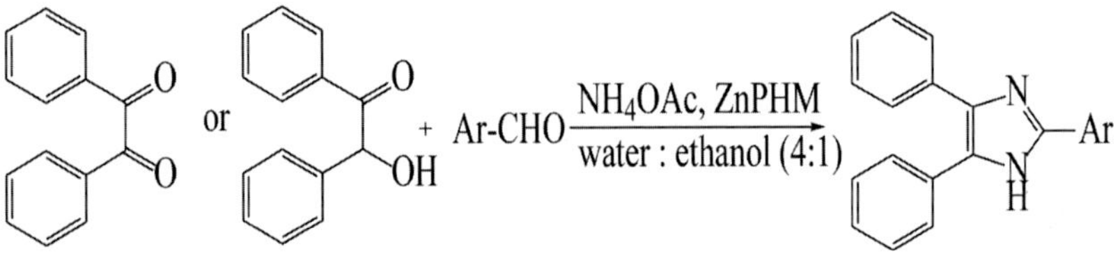
O
O
or
O
OH
+ Ar-CHO
NH4OAc, ZnPHM
water : ethanol (4:1)
N
Ar
N
H

SCHEME 9.36 2,4,5-triaryl-1*H*-imidazoles catalyzed by using zinc-proline hybrid material (ZnPHM).

Kulkarni et al. [71] was proposed the synthesis of 2,4,5-triarylimidazole from benzil, aldehydes, and ammonium acetate in the presence of wet Cyanuric chloride $SO_4^{2-}/CeO_2–ZrO_2$ as a catalyst in solvent-free condition at 100–120°C to give 65–90% yields (Scheme 9.37).

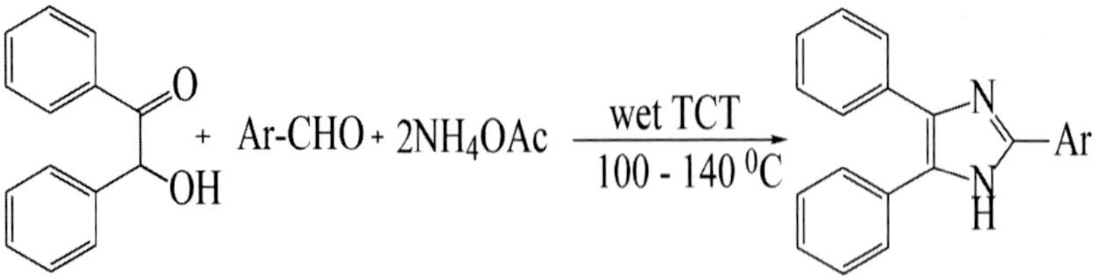
O
OH
+ Ar-CHO + 2NH4OAc
wet TCT
100 - 140 0C
N
Ar
N
H

SCHEME 9.37 The synthesis of 2,4,5-triarylimidazoles in the presence of wet TCT as a catalyst.

Najmeh et al. [72] was proposed the synthesis of Imidazoles by an one-pot three-component condensation of [9, 10]-phenanthraquinone, aryl aldehydes and ammonium acetate in presence of Silica supported La0.5 Pb0.5MnO_3 Nano Particles excellent yield under reflux, and also solvent free conditions (Scheme 9.38).

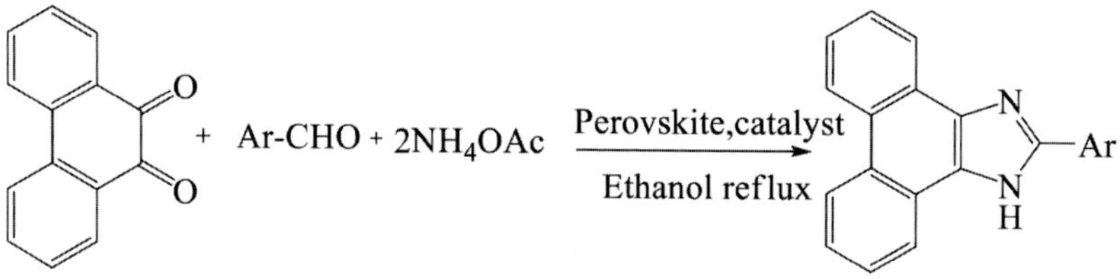

+ Ar-CHO + $2NH_4OAc$ Perovskite,catalyst / Ethanol reflux

SCHEME 9.38 Synthesis of imidazoles by using perovskite catalyst.

Ágnes et al. [73] has described the one-pot condensation reaction of 1,2-dicarbonyl compounds, benzaldehydes, and ammonium acetate in the presence of 4 A° molecular sieves-modified with titanium(IV) as an efficient heterogeneous catalyst with excellent yields of 2, 4, 5-trisubstituted imidazoles (Scheme 9.39).

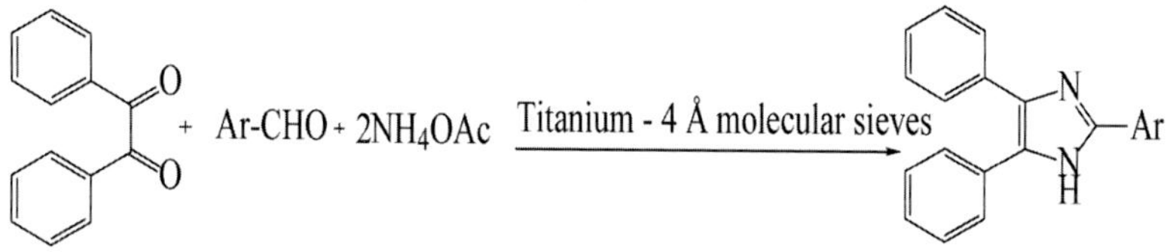

+ Ar-CHO + $2NH_4OAc$ Titanium - 4 Å molecular sieves

SCHEME 9.39 4A° molecular sieves modified with titanium(IV) as a heterogeneous catalyst for the synthesis of imidazoles.

Majid and co-worker [74] has introduced a very simple, one-pot three-component procedure for the preparation of 2,4,5-triaryl-imidazoles from the reaction of benzyl, aldehydes, and ammonium acetate, as ammonia source, catalyzed by Caro's acid-silica gel under Solvent-free condition (Scheme 9.40).

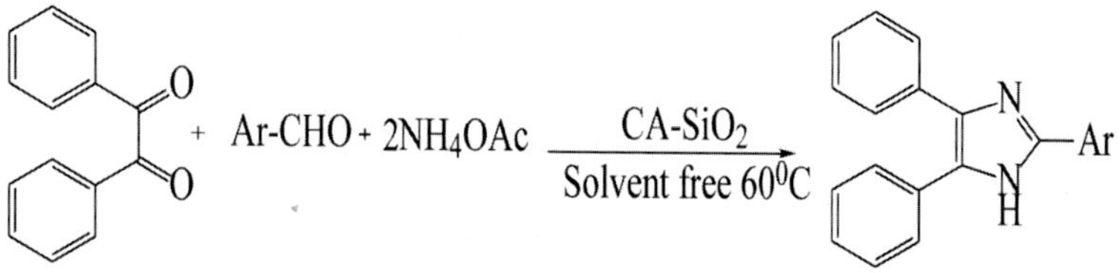

+ Ar-CHO + $2NH_4OAc$ $CA\text{-}SiO_2$ / Solvent free 60^0C

SCHEME 9.40 Preparation of 2,4,5-triaryl-imidazoles catalyzed by Caro's acid-silica gel.

Chunsheng Liu and Genxiang Luo [75] was reported Yttrium(III) trifluoroacetate is used as an efficient catalyst for the reaction of benzil, aldehydes, and ammonium acetate under mild and solvent-free conditions to afford the corresponding 2,4,5-triarylimidazoles at 100°C (Scheme 9.41).

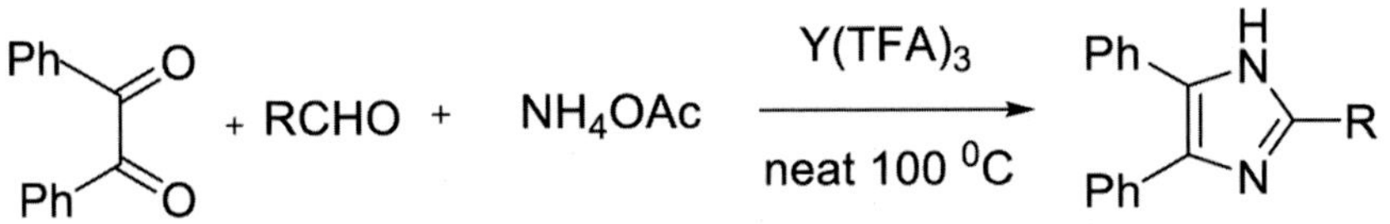

Ph, O, Ph, O + RCHO + NH_4OAc → $Y(TFA)_3$, neat 100 ^0C → Ph, H, N, Ph, N, R

SCHEME 9.41 The synthesis of 2,4,5-triarylimidazoles by using Yttrium(III) trifluoroacetate catalyst.

9.5 CONCLUSION

We have described a general and highly efficient procedure for the preparation of substituted imidazoles derivatives using commercially available inexpensive L-proline as an organocatalyst in ethanol, tannic acid as a catalyst in ethanol, Rochelle salt is also an excellent catalyst for MW-assisted organic synthesis. The remarkable advantage of this protocol is mild reaction conditions, excellent yields of product, operational, and experimental simplicity.

KEYWORDS

- **2,4,5-triaryl-1*H*-imidazole**
- **biological active heterocyclic**
- **environmental benign**
- **green chemistry**
- **multi-component reactions**
- **organocatalyst**

REFERENCES

1. Lee, J., Laydon, J., McDonnell, P., Kumar, S., Green, D., McNulty, D., et al., (1994). *Nature, 372*, 739.

2. Heeres, J., Backx, L., Mostmans, J., & Vancustem, J., (1979). *J. Med. Chem.*, *22,* 1003.
3. Wasserscheid, P., & Keim, W., (2000). *Angew. Chem. Int. Ed. Eng.*, *39,* 3772.
4. Bourissou, D., Guerret, O., Gabbai, F., & Bertrand, G., (2000). *Chem. Rev.*, *100*, 39.
5. Wauquier, A., Van, D. B. W., Verheyen, J., & Janssen, P., (1978). *J. Eur. J. Pharmacol.*, *47,* 367.
6. Tanigawara, Y., Aoyama, N., Kita, T., Shirakawa, K., Komada, F., Kasuga, M., & Okumura, K., (1999). *Clin. Pharmacol. Ther.*, *66,* 528.
7. Abrahams, S. L., Hazen, R. J., Batson, A. G., & Phillips, A. P., (1989). *J. Pharmacol. Exp. Ther.*, *249*, 359.
8. Grimmett, M., Katritzky, A., Rees, C., & Scriven, E., (1996). *Pergamon: New York*, *3,* 77.
9. Krieg, B., & Mancke, G., (1967). *Naturforschg.*, *226*, 132.
10. Mizuno, N., & Misono, M., (1998). *Chem. Rev.*, *98*, 199.
11. Poliakoff, M., & Anastas, P. T., (2001). *Nature, 413*, 257.
12. DeSimone, J. M., (2002). *Science, 297*, 799.
13. Varma, R. S., (2001). *Pure Appl. Chem.*, *73*, 193.
14. Anastas, P. T., & Warner, J. C., (1998). *Green Chemistry: Theory, and Practice*. Oxford Uni. Press, Inc. New York.
15. Matlack, A. S., (2001). *Introduction to Green Chemistry*. Marcel Dekker, Inc, New York.
16. (a) Toda, F., & Tanaka, K., (2000). *Chem. Rev.*, *100*, 1025. (b) Dolman, S. J., Satterly, E. S., Hoveyda, A. H., & Schrock, R. R., (2002). *J. Am. Chem. Soc.*, *124*, 6991.
17. (a) Poliakoff, M., Fitzpatrick, J. M., Farren, T. R., & Anastas, P. T., (2002). *Science*, *297*, 807. (b) Desimone, J. M., (2002). *Science*, *297*, 799.
18. (a) Li, C. J., & Chan, T. H., (1997). *Organic Reactions in Aqueous Media*. Wiley & Sons, New York. (b) Ten, G. J., Brink, W. C. E., & Sheldon, R. A., (2000). *Science*, *287*, 1636. (c) Lindstroem, U. M., (2002). *Chem. Rev.*, *102*, 2751.
19. (a) Luo, Z. Y., Zang, Q. S., Oderaotoshi, Y., & Curran, D. P., (2001). *Sci.*, *291,* 1766. (b) Horvath, I. T., (1998). *Acc. Chem. Res.*, *31,* 641.
20. Oakes, R. S., Clifford, A. A., & Rayner, C. M., (2001). *J. Chem. Soc. Perkin Trans, 1*, 917.
21. (a) Sheldon, R. A., (2001). *Chem. Commun.*, p. 2399.
22. Zhang, Z., Yin, L., Wang, Y., Liu, J., & Li, Y., (2004). *Green Chem.*, *6*, 563.
23. Perrier, S., Gemici, H., & Li, S., (2004). *Chem. Commun.*, p. 604.
24. Zhu, J., & Bienayme, H., (2005). *Multicomponent Reactions*. Wiley-VCH: Weinheim.
25. Krasavin, M., Tsirulnikov, S., Nikulnikov, M., Kysil, V., & Ivachtchenko, A., (2008). *Tetrahedron Letst.*, *49*, 5241.
26. Yavari, I., Sabbaghan, M., & Hossaini, Z., (2007). *Mol. Diversity*, *11*, 1.
27. Shaabani, A., Soleimani, E., Sarvary, A., & Rezayan, A. H., (2008). *Bioorg. Med. Chem. Lett.*, *18*, 3968.
28. Banfi, L., Basso, A., Cerulli, V., Guanti, G., & Riva, R., (2008). *J. Org. Chem.*, *73*, 1608.
29. (a) Russowsky, D., Lopes, F. A., Da Silva, V. S. S., Canto, K. F. S., Montes D'Oca, M. G., & Godoi, M. N., (2004). *J. Braz. Chem. Soc.*, *15*, 165. (b) Vieira, Y. W., Nakamura, J., Finelli, F. G., Brocksom, U., & Brocksom, T. J., (2007). *J. Braz. Chem. Soc.*, *18*, 448.
30. Biginelli, P., (1893). *Gazz. Chim. Ital.*, *23*, 360.
31. (a) Radziszewski, B., (1882). *Chem. Ber.*, *15,* 1493. (b) Japp, F. R., & Robinson, H. H., (1882). *Chem. Ber.*, *15*, 268.
32. Hantzsch, (1882). *Justus Liebigs Ann. Chem.*, *215*, 1.

33. Mannich, C., & Krosche, W., (1912). *Archiv der Pharmazie, 250*, 647.
34. Ugi, I., Meyr, R., Fetzer, U., & Steinbruckner, C., (1959). *Angew. Chem., 71*, 386.
35. Balalaie, S., Arabanian, A., & Hashtroudi, M. S., (2000). *Monatsh. Chem., 131*, 945.
36. Siddiqui, S. A., Narkhede, U. C., Palimkar, S. S., Daniel, T., Lahoti, R. J., & Srinivasan, K. V., (2005). *Tetrahedron, 61,* 3539.
37. Sharma, G., Jyothi, Y., & Lakshmi, P., (2006). *Synth. Commun., 36,* 2991.
38. Khodaei, M. M., Bahrami, K., & Kavianinia, I., (2007). *J. Chin. Chem. Soc. (Taipei), 54,* 829.
39. Yu, M., Lei, W., & Su, Y., (2007). *Synth. Commun., 37,* 3301.
40. Mohammed, A. F., Kokare, N. D., Sangshetti, J. N., & Shinde, D. B., (2007). *J. Korean Chem. Soc., 51*, 418.
41. Sangshetti, J. N. J., Kokare, N. D., Kotharkar, S. A., & Shinde, D. V., (2008). *Monatsh. Chem., 139*, 125.
42. Khosropour, A. R., (2008). *Can. J. Chem., 86,* 264.
43. Mohammadi, A. A., Mivechi, M., & Kefayati, H., (2008). *Monatsh. Chem., 139,* 935.
44. Shelke, K., Kakade, G., Shingat, B., & Shingare, M. S., (2008). *Rasa. J. Chem., 1*, 489.
45. Shitole, N. V., Shelke, K. F., Sonar, S. S., Sadaphal, S. A., Shingate, B. B., & Shingare, M. S., (2009). *Bull. Korean Chem. Soc., 30*, 1963.
46. Shelke, K. F., Sapkal, S. B., Sonar, S. S., Madje, B. R., Shingate, B. B., & Shingare, M. S., (2009). *Bull. Korean Chem. Soc., 30*, 1057.
47. Shelke, K. F., Sapkal, S. B., Sonar, S. S., & Shingare, M. S., (2009). *Chin. Chem. Lett., 20*, 283.
48. Shelke, K. F., Sapkal, S. B., Kakade, G. K. B. B., & Shingare, M. S., (2010). *Green Chem. Lett. and Rev., 3*, 27.
49. Nalage, S. V., Kalyankar, M. B., Patil, V. S., Bhosale, S. V., Deshmukh, S. U., & Pawar, R. P., (2010). *Open Cata. J., 3,* 58.
50. Bahador, K., Khalil, E., & Abdolmohammad, G., (2012). *Turk. J. Chem., 36*, 601.
51. Mohammadi, A., Keshvari, H., Sandaroos, R., Rouhi, H., & Sepehr, Z., (2012). *J. Chem. Sci., 124*, 717.
52. Behrooz, M., Hossein, K. S., Fereshteh, T., & Elahe, A., (2012). *Inter. J. Org. Chem., 2*, 93.
53. Dake, S. A., Khedkar, M. B., Irmale, G. S., Ukalgaonkar, S. J., Thorat, V. V., & Shintre S. A., (2012). *Synth. Comm., 42,* 323.
54. Shitole, N. V., Shitole, B. V., Kakde, G. K., & Shingare, M. S., (2013). *Orbital Elec. J. Chem., 5,* 35.
55. Adel, A. M., Vagif, M. A., Avtandil, H. T., & Shaaban, K. M., (2013). *World J. Org. Chem., 1,* 6.
56. Farhad, H., & Hadises, K., (2014). *Oriental J. Chem., 30*, 329.
57. Hangirgekar, S. P., Kumbhar, V. V., Shaikh, A. L., & Bhairuba, I. A., (2014). *Der Pharma Chemica, 6,* 164.
58. Safari, J., Naseh, S., Zarnegar, Z., & Akbari, Z. J., (2014). *Taibah Univ. Sci., 8*, 323.
59. Shaikh, S., (2014). *Res. J. Chem. Sci., 4*, 18.
60. Niralwad, K. S., Ghorade, I. B., & Shingare, M. S., (2014). *Inter. J. Scie. Rese.*, 2277.
61. Shitole, B. V., Shitole, N. V., Ade, S. B., & Kakde, G. K., (2015). *Orbital: Electron. J. Chem., 7,* 240.
62. Chen, Y., Wang, R., Ba, F., Hou, J., Ding, A., Zhoum, M., & Guo, H., (2017). *J. Saudi Chem. Soc., 21*, 76.

63. Ghodsi, M. Z., Alireza, B., Negar, L., & Zahra, F., (2016). *J. Saudi Chem. Soci.*, *20*, 419.
64. Narkhede, H. P., (2016). *Intern. J. Engine. Sci. Comp., 6,* 6061.
65. Nora, C., Taoues, B., Imane, T., Boudjemaa, B., & Abdelmadjid, D., (2016). *Der Pharma Chemica.*, *8*, 202.
66. Bhirud, J. D., & Narkhede, H. P., (2016). *J. Appli. Chem., 5,* 1075.
67. Patil, V. D., Sutar, N. R., & Patil, K. P., (2016). *J. Chem. Pharm. Rese.*, *8,* 728.
68. Mohammad, A., & Mozhgan, A., (2017). *Bull. Chem. Soc. Ethiop.*, *31*, 177.
69. Korupolu, R. B., Srividhya, M., Suri, B. M., & Nooka, R. A., (2017). *Chem. Sci. Tran.*, *6*, 428.
70. Agarwal, S., Kidwai, M., Poddar, R., & Nath, M., (2017). *Chem. Select*, *2,* 10360.
71. Kulkarni, M. S., Mamgain, R., Saravate, K., & Nagarkar, R. R., (2018). *Pharma Innov. J.*, *7,* 153.
72. Najmeh, Z., Ali, J., Mohammad, K. M., & Haman, T., (2018). *Bull. Chem. Soc. Ethiop.*, *32*, 1011.
73. Ágnes, M., & Zoltán, H., (2019). *Synlett.*, *30*, 89.
74. Majid, M. H., Narges, K., & Samira, P., (2019). *Adv. J. Chem. A, 2*, 73.
75. Chunsheng, L., & Genxiang, L., (2010). A convenient synthesis of 2,4,5-triarylimidazoles catalyzed by $Y(TFA)_3$. *Green Chemistry Letters and Reviews, 3*(2), 101–104.

CHAPTER 10

Microwave-Assisted Synthesis: Revolution in Synthetic Chemistry

VIKAS V. BORGAONKAR

Department of Chemistry, Shri Siddheshwar Mahavidyalaya, Majalgaon–431131, Maharashtra, India
E-mail: vikasborgaonkar@rediffmail.com

ABSTRACT

As the world is facing a huge problem of pollution, the research community is giving the highest priorities to green chemical synthetic ways. The microwave (MW) induced synthetic route found to be a good alternative over traditional methods. The reactions carried out by MW induced techniques have several advantages in comparison to conventional heating methods. In the last two decades, there is a dramatic surge in the methodologies using MW induced techniques as an energy source to promote chemical conversions. Many organic transformations which require several hours or even days to complete the successful reactions in a few minutes by using MW induced techniques. MW assisted synthesis have advantages of providing ecofriendly conditions such as quick reaction time, operational simplicity, enhanced yields of products.

10.1 INTRODUCTION

In this chapter, we mentioned observations and other literature evidences an attempt has been made to summarize the significance of MW technique and its use in various chemical transformations. MW-assisted reactions show in Figure 10.1.

Synthesis of new chemical entities conducts a major role in drug discovery. Conventional heating methods for various chemical syntheses are very well documented. The methods for synthesis (Heating process) of organic compounds have continuously changing with new discoveries.

Robert Bunsen invented the burner which works as energy source for heating a reaction vessel, heating mental, oil bath or hot plate are later on used for the heating purpose but the shortcoming of heating method remain the same specially regarding environment issues.

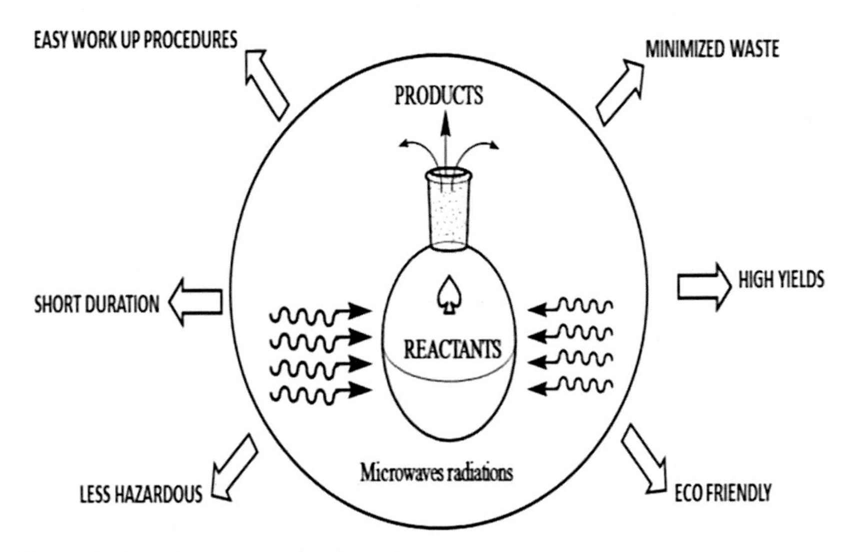

FIGURE 10.1 Microwave-assisted reactions.

To overcome this environmental issue, movement started for developing the concept of benign methodology, that is designing environmental friendly processes and chemical products. This incorporated a new concept of Green Chemistry. Green chemistry efficiently utilizes raw materials, eliminates waste, and avoids the use of toxic or hazardous reagents and solvents in the manufacture and application of chemical products.

10.2 MICROWAVE (MW) ASSISTED TECHNIQUES

The use of microwave (MW) irradiation in organic synthesis has become increasingly popular within the pharmaceutical and academic arenas, because it is a new enabling technology for drug discovery and development [1]. MW chemistry has an edge over conventional heating methods for conducting chemical reactions, and it will soon emerge as the preferred technology for performing chemical synthesis relating to lead development in pharmaceutical and biotechnology companies. Moreover, the use of MW chemistry for industrial production holds promise, since research has already been initiated to scale up MW chemistry reactions from milligrams to kilograms. The classical method is a slow and in efficient method for transferring energy into the reacting system. In MW technique, MWs couple directly with the molecules of the entire reaction mixture, leading to a rapid rise in temperature. MW irradiations will directly activate most molecules that possess a dipole or are ionic. Since energy, the transfer occurs in less than a nanosecond.

An alternative method for performing MW assisted organic reactions, termed "enhanced microwave synthesis" (EMS), has been examined [2]. By externally cooling there action vessel with compressed air, while simultaneously administering MW irradiation, more energy can be directly applied to the reaction mixture. Research published in leading organic synthesis journals supports the use of simultaneous cooling of reactions being heated by MW energy [3, 4].

The development of modern organic chemistry has been based on the fact that organic reactions are often carried out in organic solvents. According to the environmental view, chemists are concentrating on the solvent-free reactions that reduce pollution. MW heating for carrying out reactions on solids has also attracted considerable attention. For such 'dry media' reactions, solid supports such as alumina, silica, montmorillonite clay, and zeolites have been investigated.

10.3 APPLICATIONS IN CHEMICAL SYNTHESIS

Organic synthesis is the preparation of a desired organic compound from available starting materials. MW-assisted organic synthesis has been the foremost and one of the most researched applications in chemical reactions. Since then, chemists have successfully conducted a large range of organic reactions like Diels-Alder reaction, Suzuki reaction, Mannich, Wittig reaction using MW techniques.

The Wittig reaction is carried out between aromatic aldehydes and stabilized yields to generate alkenes. Since the method of generating yields is a rate-determining step, the conventional method is a slow process. However, MW heating performs the reactions in minutes [5]. In the case of organometallic compounds, it has been observed that the use of MWs in the synthesis of organo-B-metal compounds has resulted in a 40-fold increase in the rate of reaction. This enhanced rate of reaction was achieved under conditions identical to high-pressure organic synthesis [6].

Anushu Dandia et al. [7] was used MW technique to react 2-aryl-3-(3-aryl-3-oxo propen-1-yl)-indoles (3) with hydrazine hydrate/phenyl hydrazine and thiourea to obtain 2-aryl-3-(3-arylpyrazol-5-yl)indoles (4) and 2-aryl-3-(4-aryl-5,6-dihydro2(*1H*)thiooxopyrimid-ine-6-yl)-indole (5) Scheme 10.1.

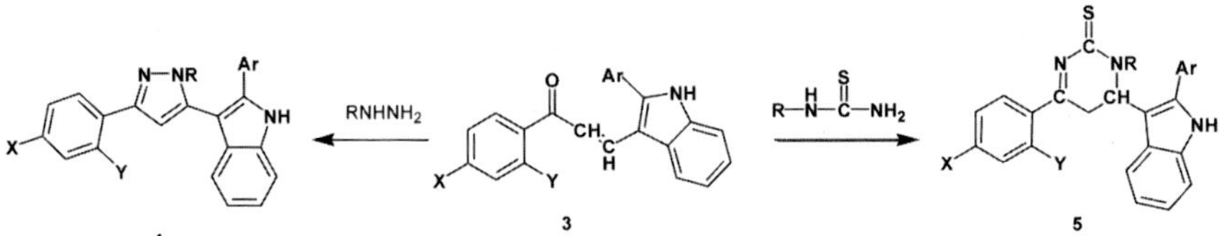

SCHEME 10.1 Synthesis of indole derivatives in microwave.

Balalaie and co-workers [8] has described the MW-assisted synthesis of 2,4,5-triarylimidazole from benzil, aldehydes, and ammonium acetate in the presence of Zeolite HY producing 54–94% yields, reduction in time and cleaner reactions are the advantages Scheme 10.2.

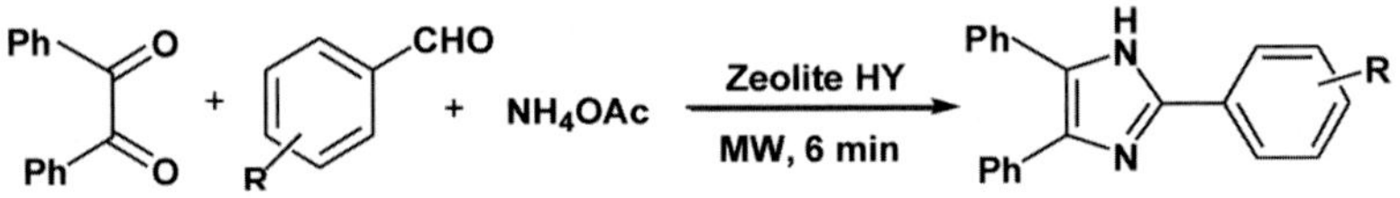

SCHEME 10.2 The microwave-assisted synthesis of 2,4,5-triarylimidazoles.

The deprotection of 2,4-dinitrophenylhydrazones to the corresponding carbonyl compounds has been reported in good yields with [BNBTS] under MW irradiation by A. Khazaei and R. Ghorbani-Vagei [9] Scheme 10.3.

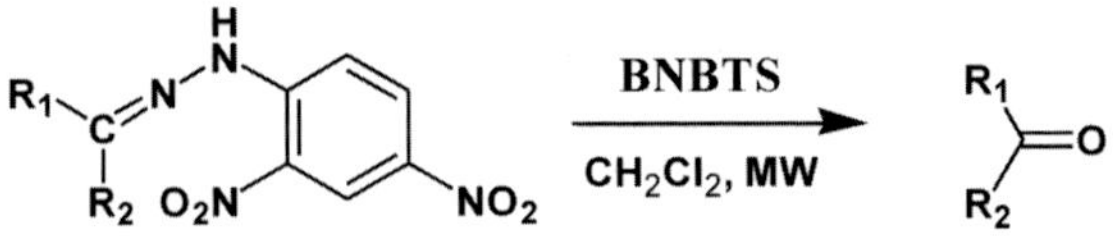

SCHEME 10.3 Synthesis of carbonyl compounds in microwave.

MW activation coupled with dry media technique as a green chemistry procedure has been applied by A. Dandia et al. [10] to synthesis of a series of some new Benzothiazepin derivatives (7). The results obtained under MW irradiation when compared with that following conventional method demonstrate the versatility of the process Scheme 10.4.

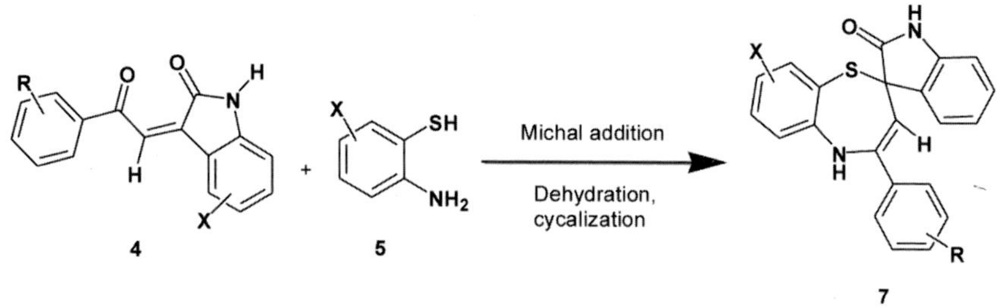

SCHEME 10.4 Synthesis of benzothiazepin derivatives in microwave.

Thanh et al. [11] were performed the synthesis of ortho-alkylated ketones in mono mode reactors. The synthesis was carried out *via* a chelation-assisted Rh(I) catalyzed ortho-alkylation reaction of aromatic imines under MW activated solvent-free conditions Scheme 10.5.

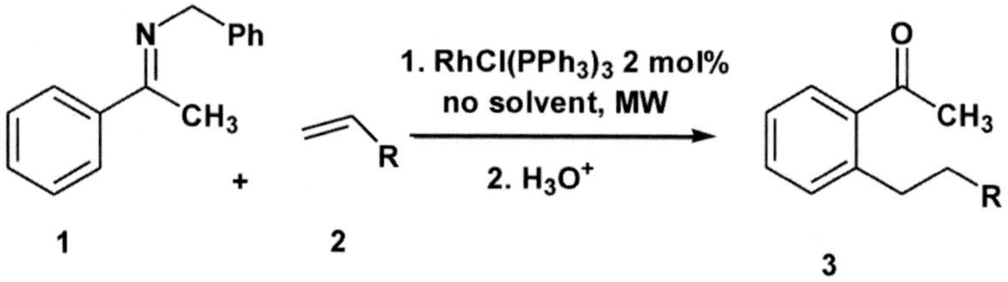

SCHEME 10.5 The synthesis of ortho-alkylated ketones under microwave condition.

Angela Rao et al. [12] were described the MW-assisted synthesis of 1*H*,3*H*-thiazolo[3,4-a]benzimidazoles, 2-aryl-1-benzylbenzimidazoles and 2,3-diaryl-1,3-thiazolidin-4-ones, which achieved reduction in reaction time, higher yields, cleaner reactions than previously described synthetic processes Scheme 10.6.

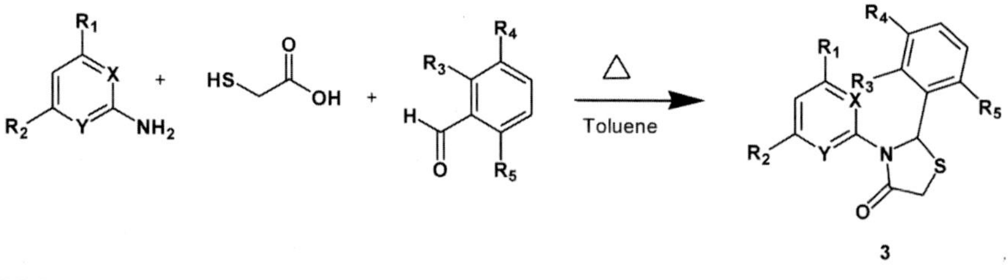

SCHEME 10.6 Synthesis of 1*H*,3*H*-thiazolo[3,4-a]benzimidazoles, 2-aryl-1-benzylbenzimidazoles and 2,3-diaryl-1,3-thiazolidin-4-ones under microwave condition.

Bolognese et al. [13] was prepared a range of 1,3-thiazolidin-4-one derivatives by the MW-assisted reaction between benzylidene-anilines and mercaptoacetic acid in benzene at 30°C in 10 min Scheme 10.7.

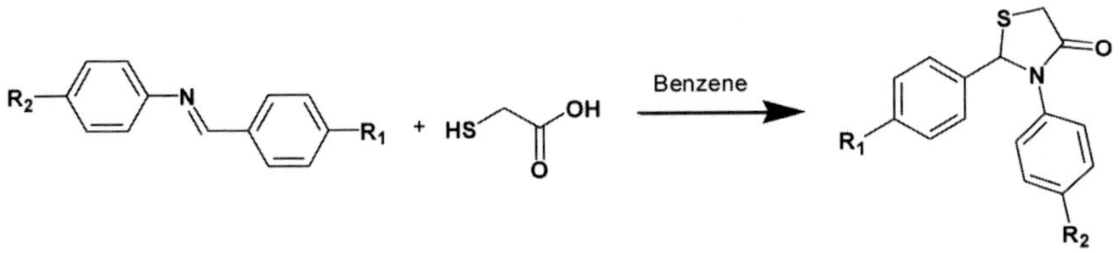

SCHEME 10.7 Preparation of 1,3-thiazolidin-4-one derivatives by the microwave-assisted reaction.

Spiro[3H-indole-3,20-[4H] pyrido[3,2-e]-1,3-thiazine]-2,40(1H)diones does not form under conventional conditions, can be prepared by treatment of '*in situ*' generated 3-indolylimine derivatives with 2-mercaptonicotinic acid under MW irradiation in absence of any solvent or solid support. The facile one-pot reaction is generalized for a variety of ketones and amines to give pure pyrido [3,2-e] thiazine derivatives, which do not require further purification processes was developed by A. Dandia et al. [14] Scheme 10.8.

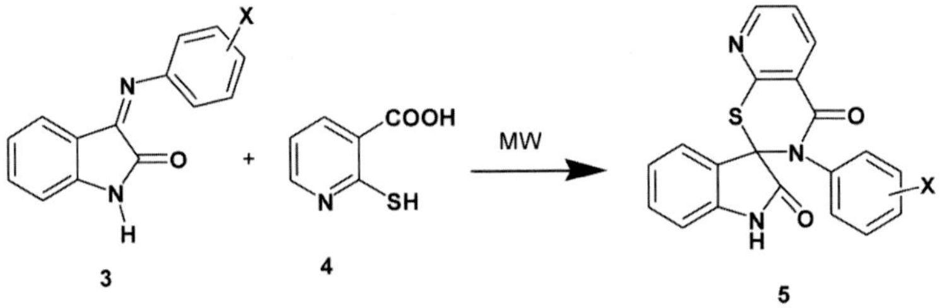

SCHEME 10.8 Synthesis of Spiro [3H-indole-3,20-[4H] pyrido[3,2-e]-1,3-thiazine]-2,40 (1H)diones.

Hazarkhani and Karimi [15] have been reported as catalyst for the preparation of dihydropyrimidinones under MW irradiation by using NBS Scheme 10.9.

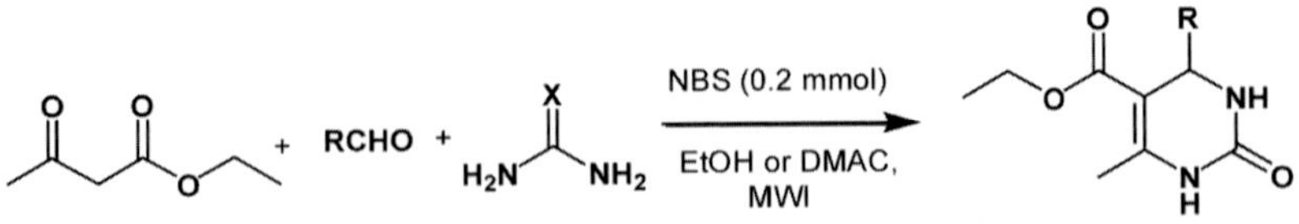

SCHEME 10.9 Preparation of dihydropyrimidinones under microwave irradiation.

Kidwai et al. [16] have developed convenient synthetic procedure for the preparation of benzodiazepines by coupling MWs with the solvent-free technique. Through modernization and simplification of the classical procedure, and avoidance of volatile and toxic solvents and external bases. All the reactions were also pursued under MWI in solution (method B) as well as the conventional method (method A) for a comparative study. Solid-supported reactions gave improved yield and were completed within seconds of MWI in comparison to hours and minutes in conventional and MW assisted syntheses in solution, respectively Scheme 10.10.

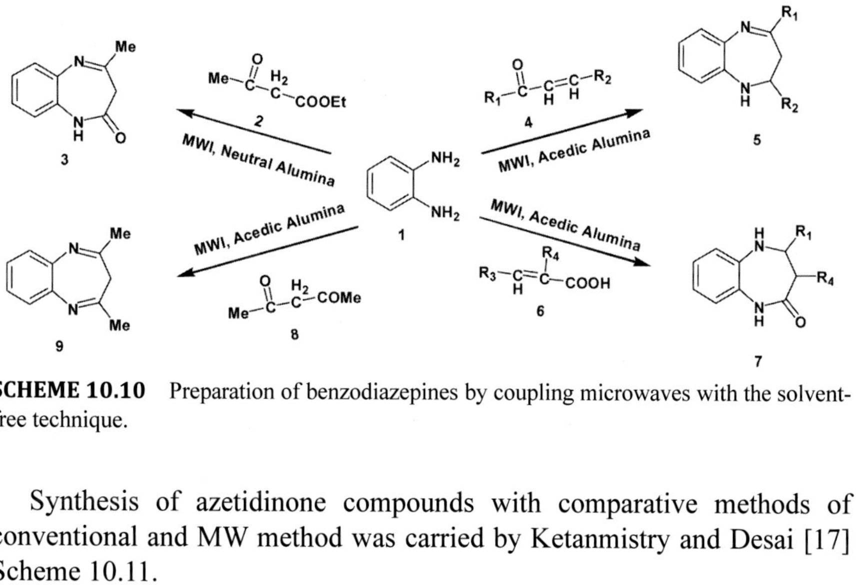

SCHEME 10.10 Preparation of benzodiazepines by coupling microwaves with the solvent-free technique.

Synthesis of azetidinone compounds with comparative methods of conventional and MW method was carried by Ketanmistry and Desai [17] Scheme 10.11.

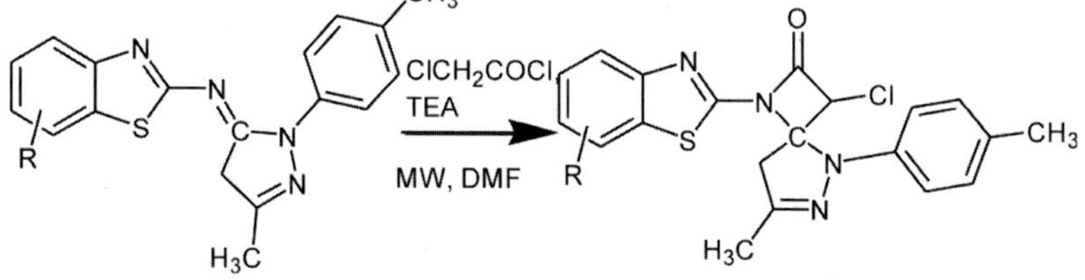

SCHEME 10.11 Synthesis of azetidinone compounds under microwave irradiation conditions.

A. Dandia et al. [18] have developed a new, economical, safe, environmentally benign one-pot synthesis of novel spiro[indole-pyrido(2,3-d) pyrimidines] under MW irradiation. The synthesis gives excellent yields of the required products 6 and 7 (85–89%) in 11–13 min Scheme 10.12.

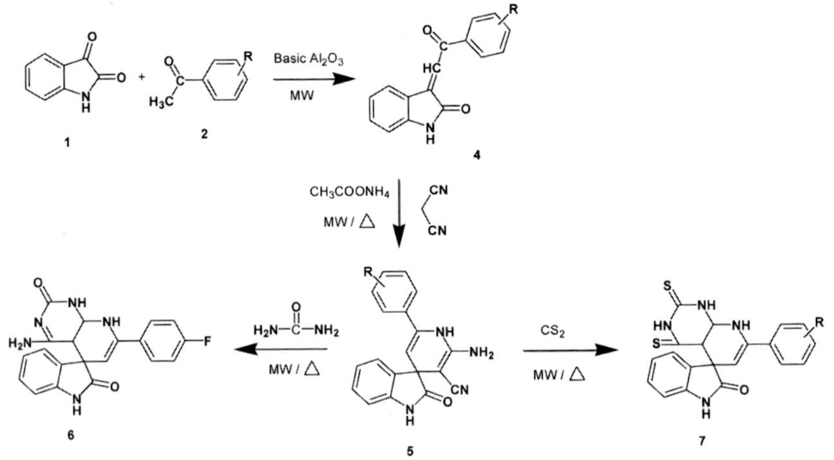

SCHEME 10.12 Synthesis of spiro[indole-pyrido(2,3-d)pyrimidines] under microwave irradiation conditions.

An effective method for the exclusive one-pot synthesis of novel benzothiazepines derivatives (7) having a spiro-3H-indoline nucleus was described by Anshu Dandia et al. [19]. Conventionally, 3 did not undergo reactions with 5 even under drastic conditions of prolonged reflux using strong acidic or basic catalysts in high boiling organic solvents for many days. However, the exclusive formation of the title products in a satisfactory yield (71%) was succeeded under MW irradiation coupled with various inorganic supports revealing a very strong specific MW effect Scheme 10.13.

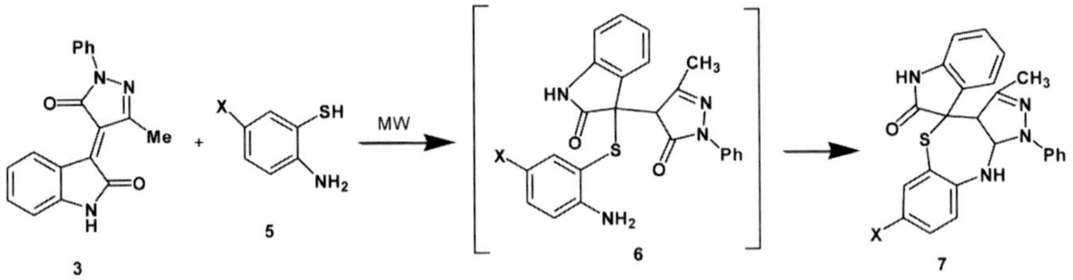

SCHEME 10.13 Synthesis of spiro-3H-indoline under microwave irradiation conditions.

Mistry and Desai [20] was applied MW technique for synthesis of pharmacologically active Azetidinones (4) and Thiazolidinones (5) from Schiff bases. A comparative study of conventional and MW method was carried Scheme 10.14.

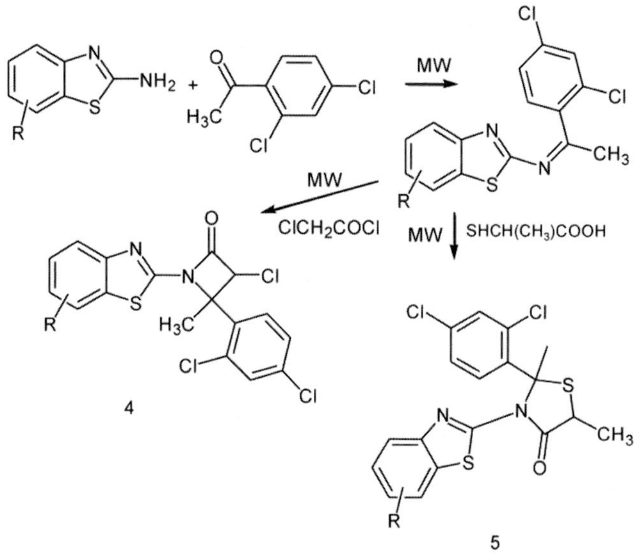

SCHEME 10.14 MW technique for synthesis of azetidinones and thiazolidinones.

Synthesis of triazolo[4,3-a]pyrimidines, an important class of building blocks in herbicidal drugs and pharmaceuticals, has been developed by A. Dandia et al. [21] *via* a multi-component condensation reaction under MW irradiation conditions in an aqueous medium. The simplicity of this short procedure and enhanced yields render this method particularly attractive for the rapid synthesis of triazolo[4,3-a]pyrimidines Scheme 10.15.

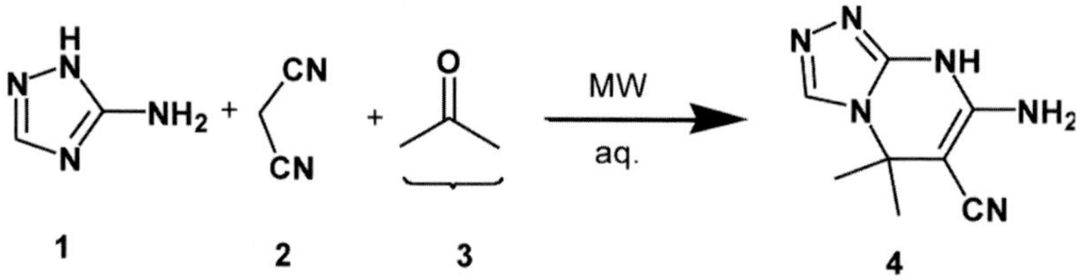

SCHEME 10.15 Synthesis of triazolo[4,3-a]pyrimidines under microwave irradiation conditions.

NBP was used by Khazaei et al. [22] for the facile oxidation of thiols to symmetrical disulfides in a mixture of acetone-water under MW irradiation.

Both aromatic and aliphatic thiols were selectively oxidized to disulfides (sulfoxides, sulfones, sulfonic acid were not formed) in good to excellent yields Scheme 10.16.

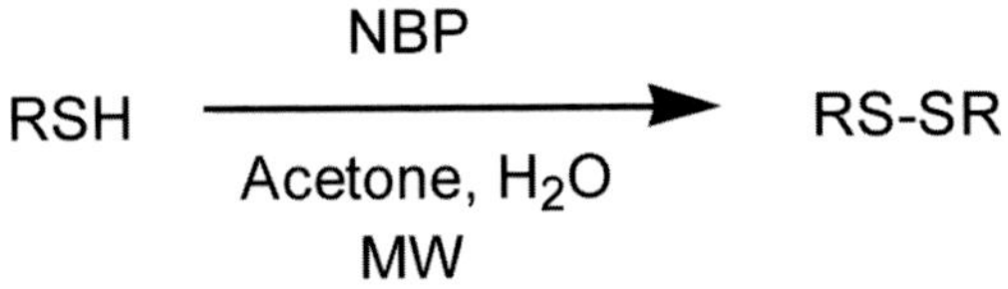

$$\text{RSH} \xrightarrow[\text{Acetone, } H_2O \text{ MW}]{\text{NBP}} \text{RS-SR}$$

SCHEME 10.16 Synthesis of symmetrical disulfides under microwave irradiation.

Mu [23] was developed a solvent-free and catalyst-free method for the synthesis of α-aminophosphonates by a MW-assisted three-component Kabachnik-Fields reaction containing aldehyde, amine, and dimethyl phosphite Scheme 10.17.

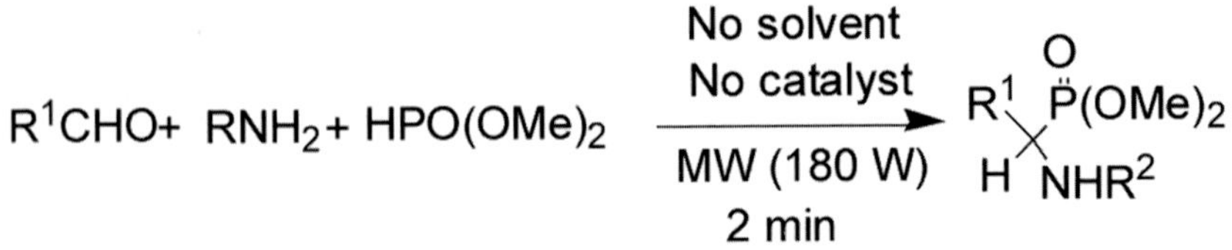

$$R^1CHO + RNH_2 + HPO(OMe)_2 \xrightarrow[\text{MW (180 W) 2 min}]{\text{No solvent, No catalyst}} R^1CH(NHR^2)P(O)(OMe)_2$$

SCHEME 10.17 The synthesis of α-aminophosphonates by a microwave-assisted condition.

Xu et al. [24] has been developed a simple method for the synthesis of various α-aminophosphonate derivatives under MW irradiation Scheme 10.18.

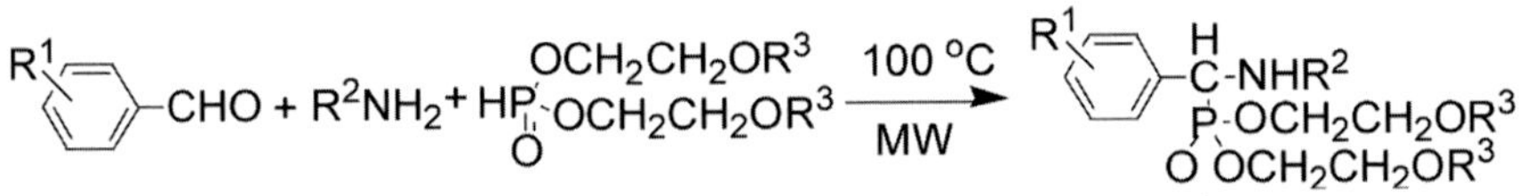

$$R^1C_6H_4\text{-CHO} + R^2NH_2 + HP(O)(OCH_2CH_2OR^3)_2 \xrightarrow[\text{MW}]{100\ ^\circ C} R^1C_6H_4\text{-CH(NHR}^2)\text{-P(O)(OCH}_2CH_2OR^3)_2$$

SCHEME 10.18 Synthesis of various α-aminophosphonate derivatives under microwave irradiation.

Ghorbani-Vaghei et al. [25] were reported methodology for the oxidation of primary and secondary alcohols to the corresponding carbonyl functions with *N,N,N,N*-tetrabromobenzene-1,3-disulfonamide and poly(*N*-bromobenzene-1,3-disulfonamide) using MW irradiation under solvent-free conditions. Aliphatic, benzylic, and allylic alcohols are rapidly oxidized to aldehydes without over-oxidation to carboxylic acids. Secondary carbinols are slowly oxidized so that the reaction is highly chemoselective Scheme 10.19.

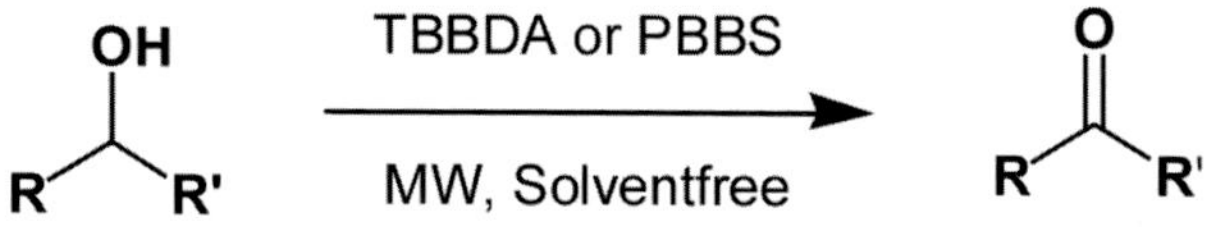

SCHEME 10.19 Oxidation of primary and secondary alcohols to the corresponding carbonyl functions.

Desai et al. [26] was synthesized a series of azetidinone and thiazolidinone using MW irradiation method as well as conventional method Scheme 10.20.

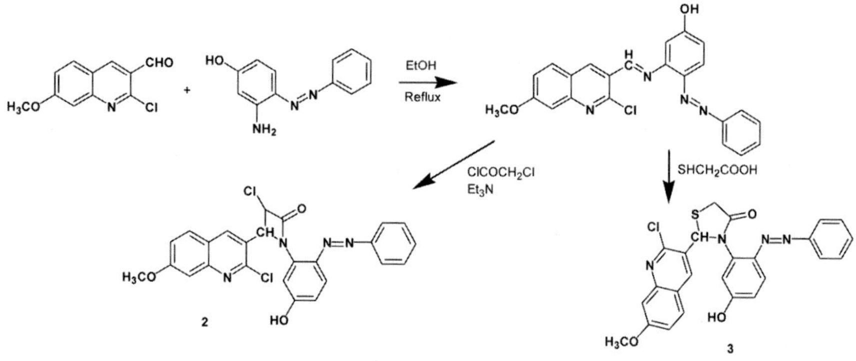

SCHEME 10.20 Synthesis of azetidinone and thiazolidinone using microwave irradiation method.

MW irradiation technique was used by Tsoleridis et al. [27] for synthesis of 4-phenyl-1H-1,5-benzodiazepines. The facile method produced the target molecule in excellent yield from *o*-phenylenediamine and 1,3-diketones in presence of a catalytic amount of acetic acid Scheme 10.21.

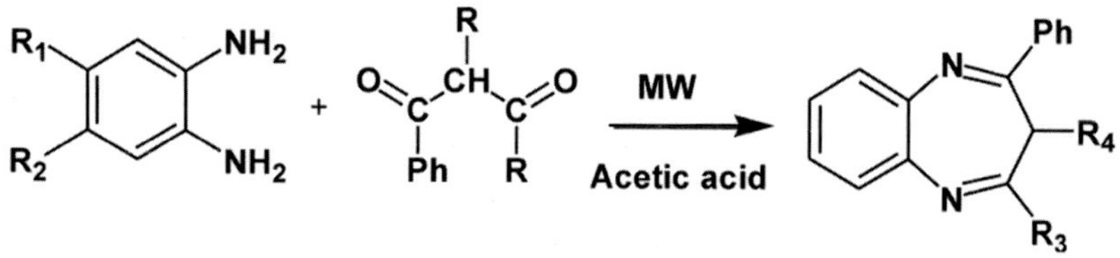

SCHEME 10.21 Synthesis of 4-phenyl-1H-1,5-benzodiazepines under MW.

Bhattacharya et al. [28] developed three-component reaction of amine, aldehyde or ketone and diethyl phosphite catalyzed by Amberlite-IR 120 resin producing α-aminophosphonates Scheme 10.22.

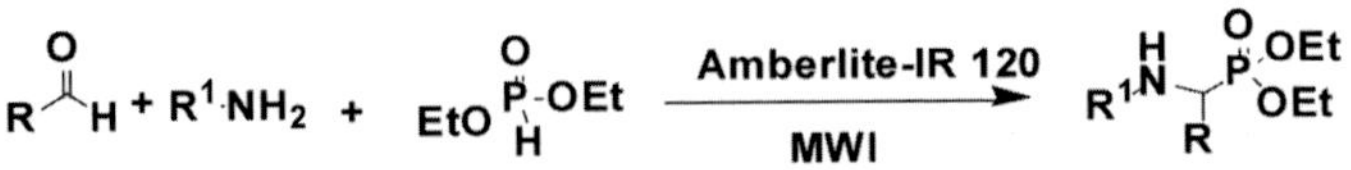

SCHEME 10.22 Synthesis of α-aminophosphonates catalyzed by amberlite-IR 120 resin.

New *ortho*-hydroxyketimines were synthesized by conventional as well as MW method by A. Kundu et al. [29] Scheme 10.23.

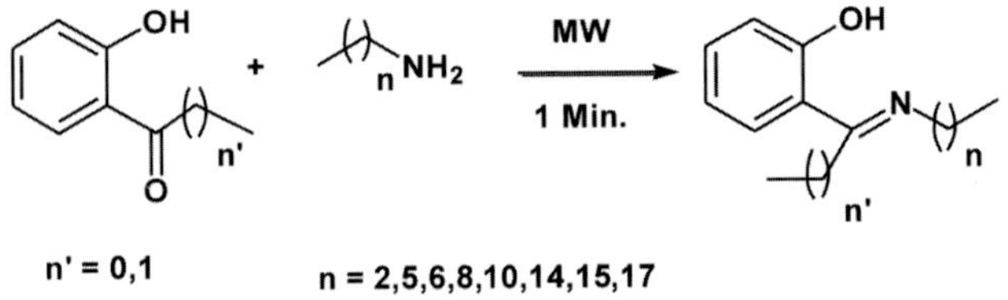

SCHEME 10.23 *Ortho*-hydroxyketimines were synthesized by microwave method.

Raval et al. [30] was reported efficient and fast procedure for the synthesis of 3-chloro-4-(4-(diethylamino)-2-hydroxyphenyl)-1-(substitutedphenyl)-azetidin-2-one using 4-diethylamino-2-hydroxybenzaldehyde, various amines and chloroacetyl chloride in presence of triethyl amine under MW irradiation and similarly, conventional methods are used for comparison Scheme 10.24.

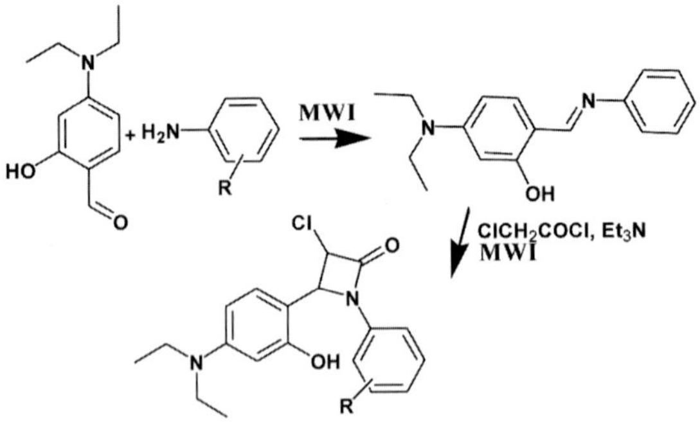

SCHEME 10.24 Synthesis of 3-chloro-4-(4-(diethylamino)-2-hydroxyphenyl)-1-(substitutedphenyl)-azetidin-2-ones.

Vibhute et al. [31] was demonstrated the use of MW technique for synthesis of Schiff bases by offering advantages such as decrease in the reaction time, outstanding yields with without formation of unwanted side products, operational simplicity and easy work-up Scheme 10.25 & 10.26.

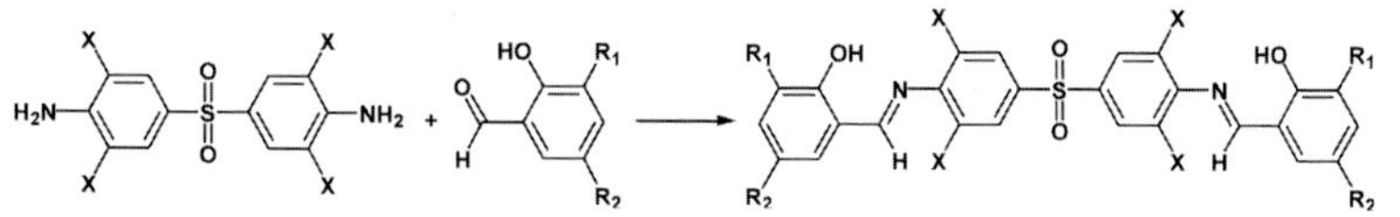

SCHEME 10.25 Synthesis of Schiff bases by using microwave technique.

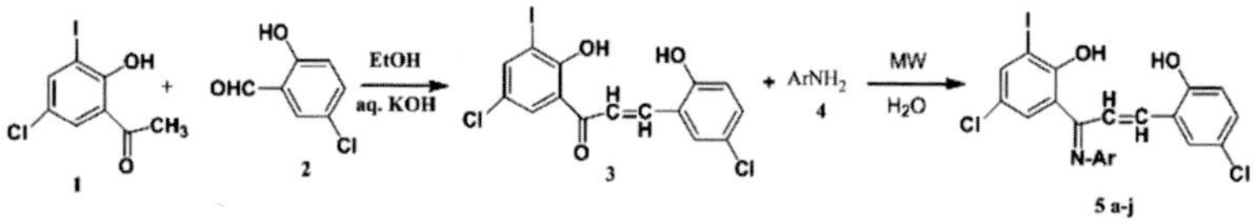

SCHEME 10.26 Synthesis of α, β-unsaturated ketimines by using microwave technique.

Kondapalli et al. [32] was reported the synthesis of thiazolidinones was achieved by using MW assisted solvent-free condition technique. It provides advantages such as shorter reaction time, solvent-free conditions, and minimal purification of the products Scheme 10.27.

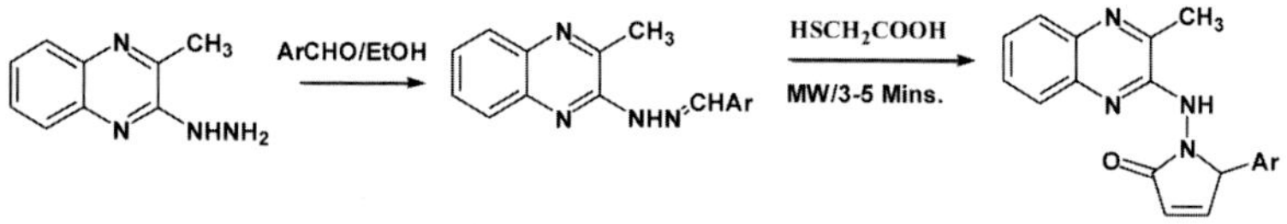

SCHEME 10.27 Synthesis of thiazolidinones by using microwave-assisted solvent-free condition.

Meshram et al. [33] was developed a zeolite-catalyzed the synthesis of 2-(2-chloroquinoline-3-yl)-3-substituted phenyl thiazolidin-4-ones. This reaction is scalable to milligram scale. These compounds have been synthesized in high yield by using zeolite 5A° and avoiding the use of any solvent under MWs Scheme 10.28.

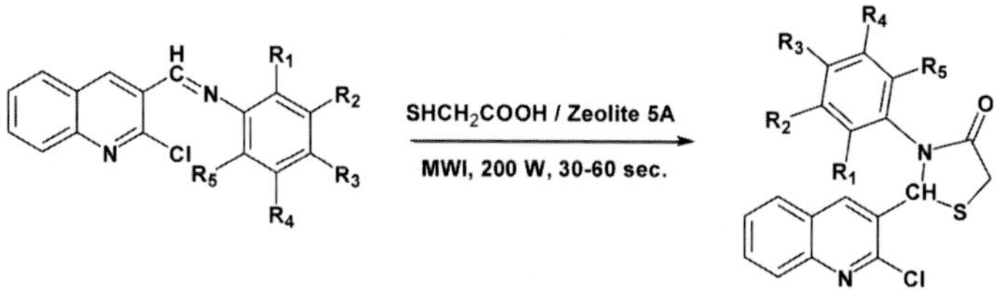

SCHEME 10.28 Synthesis of 2-(2-chloroquinoline-3-yl)-3-substituted phenyl thiazolidin-4-ones.

Aliphatic and aromatic nitro compounds were selectively reduced to their corresponding amino derivatives in good yields using formic acid and CeY zeolite under MW by K. Arya et al. [34]. Thus, the reduction of nitro compound can be accomplished with CeY zeolite instead of expensive Ni, Pt, Pd, etc., without affecting the reduction of any reducible substituents including halogen and carbonyl compounds (Scheme 10.29).

$$\underset{1}{X\text{-}NO_2} + \underset{2}{HCOOH/HCOONH_4} \xrightarrow[\text{MW, 3 - 10 min, 80 - 90\%}]{\text{CeY Zeolite}} \underset{3}{X\text{-}NH2}$$

SCHEME 10.29 Synthesis of amino derivatives by using formic acid and ceY zeolite under MWI.

Nalage and co-workers [35] was developed an efficient and green synthetic route for the synthesis of 2, 4, 6-triaryl-1*H*-imidazole in polyethylene glycol under MW irradiation in excellent yield Scheme 10.30.

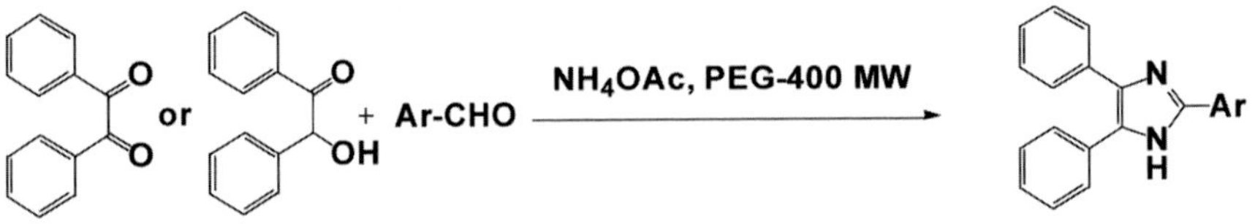

SCHEME 10.30 Synthesis of 2,4,6-triaryl-1*H*-imidazole in polyethylene glycol under microwave irradiation.

MW assisted, the three-component technique was applied for a mixture of an aromatic aldehyde, 2-amino-4*H*-benzothiopyrano[4,3-*d*]thiazole and meraptoacetic acid for the synthesis of thiazolidinones by Ma et al. [36] Scheme 10.31.

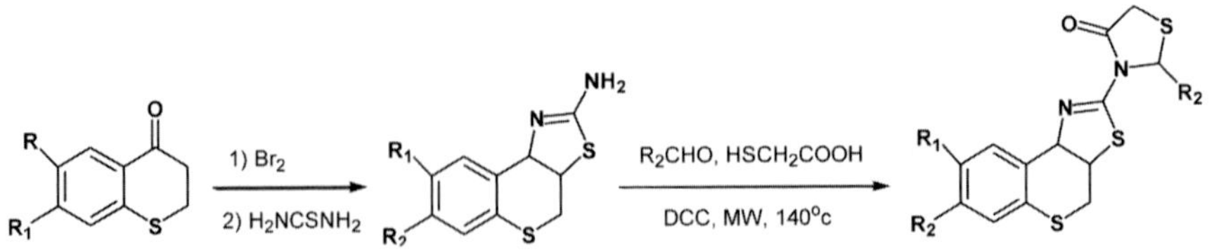

SCHEME 10.31 The synthesis of thiazolidinones by using microwave-assisted technique.

A series of thaizolidinones, 2-(arylphenyl)-3-(4-(4-(2-hydroxyphenyl)-2,3,5,6-tetramethyl1,7-diphenyl-1,2,7,7a-tetrahydro-dipyrazolo[3,4-b:4,'3'-e] pyridine8(4H,6H, 8aH) yl)phenyl)thiazolidin-4-one was synthesized by J. S. Meshram et al. [37]. These compounds were synthesized from new Schiff bases, under MW and conventional methods Scheme 10.32.

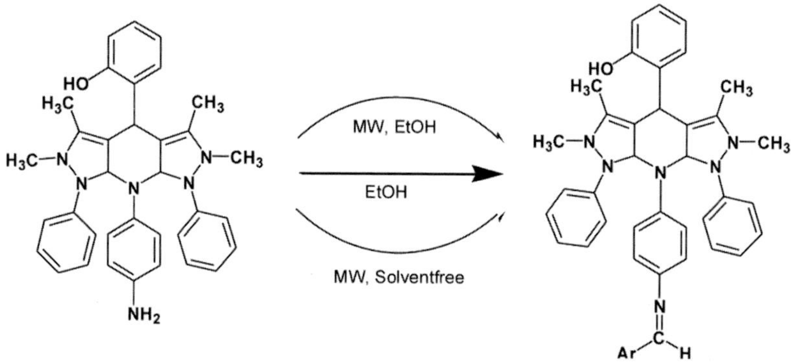

SCHEME 10.32 Synthesis of series of thiazolidinones under microwave method.

A series of 2-azetidinones and 4-thiazolidinones were successfully synthesized from 4,4'-sulphonyldianiline by Vibhute et al. [38]. Reaction showed that MW technique is remarkably successful in higher yield in less reaction time compared to conventional heating method Scheme 10.33.

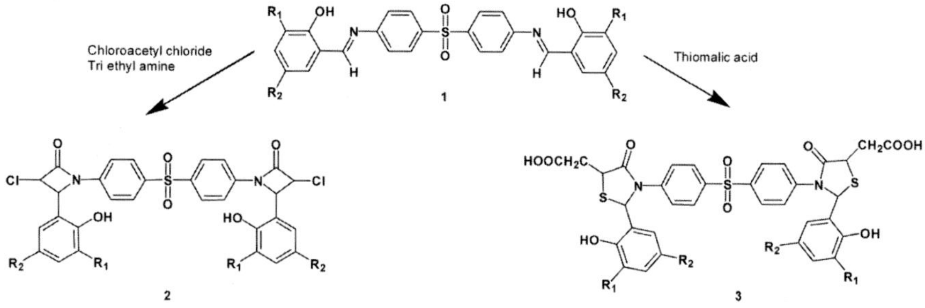

SCHEME 10.33 Synthesis of 2-azetidinones and 4-thiazolidinones under microwave technique.

Srivastava et al. [39] was reported synthesis of azetidinone derivatives using MW techniques Scheme 10.34.

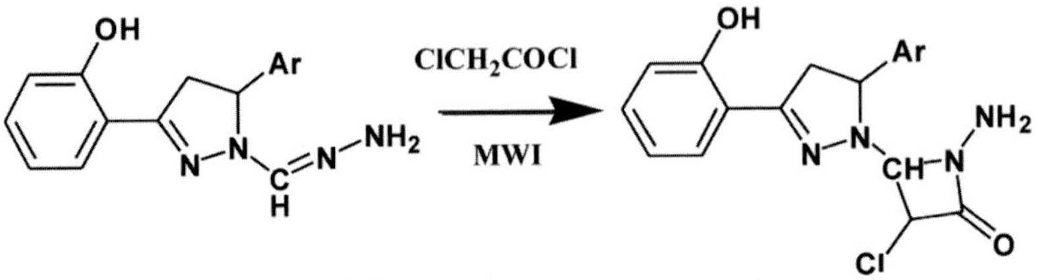

SCHEME 10.34 Synthesis of azetidinone derivatives using microwave techniques.

Dinnimath et al. [40] was synthesized series of Chloro, Fluoro, Phenyl substituted Azetidin-2-ones with different substituent's both by MW technique and conventional method Scheme 10.35.

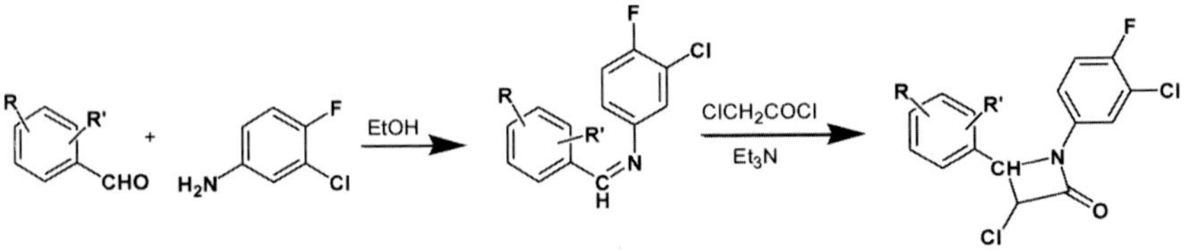

SCHEME 10.35 Synthesized series of chloro, fluoro, phenyl substituted azetidin-2-ones.

Zangade et al. [41] was reported the synthesis of 1,5-benzodiazepines by used MW irradiation technique was also adopted as an ecofriendly synthetic route. Clean reaction conditions, easy work-up, time-saving and higher yields are noteworthy advantages of used method Scheme 10.36.

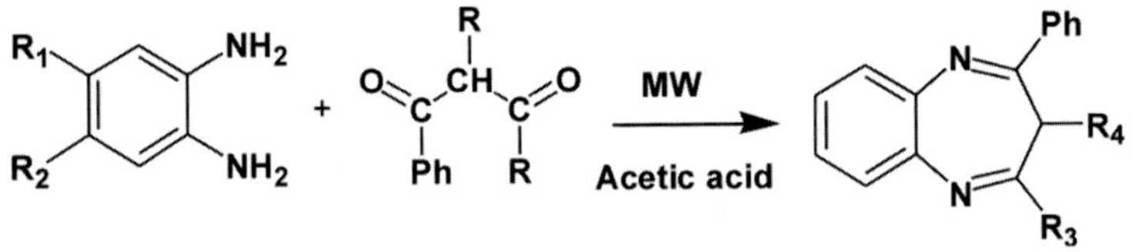

SCHEME 10.36 Synthesis of 1,5-benzodiazepines by used Microwave irradiation technique.

S. Das et al. [42] was applied MW technique catalyzed by *p*-toluene sulphonic acid effectively for preparation of symmetrical bis-imines from the reaction between dialdehydes and mono amines or diamines and mono aldehydes. Reduced reaction time, good yields, and simple workup are advantages of the methodolgy Scheme 10.37.

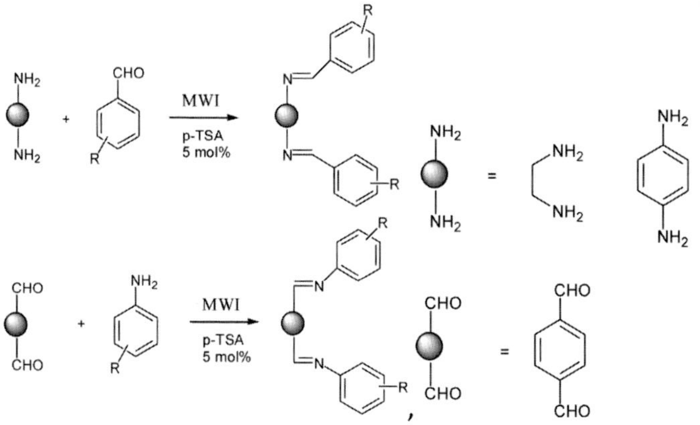

SCHEME 10.37 Preparation of symmetrical bis-imines catalyzed by *p*-toluene sulphonic acid.

Shinde et al. [43] were synthesized series of bis-Schiff bases starting with propane-1,3-diamine and different halogen substituted benzaldehydes under MW irradiation. Structures of the new synthesized bis-Schiff bases Scheme 10.38.

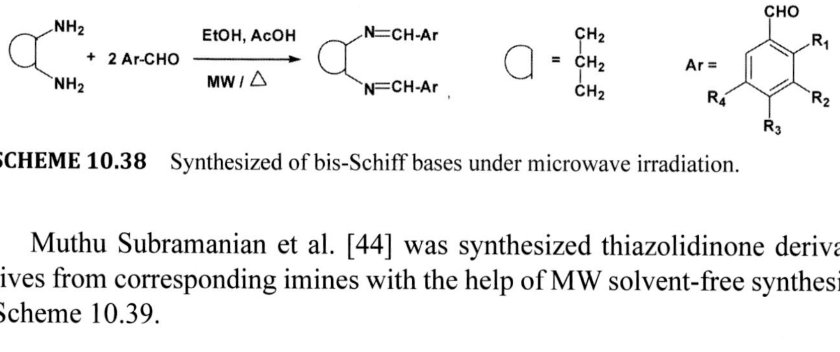

SCHEME 10.38 Synthesized of bis-Schiff bases under microwave irradiation.

Muthu Subramanian et al. [44] was synthesized thiazolidinone derivatives from corresponding imines with the help of MW solvent-free synthesis Scheme 10.39.

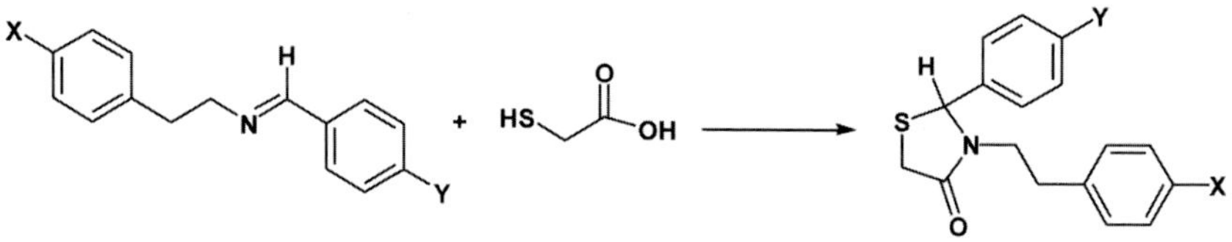

SCHEME 10.39 Synthesized thiazolidinone derivatives under microwave solvent-free.

Ethylenediamine based bioactive bis-ketiminies recently entered in MW assisted solvent-free synthetic route by Vikas and Bhagwan [45] Scheme 10.40.

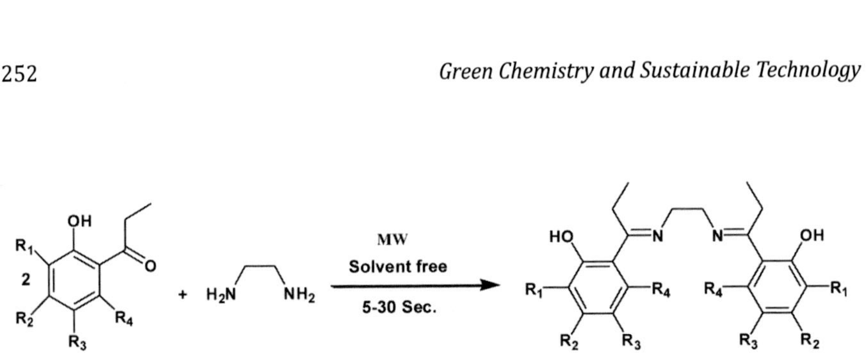

SCHEME 10.40 Synthesized of bioactive bis-ketiminies under microwave-assisted condition.

Borgaonkar et al. [46] were reported synthesis of novel ketimines using MW assisted solvent-free conditions which provide advantages such as shorter reaction time, solvent-free conditions, enhanced yields of products and ecofriendly one Scheme 10.41.

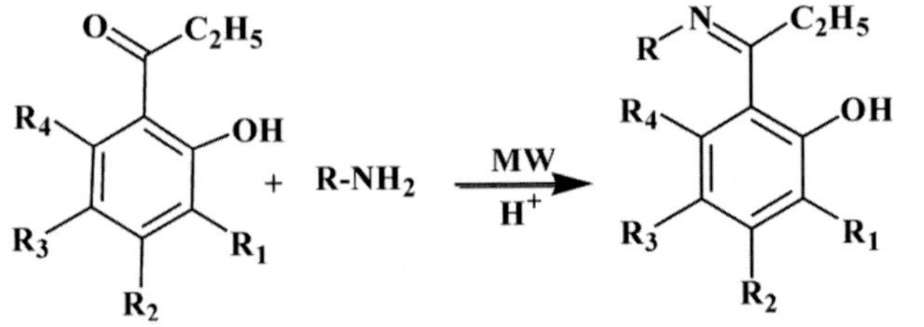

SCHEME 10.41 Synthesis of novel ketimines using microwave-assisted solvent-free conditions.

Reddy et al. [47] has been developed environmentally benign method to afford α-aminophosphonates using MW irradiation reaction with an amine, an aldehyde and diethyl phosphite catalyzed by Amberlyst-15 under solvent-free conditions Scheme 10.42.

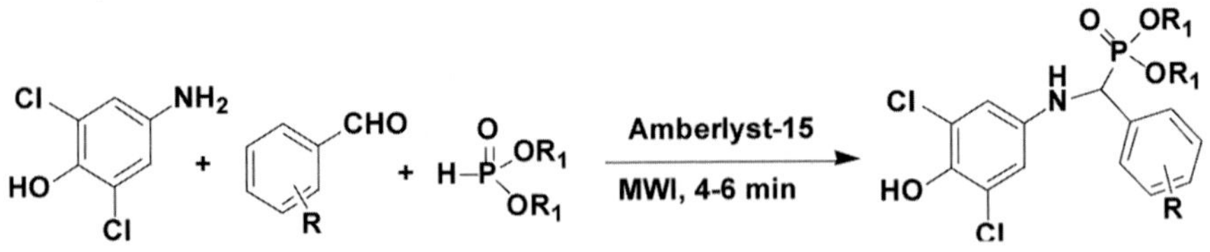

SCHEME 10.42 Synthesis of α-aminophosphonates using microwave irradiation.

Hangirgekar et al. [48] was described the synthesis of 2,4,5-triarylimidazole from benzil, aldehydes, and ammonium acetate in the presence of cupric chloride as a catalyst in a MW oven to give 85–91% yields Scheme 10.43.

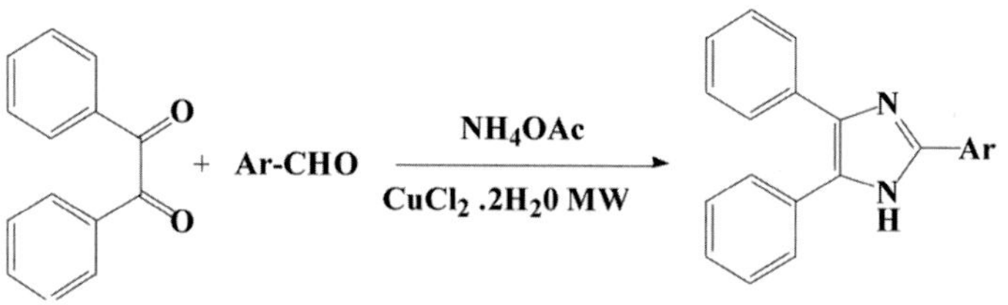

SCHEME 10.43 Synthesis of 2,4,5-triarylimidazoles in the presence of cupric chloride as a catalyst.

Safari et al. [49] were explored cyclo-condensation of benzil, aldehydes, and ammonium to synthesize 2,4,5-trisubstituted-1*H*-imidazole derivatives under MW irradiation with silica-supported $SbCl_3$ ($SbCl_3/SiO_2$) as a heterogeneous catalyst to gives 77–97% of yields Scheme 10.44.

SCHEME 10.44 Synthesize 2,4,5-trisubstituted-1*H*-imidazole derivatives under microwave irradiation.

Niralwad et al. [50] was proposed the silica sulfuric acid was found to be an efficient catalyst for the green synthesis of 2,4,5-triaryl-1*H*-imidazoles by the coupling of benzil/benzoin, aldehyde, and ammonium acetate under MW-irradiation at ambient temperature for appropriate time to furnish to excellent yield Scheme 10.45.

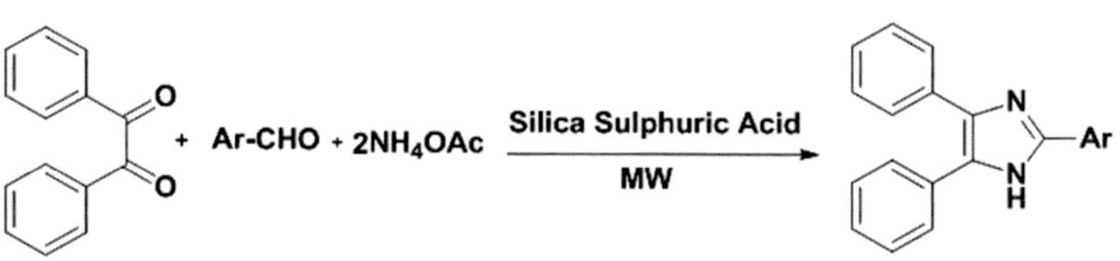

SCHEME 10.45 Synthesis of 2,4,5-triaryl-1*H*-imidazoles by using silica sulphuric acid catalyst under microwave-irradiation

Borgaonkar et al. [51] was synthesized novel thiazolidinones from ketimines using MW accelerated conditions Scheme 10.46.

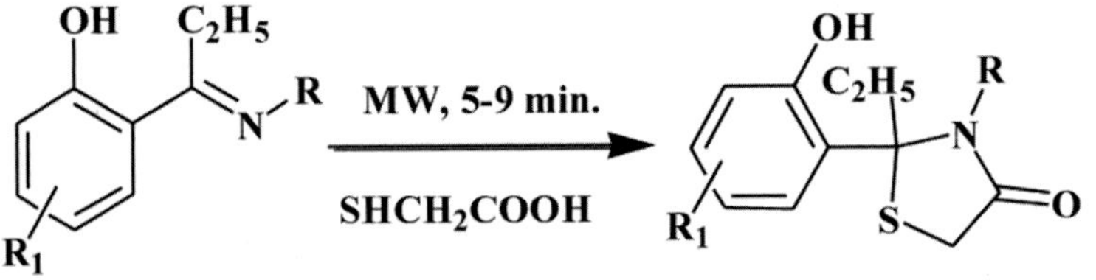

SCHEME 10.46 Synthesis of thiazolidinones from ketimines using microwave accelerated conditions.

Vikas and Bhagwan [52] was described the synthesis biologically eminent 1,5-benzodiazepines in good yields from hydroxypropiophenones and *o*-phenylenediamine by MW irradiation method Scheme 10.47.

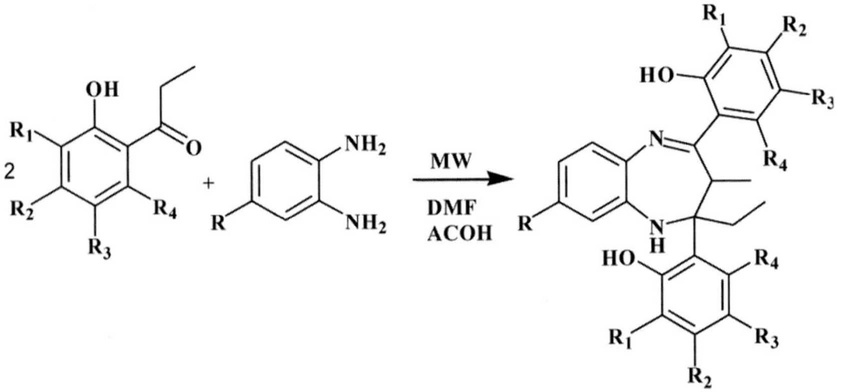

SCHEME 10.47 Synthesis of 1,5-benzodiazepines by using microwave irradiation method.

Radai et al. [53] has been accomplished a new, environmental friendly method to afford the corresponding α-hydroxyphosphonates by using Pudovik reaction between substituted benzaldehydes and dialkylphosphites Scheme 10.48.

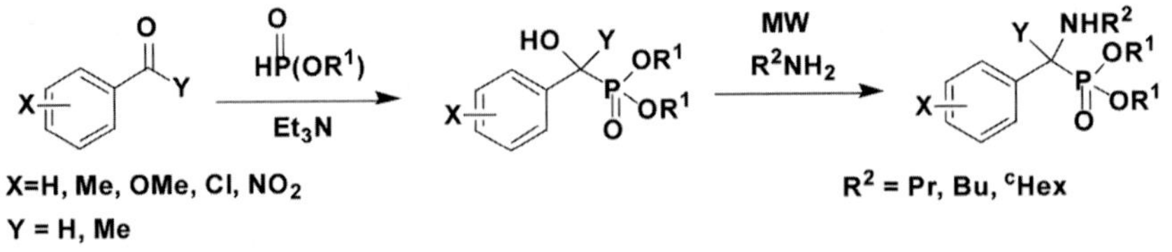

SCHEME 10.48 Synthesis of α-hydroxyphosphonates by using Pudovik reaction under MW.

Liu et al. [54] was reported non-fluorescent ND can be easily converted into FND via surface modified with aggregation-induced emission fluorophore having luminescent polymers through the MW assisted typical Diels-Alder (D-A) reaction.

Balint's and coworkers [55] were offered the synthesis and application of optically active α-aminophosphonate derivatives by Kabachnik-Fields reaction in the MW-induced condensation with paraformaldehyde and >P(O)H species, such as dialkylphosphites, ethyl phenyl-H-phosphinate and diphenylphosphine oxide to offer optically active α-aminophosphonates, α-aminophosphinate and α-amino phosphine oxide, or using a two equivalent quantity of paraformaldehyde and $(RO)_2P(O)H$ or $Ph_2P(O)H$, bis(phosphono methyl)amines and bis(phosphinoylmethyl) amine, respectively. The bis(phosphinoyl methyl) amine served as a precursor for an optically active bidentate P-ligand in the synthesis of a chiral platinum complex Scheme 10.49.

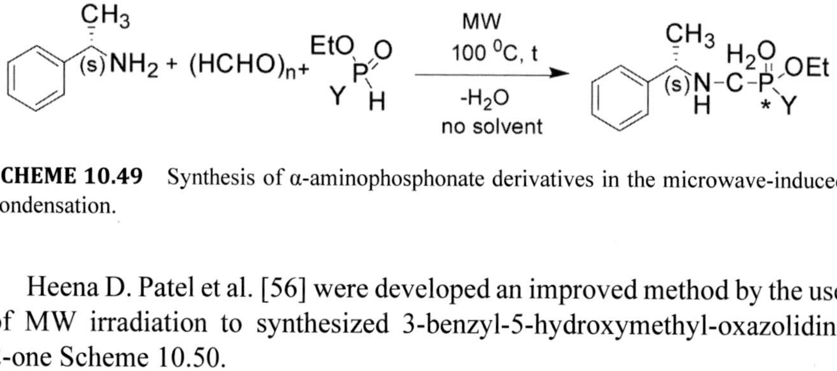

$$\text{(S)-PhCH(CH}_3\text{)NH}_2 + (HCHO)_n + \text{EtO(Y)P(O)H} \xrightarrow[\text{-H}_2\text{O, no solvent}]{\text{MW, 100 }^0\text{C, t}} \text{(S)-PhCH(CH}_3\text{)NH-CH}_2\text{-P}^*\text{(O)(OEt)Y}$$

SCHEME 10.49 Synthesis of α-aminophosphonate derivatives in the microwave-induced condensation.

Heena D. Patel et al. [56] were developed an improved method by the use of MW irradiation to synthesized 3-benzyl-5-hydroxymethyl-oxazolidin-2-one Scheme 10.50.

Benzyl amine + Epichlorohydrin → 3-benzyl-5-hydroxymethyl-1,3oxazolidin-2-one

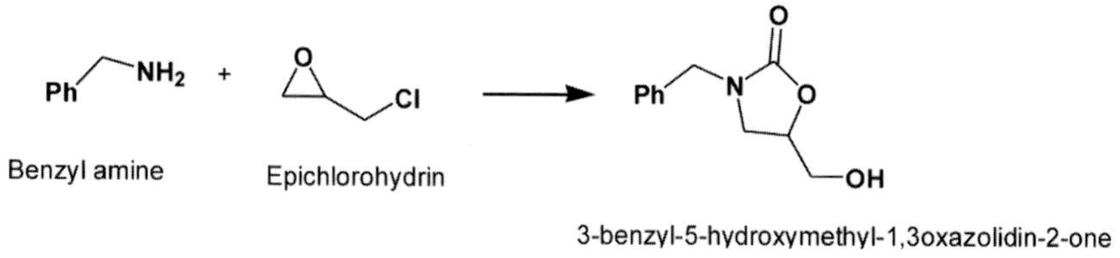

SCHEME 10.50 Synthesis of 3-benzyl-5-hydroxymethyl-oxazolidin-2-one by using microwave irradiation.

The Michael type addition of pyrroles and indoles bearing strongly electron-withdrawing N-protecting groups (e.g., arylsulfonyl-, benzoyl-) to methyl vinyl ketone have been reported by Daniel M. Ketcha et al. [57] in low to moderate yields using bismuth triflate as catalyst in conjunction with MW irradiation Scheme 10.51.

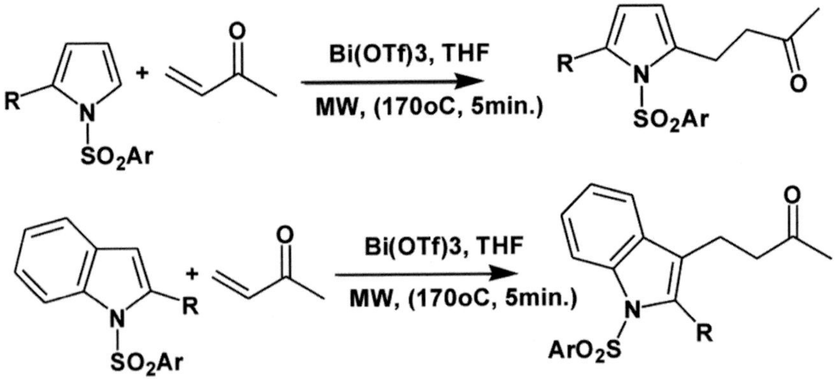

SCHEME 10.51 Synthesis of methyl vinyl ketone bismuth triflate as catalyst in microwave irradiation.

G. Shrivastava et al. [58] was synthesized New Schiff base (2-[(1*H*-benzimidazol-2-ylimino) methyl]-4,6-diiodophenol) by the condensation of aryl/hetero aromatic aldehyde (3,5-diiodosalicylaldehyde) with 2-amino benzimidazole under conventional and MW conditions Scheme 10.52.

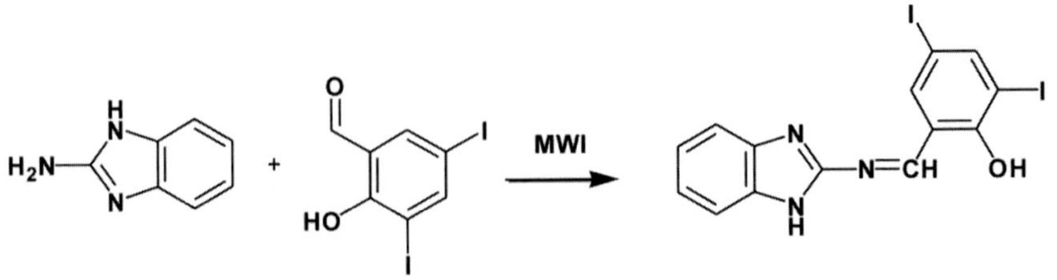

SCHEME 10.52 Synthesis of Schiff base (2-[(1*H*-benzimidazol-2-ylimino) methyl]-4,6-diiodophenol) derivatives.

Hosamani et al. [59] was synthesized coumarin-purine hybrids (3) under MW irradiation method. The MW approach has been demonstrated to be extremely fast, with enhanced reaction rate and shorter time, providing desired compounds in good to excellent yields and in high purity as compared to the conventional method Scheme 10.53.

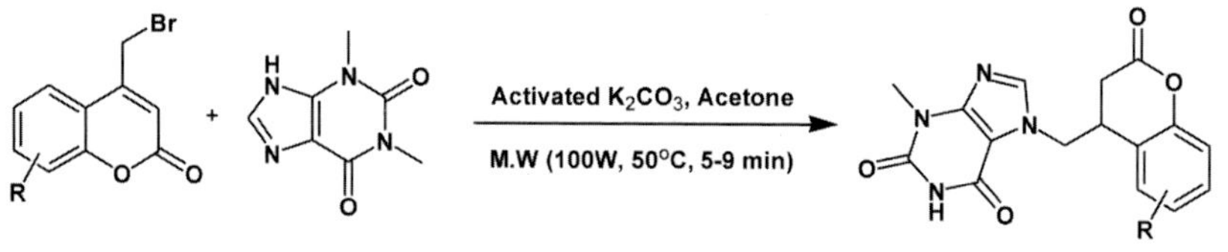

SCHEME 10.53 Synthesized coumarin-purine hybrids under microwave irradiation method.

10.4 CONCLUSION

Drug discovery is a challenging and demanding symphony, no one technique alone can whistle it. It was required to take a whole orchestra of strategies and techniques to play it well, in combination with strength. MW Chemistry has already been constantly combined with other enabling technologies and strategies such as multi-component reactions (MCRs), solid-phase organic synthesis, or combinatorial chemistry. The combination of a multidisciplinary approach with MW heating encourages scientists to initiate new and unexplored areas of complex pharmaceutical systems. This enabling technique has changed from the 'last sort' in early days to the 'first choice' nowadays for carrying out synthetic transformations requiring heat.

MW assisted green synthesis is a very good technique in the field of green chemistry which governs a flexible platform for many named organic reactions. This compiled review certainly provides an idea on MW assisted chemical synthesis, which will be beneficial for researchers.

KEYWORDS

- **chemical transformations**
- **energy source**
- **green chemistry**
- **microwave assisted techniques**
- **microwave technique**
- **synthetic chemistry**

REFERENCES

1. (a) Lidström, P., Tierney, J., Wathey, B., & Westman, J., (2001). *Tetrahedron, 57*, 9225. b) Larhed, M., & Hallberg, A., (2001). *Drug Discovery Today, 6*, 406.
2. Hayes, B. L., & Collins, M. J., (2004). *World Patent WO 04002617*.
3. Chen, J. J., & Deshpande, S. V., (2003). *Tetrahedron Lett., 44*, 8873.
4. Humphrey, C. E., Easson, M., A. M., Tierney, J. P., & Turner, N. J., (2003). *Org. Lett., 5*, 849.
5. Frattini, S., Quai, M., & Cereda, E., (2001). *Tetrahedron Letters*.
6. Ali, M., Bond, S. P., Mbogo, S. A., McWhinnie, W. R., & Watts, P. M., (1989). *J. Organometallic Chem., 371,* 11.
7. Dandia, A., Rani, B., & Saha, M., (1998). *Indian Journal of Chemical Technology, 5*, 159–162.
8. Balalaie, S., Arabanian, A., & Hashtroudi, M. S., (2000). *Monatsh. Chem., 131*, 945.
9. Khazaei, A., & Ghorbani-Vagei, R., (2002). *Molecules, 7*, 465.
10. Dandia, A., Sati, M., Arya, K., Sharma, R., & Loupy, A., (2003). *Chem. Pharm. Bull., 51*(10), 1137–1141.
11. Thanh, G. V., Lahrache, H., Loupy, A., Kim, I., Chang, D., & Junb, C., (2004). *Tetrahedron, 60*, 5539–5543.
12. Rao, A., Chimirri, A., Ferro, S., Monforte, A. M., Monforte, P., & Zappala, M., (2004). *Arkivoc, V*, 147–155.
13. Bolognese, A., Correale, G., Manfra, M., Lavecchia, A., Novellino, E., & Barone, V., (2004). *Org. Biomol. Chem., 2*, 2809.
14. Dandia, A., Arya, K., Sati, M., & Gautam, S., (2004). *Tetrahedron, 60*, 5253–5258.
15. Hazarkhani, H., & Karimi, B., (2004). *Synthesis*, p. 1239.
16. Kidwai, R. M., & Venkataramanan, R., (2004). *Chemistry of Heterocyclic Compounds, 40*(5), 631–634.
17. Mistry, K., & Desai, K. R., (2005). *Indian Journal of Chemistry, 44B*, 1452–1455.
18. Dandia, A., Arya, K., Khaturia, S., & Yadav, P., (2005). *Arkivoc., XIII*, pp. 80–88.
19. Dandia, A., Sati, M., Arya, K., Sarawgi, P., & Loupy, A., (2005). *Arkivoc., I*, pp. 105–113.
20. Mistry, K., & Desai, K. R., (2006). *Indian Journal of Chemistry, 45B*, 1762–1766.
21. Dandia, A., Sarawgi, P., Arya, K., & Khaturia, S., (2006). *Arkivoc., XVI*, pp. 83–92.
22. Khazaei, A., Aminmanesh, A., & Rostami, A., (2006). *J. Chem. Res., S*, p. 391.
23. Mu, X. J., Lei, M. Y., Zoua, J. P., & Zhang, W., (2006). *Tetrahedron Let., 47,* 1125–1127.
24. Xu, Y., Yan, K., Yan, B., Xu, Y., Yang, S., Xue, W., Hu, D., Lu, P., Ouyang, G., & Zhuo, C., (2006). *Molecules, 11,* 666–676.
25. Ghorbani-Vaghei, R., Veisi, H., & Amiri, M., (2007). *J. Chin. Chem. Soc., 54*, 1257.
26. Patel, J. A., Mistry, B. D., & Desai, K. R., (2008). *Indian Journal of Chemistry, 47B*, 1695–1700.
27. Tsoleridis, C. A., Pozarentzi, M., Mitkidou, S., & Stephanatou, S., (2008). *Arkivoc., XV*, pp. 193–209.
28. Bhattacharya, A. K., & Rana, K. C., (2008). *Tetrahedron Lett., 49,* 2598–2601.
29. Kundu, A., Shakil, N. A., Saxena, D. B., Pankaj, J. K., & Walia, S., (2009). *J. of Envir. Sci. and Health Part B, 44*, 428–434.
30. Raval, J. P., Patel, H. V., Patel, P. S., Patel, N. H., & Desai, K. R., (2009). *Asian J. Research Chem., 2*(2), 171–177.

31a. Bhusnure, O. G., Zangade, S. B., Chavan, S. B., & Vibhute, Y. B., (2010). *J. Chem. Pharm. Res., 2*(6), 234–243.
31b. S. M. Lonkar, S. S. Mokle, A. Y. Vibhute, Y. B. Vibhute, *Der Chemica Sinica,* **2010**, *1*(2), 119–124.
32. Kondapalli, G. C., Rao, V. S., Reddy, A. S., Sunandini, R., & Kumar, V., S. A., (2010). *Bull. Korean Chem. Soc., 31*(5), 1219–1222.
33. Tiwari, V., Meshram, J., & Ali, P., (2010). *Der Pharma Chemica, 2*(3), 187–195.
34. Arya, K., & Dandia, A., (2010). *Journal of the Korean Chemical Society, 54*(1), 55–58.
35. Nalage, S. V., Kalyankar, M. B., Patil, V. S., Bhosale, S. V., Deshmukh, S. U., & Pawar, R. P., (2010). *Open Cata. J., 3,* 58.
36. Ma, Z., Zhang, X., Bai, L., Zheng, Y., & Yang, G., (2011). *Modern Applied Science, 5*(3), 207–212.
37. Meshram, J. S., Chopde, H. N., Pagadala, R., & Jetti, V., (2011). *Int. J. of Pharma and Bio. Sci., 2*(1), 667–676.
38. Bhusnure, O. G., Mokle, S. S., Nalwar, Y. S., & Vibhute, Y. B., (2011). *Journal of Pharmaceutical and Biomedical Sciences, 06*(08), 1–7.
39. Malhotra, G., Gothwal, P., & Srivastava, Y., K., (2011). *Der. Chemica. Sinica., 2*(3), 47–50.
40. Dinnimath, B. M., Hipparagi, S. M., & Gowda, M., (2011). *International Journal of Pharmacy and Technology, 3*(4), 3792–3801.
41. Zangade, S., Mokle, S., Chavan, S., & Vibhute, Y. B., (2011). *Orbital Elec. J. Chem., 3*(3), 144–149.
42. Das, S., Das, V. K., Saikia, L., & Thakur, A. J., (2012). *Green Chemistry Letters and Reviews, 5*(3), 457–474.
43. Shinde, A., Zangade, S., Chavan, S., & Vibhute, Y., (2014). *Org. Commun., 7*(2), 60–67.
44. Amutha, C., Saravaranan, S., & Muthusubramanian, S., (2014). *Indian Journal of Chemistry, 53B*, 377–383.
45. Borgaonkar, V. V., & Patil, B. R., (2014). *Der Pharma Chemica, 6*(1), 120–124.
46. Borgaonkar, V. V., & Patil, B. R., (2014). *Int. J. of Chem. Tech Res., 6*(7), 3535–3539.
47. Reddy, G. S., Rao, K. U. M., Sundar, C. S., Sudha, S. S., Haritha, B., Swapna, S., & Reddy, C. S., (2014). *Arabian Journal of Chemistry, 7,* 833–838.
48. Hangirgekar, S. P., Kumbhar, V. V., Shaikh, A. L., & Bhairuba, I. A., (2014). *Der. Pharma. Chemica., 6,* 164.
49. Naseh, S. J., Zarnegar, Z., & Akbari, Z. J., (2014). *Taibah Univ. Sci., 8*, 323.
50. Niralwad, K. S., Ghorade, I. B., & Shingare, M. S., (2014). *Inter. J. Scie. Rese.,* 2277.
51. Borgaonkar, V. V., & Patil, B. R., (2015). *Am. J. Pharm. Tech Res., 5*(2), 488–495.
52. Borgaonkar, V. V., & Patil, B. R., (2015). *Heterocyclic Letters, 5*(2), 261–268.
53. Radai, Z., Kiss, N. Z., Mucsi, Z., & Keglevich, G., (2016). *21st International Conference on Phosphorus Chemistry (ICPC-2016)* (Vol. 191, pp. 11–12). doi: 10.1080/10426507.2016.1213261.
54. Liu, X., Wan, Q., Zhao, Z., Liu, J., Zhang, Z., Deng, F., Liu, M., Wen, Y., & Zhang, X., (2017). *Materials Chemistry and Physics, 197*, 256–265.
55. Balint, E., Tajti, A., Kalocsai, D., Matravolgyi, B., Karaghiosoff, K., Czugler, M., & Keglevich, G., (2017). *Tetrahedron, 73*, 5659–5667.
56. Patel, H. D., Al Shahwan, M., Ashames, A., Islam, M. W., & Bhandare, R., (2018). *J. Pharm. Sci. and Res., 10*(11), 2814–2817.
57. Miles, K. C., Kohane, B. J., Southerland, B. K., & Ketcha, D. M., (2018). *Arkivoc, Part IV*, pp. 149–157.

58. Shrivastava, G., & Shrivastava, M., (2018). *International Journal of Pharmaceutical Sciences and Drug Research, 10*(4), 293–296.
59. Mangasuli, S. N., Hosamani, K. M., & Managutti, P. B., (2019). *Heliyon, 5*, e01131. doi: 10.1016/j.heliyon.2019.e01131.

CHAPTER 11

Green Synthesis, Characterization, and Biological Studies of 1,3,4-Thiadiazole Derived Schiff Base Complexes

AJAY M. PATIL,[1] RAVINDRA S. SHINDE,[2] B. R. SHARMA,[3] and SUNIL R. MIRGANE[4]

[1]*Department of Chemistry, Pratishthan Mahavidyalaya, Paithan–431107, Maharashtra, India*

[2]*Department of Chemistry, Dayanand Science College, Latur–413512, Maharashtra, India*

[3]*Department of Physics, Pratishthan Mahavidyalaya, Paithan–431107, Maharashtra, India*

[4]*Department of Chemistry, J.E.S. College, Jalna–431203, Maharashtra, India*

ABSTRACT

A green route for the synthesis of heterocyclic ligand and their Schiff base metal complexes are followed to get minimization of time and solvent and high yield. Metal complexes derivatives of 2,4-diiodo-6-(((5-mercapto-1,3,4-thiadiazol-2-yl)imino)methyl)phenol, HL with the metal ions Cu(II), Zn(II) and Cd (II) have been successfully synthesized by using microwave (MW) irradiation method. The complexes formed by the MW irradiation method are characterized quantitatively and qualitatively by using the following technique, i.e., elemental analysis, UV-Vis spectroscopy, FT-IR, mass spectroscopy, ^{1}H & ^{13}C-NMR, magnetic susceptibility, and molar conductance. The spectral study, all the complexes formed as a dimeric in structure and the center metal are six-coordinated with octahedral geometry. The preliminary *in vitro* antibacterial and antifungal screening activity explains that most of the metal complexes showed moderate to excellent activity against tested bacterial strains *S. aureus* and *B. subtilis* and fungal strains *A. Niger* and *F. Oxysporum* higher compared to the ligand, HL by using Kirby-Bauer disc diffusion method.

11.1 INTRODUCTION

The fundamentals of microwave (MW) chemistry were studied and reported in the post-war age for the applications such as decomposing simple organic compounds in a MW discharge, wet ashing techniques, and dehydrating inorganic salts, of biological and organic materials [1–5]. On MW heating technique, first two studies of organic synthesis were reported by Geyde et al. and Giguere et al. in 1986 [6, 7]. These reports mainly focused on the significant rate acceleration for reactions; MW heating reactions are still attracting the consideration of synthetic chemists. MW irradiation results in the dielectric heating, which is differed from classical thermal heating. Presently, the basic principles and theory of dielectric heating, the characteristic "superheating effect," and the experimental results, i.e., considerably reduce reaction time, are extensively studied, and are known characteristics of MW-assisted synthesis [8–11]. Several transient-metal sandwich compounds were synthesized by MW irradiation technique in a closed system at high temperature and pressure. Ali et al. reported that ligand exchange reaction by MW irradiation technique in that same year [12]. Thereafter, many efforts have been made to synthesize metal complexes by MW irradiation. Research areas and studies on metal complexes are increasing significantly, due to their attractive prospect of playing a key role in functional materials Chemistry. Metal complexes as a photosensitizers and photocatalysts in photofunctional materials are now contributing to providing solutions to energy and environmental problems [13, 14]. Asymmetric synthesis and Coupling reactions are promoted by organometallic compounds [15, 16]. Organic light-emitting device materials, which are among the most well-known subjects in materials science in the first decade of the 21st century, and metal complexes are the candidate materials for practical applications [17]. Schiff bases obtain from an amino ($-NH_2$) and carbonyl (>C=O) groups are an important class of ligands that coordinate to metal ions through azomethine nitrogen and have been studied largely in the field of chemistry. In azomethine derivatives, the C=N linkage is important group for the biological activity, different azomethine has been indicated that remarkable anticancer antibacterial, antimalarial, and antifungal activities [18] 1,3,4-Thiadiazole derivatives having strong aromaticity of this ring system due to that shows beneficial and interesting biological activity, which leads to better in vivo stability and a lack of toxicity for higher vertebrates, including humans beings. When distinct functional groups present on the Thiadiazole ring that interacts with biological receptors possessing magnificent properties. Except for Few antibacterial sulfonamides, no longer used clinically, but that possessed historical importance, the most interesting and important examples are given by 5-amino-l,3,4-thiadiazole-derivatives [19]. Further addition, the chemistry and the applications of these new Schiff bases thiadiazoles ring containing moieties derivatives could be extensively studied by coordinating to diverse metal ions. As a result, the structural activity relationship study of 1,3,4-thiadiazoles ring might be enlarged in the future [20–26]. The prominent features of MW approach are shorter reaction times, simple reaction conditions, and increases in yield. The application of MW oven for the synthesis of organic compounds has proven to be an effectual, safe, and eco-friendly method [27].

Present study the synthesis and characterization of new complexes Cu (II), Zn (II) and Cd (II) of 2,4-diiodo-6-(((5-mercapto-1,3,4-thiadiazol-2-yl) imino)methyl)phenol. Moreover, the preliminary *in vitro* antibacterial and antifungal screening activities of the metal complexes formed are carried out and the results are reported herein. The MW technique would be one of the most powerful tools in the synthesis of these compounds.

11.2 PRESENT WORK

The present scheme involves the synthesis of ligand 2,4-diiodo-6-(((5-mercapto-1,3,4-thiadiazol-2-yl)imino)methyl)phenol illustrated in Scheme 11.1 and its Metal Complexes of Cu(II), Zn(II), Cd(II) shown in Figure 11.1.

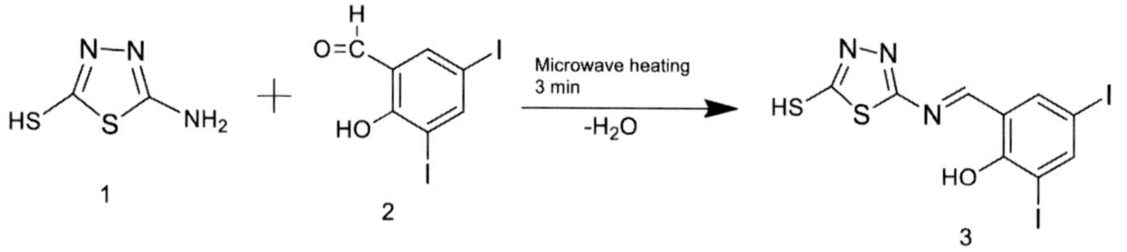

SCHEME 11.1 2,4-diiodo-6-(((5-mercapto-1,3,4-thiadiazol-2-yl)imino)methyl)phenol.

11.3 EXPERIMENTAL PART

All the chemicals of the analytical grade. All salts are metal nitrates, i.e., $Cu(NO_3)_2.3H_2O$, $Zn(NO_3)_2.6H_2O$, $Cd(NO_3)_2.4H_2O$ were purchase from Sigma-

Aldrich and used without further purification. The 3,5-diiodo-2-hydroxybenzaldehyde and 5-amino-1,3,4-thiadiazole-2-thiol from Sigma-Aldrich and Alfa Aesar used without further purification. The dist.ethanol used for the synthesis of metal complexes and ligand (Sigma-Aldrich). The IR Spectra recorded on Perkin Elmer Spectrometer in range 4000–400 cm^{-1} KBr pellets. ^{1}H and ^{13}CNMR Spectra were recorded on BRUKER AVANCE III HD NMR 500 MHz spectrophotometer. The room temperature (RT) magnetic moments by Guoy's method in B.M. Electronic Spectra using DMSO on Varian Carry 5000 Spectrometer. The molar Conductance measurements in DMSO having 1×10^{-3} concentration on Systronics conductivity bridge at RT. Mass Spectra were recorded on Bruker IMPACT HD. Elemental analysis (C,H,N) were carried out by using perkin Elmer 2400 elemental analyzer.

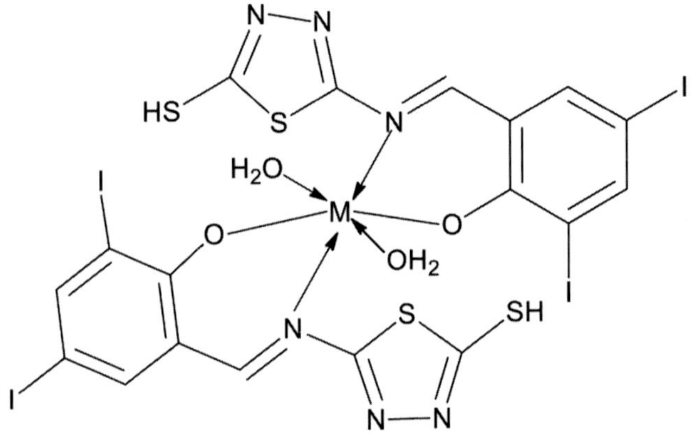

FIGURE 11.1 Proposed Structures of metal complexes M: Cu (II), Zn (II) and Cd (II).

11.3.1 BIOLOGICAL ACTIVITY

The Schiff base and their metal complexes assessed in vitro for their antibacterial activity against two Gram-Positive bacteria, viz., *B. Subtilis*; *S. aureus,* two fungal strains *A. niger* and *F. Oxysporum* by Kirby-Bauer disc diffusion method [28]. The fungal and bacterial strains sub-cultured on Potato Dextrose Agar and Nutrient Agar. The stock solution (1 mg mL^{-1}) was prepared in DMSO solution. The stock solution again diluted by using sterilized water to dilution in 500 ppm. The bacteria were subculture in agar medium and disc were kept incubated for 37°C at 24 hrs. The standard antibacterial drug Miconazole and Ciprofloxacin was screen under the same condition for comparison. The activity was measure and calculated by a zone of inhibition (mm) surrounding discs. The experimental value compares with standard drug value Miconazole for the Antifungal activity and Ciprofloxacin for the antibacterial activity of ligand and Metal Complexes.

11.3.2 MICROWAVE (MW) SYNTHESIS OF 2,4-DIIODO-6-(((5-MERCAPTO-1,3,4-THIADIAZOL-2-YL)IMINO) METHYL) PHENOL (HL_1)

The target compounds were prepared in the following steps: The equimolar (1:1) ratio of 3,5-diiodo-2-hydroxybenzaldehyde (1) (1.91 g, 0.01 mol) with 5-amino-1,3,4-thiadiazole-2-thiol (2) (1.33 g, 0.01 mol) mixed thoroughly in a grinder. The reaction mixture was then irradiated by the MW oven by taking 3–4 ml of dry ethanol as a solvent. The reaction was completed in a short time (3–4 min) with higher yields. The resulting product was then recrystallized with Ethanol and Ether then dried in air. The progress of the reaction and purity of the product was monitored by TLC using silica gel method (3) Scheme 11.1.

Yield: 85% M.P: 208°C
IR(KBr cm^{-1}): 3320 (νOH/H_2O-stretch), 1267 (ν C-O), 1651 (ν C=N), 751 νC-S-C,1479(ν -C=N-N=C).
^1H NMR (500 MHz, DMSO-d_6): δ 11.50 (s,1H,Ar-OH), 8.92 (s,1H,CH=N),7.18–8.10 (s,2H,Ar-CH), 13.21 (s,1H,SH).
MS(*ESI*) *m/z*: 489 $[M+1]^+$.
Anal. Data: Calcd. for $C_9H_5Cl_2N_3OS_{2:}$ C, 22.10; H, 1.03; N, 8.59;S, 13.11%, I; 51.89%.
Found: C, 23.50; H, 1.60; N, 7.98;S, 14.02%, I; 52.03%.

11.3.3 MICROWAVE (MW) SYNTHESIS OF METAL COMPLEXES

The metal salts copper(II) nitrate trihydrate ($Cu(NO_3)_2.3H_2O$) (0.241 g, 0.001 mol) and The ligand HL (0.612 g, 0.001 mol) were mixed in 1:2 (metal:- ligand) ratio in a grinder. The reaction mixture was then irradiated by the MW oven by taking 3–5 ml of dry ethanol as a solvent. The reaction was completed in a short time (5–6 min) with higher yields. A colored product obtain washed with ethanol, filtered, and recrystalized with ethanol and ether. Similarly, Zn $(NO_3)_2.6H_2O$, $Cd(NO_3)_2.4H_2O$ metal complexes was prepared by a similar method (Figure 11.1). The progress of the reaction and purity of the product was monitored by TLC using silica gel G.

11.4 RESULTS AND DISCUSSION

The ligand Scheme 11.1 and its transition metal complexes (Figure 11.1) of 2,4-diiodo-6-(5-mercapto-1,3,4-thiadiazol-2-yl)imino methyl phenol are stable at RT in solid-state. The ligand is soluble in organic solvent DMF, DMSO, and metal complexes are easily soluble in DMSO. The synthesized complexes having 1:2 (metal to ligand) stoichiometric ratios. The physical and analytical data shown in Table 11.1. The Spectral analysis shows the formation of ligand and its metal complexes.

11.4.1 IR SPECTRA

The IR spectra of 2,4-diiodo-6-(((5-mercapto-1,3,4-thiadiazol-2-yl)imino) methyl)phenol (HL) SB ligand and its complexes are listed in Table 11.2. The IR data of HL and complexes as compared to know the coordination site of ligand to metal in the formed chelate ring. SB ligand having the most characteristic bands at 3310–3325 cm^{-1} ν(O-H), 1620–1670 cm^{-1} ν(C=N, azomethine) and 1230–1290 cm^{-1} ν(C-O). The ligand showed bands at 3312–3322 and 1336–1348 cm^{-1} because of the stretching and deformation of the phenolic –OH [29, 30] these are missing in the spectra of the metal complexes indicates that deprotonation of the hydroxyl group (-OH) and co-ordination through phenolic oxygen of Aromatic ring of SB Ligand. The band 1,641–1,650 cm^{-1} due to the azomethine group of the Schiff bases ligand have shifted to lower frequency (1,610–1,634 cm^{-1}) after complexes formation, shows that donation of electron from nitrogen of the azomethine group to the empty vacant d-orbital metal ion [31, 32]. The phenolic λ (C-O) stretching vibration frequency at 1259–1265 cm^{-1} in SB Ligand shifted to higher frequencies (18–32 cm^{-1}) in the metal complexes. This shift proved that involvement of oxygen in the C-O-M bond in metal complexes. The occurrence of broad bands around at (3,372–3,450 cm^{-1}) in the spectra of complexes may be due to water molecules attached to metal complexes [33]. Two new bands appearing in the low frequency range 518–580 cm^{-1} and 458–490 cm^{-1} are due to ν(M-O) and ν(M-N) respectively. The ν (C-S-C) at 750–754 cm^{-1} of the Thiadiazole ring remain intact Thiadiazole ring of SB Ligand directly does not take part in the donation of electron to the metal in metal complexes [34].

TABLE 11.1 Analytical Data and Physical Properties of Ligand and its Metal Complexes

Compounds	Empirical Formula	Mol. Wt.	Color	M.P (°C)	Yield (%)	C	H	N	S	I	M
Ligand (HL)	Ligand (HL)	489	Light Yellow	208 $_o$C	82%	23.50	1.60	7.98	14.02	52.03	–
						(22.10)	(1.03)	(8.59)	(13.11)	(51.89)	
Cu(II) Complex	Cu(II) Complex	1075	Green	>300	88%	19.02	1.33	7.92	11.62	47.0	5.22
						(20.10)	(1.12)	(7.81)	(11.92)	(47.19)	(5.91)
Zn(II) Complex	Zn(II) Complex	1077	Lemon Yellow	>300	86%	20.20	1.08	7.98	11.83	47.25	5.97
						(20.06)	(1.12)	(7.80)	(11.90)	(47.11)	(6.07)
Cd(II) Complex	Cd(II) Complex	1124	Gray	>300	89%	19.08	1.11	7.96	11.81	47.56	6.02
						(19.22)	(1.08)	(7.47)	(11.40)	(45.14)	(5.69)

TABLE 11.2 Infrared Spectra of the Schiff Base and Complexes in Cm^{-1}

Compounds	νOH/H_2O	νC-O	νC=N	νM-N	νM-O	νC-S-C	ν-C=N-N=C	νN-N
Ligand	3320	1267	1651	—	—	751	1479	1028
Cu(II) Complex	3471	1271	1623	484	584	757	1470	1039
Zn(II) Complex	3445	1291	1624	481	575	754	1487	1033
Cd(II) Complex	3444	1315	1604	484	549	753	1453	1026

11.4.2 ^{1}H NMR AND ^{13}C NMR SPECTRA

The spectra of ligands singlet at δ 7.18–8.10 ppm due to aromatic proton while azomethine (>C=N-) proton resonate at singlet δ 8.92 ppm the phenolic -OH has singlet at δ 11.50 ppm and Thiadiazole containing -SH group singlet at δ 13.21 ppm [35, 36].

^{13}C NMR of SB Ligand, peak at δ159–165 ppm imine peak (>C=N-) 183 ppm Due to C-SH bonding in Thiadiazole.123.96–141.75 ppm due to aromatic carbon, 155–170 ppm peak due to Ar-OH group [37] shown in Table 11.3.

TABLE 11.3 ^{1}H NMR Signals (δ, ppm) and Their Assignments

Compound	^{1}H NMR Signals (δ, ppm) and their assignments
Ligand (HL)	11.50 (s,1H,Ar-OH), 8.92 (s,1H,CH=N),7.18–8.10 (s,2H,Ar-CH), 13.21 (s,1H,SH)

11.4.3 MASS SPECTRA

SB ligand peak at m/z 489 is M+H peak at 100% intensity this peak support to the structure Confirm the formation of ligand [29].

11.5 MAGNETIC SUSCEPTIBILITY AND MOLAR CONDUCTANCE

The magnetic susceptibility measure at r.t. Metal complexes of Zinc(II) and Cadmium (II) are Diamagnetic in nature. while Copper Cu (II) is paramagnetic in nature. Molar conductance of metal complexes was observed at RT in 1×10^{-3}M DMSO Sol. Show negligible molar conductance in range 8–12 $ohm^{-1}cm^{2}mol^{-1}$ results show in Table 11.4. All Synthesize metal complexes are non-electrolytic in nature [38, 39].

11.5.1 ELECTRONIC ABSORPTION SPECTRA

The electronic spectral data of the ligands and metal complexes in DMSO shown in Table 11.4. Geometry and Nature of the ligand field around the metal ion has now been concluding from the electronic spectral data of metal complexes and SB ligand. The band at 220–315 nm is because of transition in the benzene ring of the ligand. Band of free ligands 320–382 nm due to transition for phenolic -OH and azomethine groups (-C=N-). These band shifts longer wavelength shows formation of ligand to metal complexes [39, 40]. The spectral data of the complexes band at 422–500 nm assigned to charge transfer transition (L→M) from ligands to metal [17, 35]. The M.M value for Cu (II) complexes is 1.82 B.M is close to octahedral spectra of complex shows two bands at 360 nm and 560 nm shows that octahedral geometry of Cu (II) complex [41]. Spectra of Zn (II) complexes show band 266 nm, 370–432 nm not showing d-d transition suggest that octahedral geometry [42]. Spectra of Cd (II) shows two peaks at 326 nm and 306–360 ligand to metal donation with diamagnetic suggest that octahedral geometry [43].

TABLE 11.4 Electronic Spectral Magnetic and Molar Conductance Data

Compounds	Wavelength in nm	Magnetic Moment μ_{eff} (BM)	Molar Conductance (ohm^{-1} cm^2 mol^{-1})
Ligands (HL)	260, 370	–	6.68
$C_{18}H_{12}Cl_4CuN_6O_4S_4$	270–320, 360, 560	1.82	8.1
$C_{18}H_{12}Cl_4ZnN_6O_4S_4$	266, 370–432	Diamagnetic	10.1
C18H12Cl4CdN6O4S4	265, 306–360	Diamagnetic	12

11.6 ANTIMICROBIAL ACTIVITY

Antimicrobial activity In vitro of the SB ligand and their corresponding metal complexes on two gram-positive *bacteria S. aureus and B. Subtlis* two fungi *A. niger* and *F. Oxysporum* was screened shown in Table 11.5. The observation shows that Cu (II) Zn (II) and Cd (II) Complexes shows more bactericidal and fungicidal activity as compare to ligand hence activity of metal complexes increases due to chelation. Chelation enhances the penetration of metal complexes in lipid membrane of microbes and blocks the binding site enzymes of microorganism there are some other factors increases the activity of complexes, i.e., M-L bond length solubility, lipophilicity/hydrophilicity, and Conductivity [34, 44–48].

TABLE 11.5 Antimicrobial Activity of Ligand and its Metal Complexes

Compounds	Antibacterial Activity				Antifungal Activity			
	S. aureus		B. subtilis		A. niger		F. oxysporum	
	Diameter of Inhibition Zone in mm	% Activity Index	Diameter of Inhibition Zone in mm	% Activity Index	Diameter of Inhibition Zone in mm	% Activity Index	Diameter of Inhibition Zone in mm	% Activity Index
	500 ppm	500 ppm	500 ppm	500 ppm	500 ppm	500 ppm	500 ppm	500 ppm
Ligands (HL)	500 ppm	500 ppm	500 ppm	500 ppm	500 ppm	500 ppm	500 ppm	500 ppm
Cu Complex	23	68	22	67	23	74	19	70
Zn Complex	26	76	26	79	26	84	21	78
Cd Complex	22	64	22	67	19	61	18	67
Ciprofloxacin (Standard)	19	56	21	64	21	68	12	44
Miconazole (Standard)	34	100	33	100	–	–	–	–

11.7 CONCLUSION

In the present research studies, our efforts were to synthesize and characterize some compounds from the MW methods. MW method has been considered a green chemical route. These synthesized compounds were characterized by various physicochemical and spectral analyses. In the course of MW assisted synthesis, it was observed that the reaction time decreased from hours to minutes and availability of the product with better yield as compared to the conventional method. The use of solvent is also minimized. The ligand, 2,4-diiodo-6-(((5-mercapto-1,3,4-thiadiazol-2-yl)imino)methyl) phenol HL was coordinated to three different metal ions via oxygen and nitrogen atoms to afford the corresponding complexes. All the complexes were six coordinated and exhibited octahedral geometry. Preliminary *in vitro* antibacterial study indicated that all the complexes obtained showed moderate to excellent activity against the tested bacterial strains and slightly higher activity compared to the ligand, HL.

KEYWORDS

- **1,3,4-thiadiazole antifungal activity**
- **antibacterial activity**
- **electronic absorption spectra**
- **magnetic susceptibility**
- **metal complexes**
- **molar conductance**

REFERENCES

1. Streitwieser, A., & Ward, H. R., (1963). Organic compounds in microwave discharge. II. initial studies with toluene and related hydrocarbons. *J. Am. Chem. Soc., 85*(5), 539–542.
2. Liu, S. W., & Wightman, J. P., (1971). Decomposition of simple alcohols, ethers and ketones in a microwave discharge. *J. Appl. Chem. Biotechnol., 21*(6), 168–172.
3. Hesek, J. A., & Wilson, R. C., (1974). Use of microwave oven in in-process control. *Anal. Chem., 46*, 1160.
4. Abu-Samra, A., Morris, J. S., & Koirtyohann, S. R., (1975). Wet ashing of some biological samples in a microwave oven. *Anal. Chem., 47*(8), 1475–1477.

5. Barrett, P., Davidowski, L. J., Penaro, K. W., & Copeland, T. R., (1978). Microwave oven-based wet digestion technique. *Anal. Chem.*, *50*(7), 1021–1023.
6. Gedye, R., Smith, F., Westaway, K., Ali, H., Baldisera, L., Laberge, L., & Rousell, J., (1986). The use of microwave ovens for rapid organic synthesis. *Tetrahedron Lett.*, *27*(3), 279–282.
7. Giguere, R. J., Bray, T. L., Duncan, S. M., & Majetich, G., (1986). Application of commercial microwave ovens to organic synthesis. *Tetrahedron Lett.*, *27*(41), 4945–4948.
8. Mingos, D. M. P., & Baghurst, D. R., (1991). Applications of microwave dielectric heating effects to synthetic problems in chemistry. *Chem. Soc. Rev.*, *20*(1), 1–47.
9. Galema, S. A., (1997). Microwave chemistry. *Chem. Soc. Rev.*, *26*(3), 233–238.
10. Gabriel, C., Gabriel, S., Grant, E. H., Halstead, B. S. J., & Mingos, D. M. P., (1998). Dielectric parameters relevant to microwave dielectric heating. *Chem. Soc. Rev.*, *27*(3), 213–223.
11. de la Hoz, A., Díaz-Ortiz, Á., & Moreno, A., (2005). Microwaves in organic synthesis. Thermal and non-thermal microwave effects. *Chem. Soc. Rev.*, *34*(2), 164–178.
12. Ali, M., Bond, S. P., Mbogo, S. A., McWhinnie, W. R., & Watts, P. M., (1989). Use of a domestic microwave oven in organometallic chemistry. *J. Organomet. Chem.*, *371*, 11–13.
13. O'Regan, B., & Grätzel, M., (1991). A low-cost, high-efficiency solar cell based on dye-sensitized colloidal TiO_2 films. *Nature*, *353*, 737–740.
14. Konno, H., Ishii, Y., Sakamoto, K., & Ishitani, O., (2002). Synthesis, spectroscopic characterization, electrochemical and photochemical properties of ruthenium (II) polypyridyl complexes with a tertiary amine ligand. *Polyhedron*, *21*(1), 61–68.
15. Kitamura, M., Ohkuma, T., Inoue, S., Sayo, N., Kumobayashi, H., Akutagawa, S., Ohta, T., Takaya, H., & Noyori, R., (1988). Homogeneous asymmetric hydrogenation of functionalized ketones. *J. Am. Chem. Soc.*, *110*(2), 629–631.
16. Miyaura, N., & Suzuki, A., (1995). Palladium-catalyzed cross-coupling reactions of organoboron compounds. *Chem. Rev.*, *95*(7), 2457–2483.
17. Baldo, M. A., O'Brien, D. F., You, Y., Shoustikov, A., Sibley, S., & Thompson, M. E., (1998). Highly efficient phosphorescent emission from organic electroluminescent devices. *Nature, 395*(6698), 151–154.
18. Annapoorani, S., & Krishnan, C., (2013). Synthesis and spectroscopic studies of trinuclear n4schiff base complexes. *International, J. Chem. Tech. Res.*, *5*(1), 180–185.
19. Kornis, G., (1984). l,3,4-Thiadiazoles in comprehensive heterocyclic chemistry. In: Katritzky, R., (ed.), *Pergamon. Press*, *6*(Part 4B), 545–578.
20. Elzahany, E., Hegab, K., Khalil, S., & Youssef, N., (2008). Characterization and biological activity of some transition metal complexes with Schiff bases derived from 2-formylindole, salicylaldehyde, and N-amino rhodanine. *Aust. J. Basic Appl. Sci.*, *2*(2), 210–220.
21. Gaber, M., Mabrouk, H. E., & Al-Shihry, S. S., (2001). Complexing behavior of naphthylidene sulfamethazine Schiff base ligand towards some metal ions. *Egypt. J. Chem.*, *44*(4–6), 191–200.
22. Hadizadeh, F., & Vosoogh, R., (2008). Synthesis of α-[5-(5-Amino-1,3,4-thiadiazol-2-yl)-2-imidazolylthio]acetic acids. *J. Heterocyclic Chem.*, *45*, 1–3.
23. Jarrahpour, A., Khalili, D., DeClercq, E., Salmi, C., & Brunel, J. M., (2007). Synthesis, antibacterial, antifungal, and antiviral activity evaluation of some new bis-Schiff bases of Isatin and their derivatives. *Molecules*, *12*, 1720–1730.
24. Taggi, A., Hafez, A., Wack, H., Young, B., Ferraris, D., & Lectka, T., (2002). A practical methodology for the catalytic, asymmetric synthesis of β-lactams results from the

development of a catalyzed reaction of ketenes (or their derived zwitterionic enolates) and imines using benzoylquinine as chiral catalyst and proton sponge as the stoichiometric base. *J. Am. Chem. Soc.*, *124*, 6626–6637.
25. Salimon, J., Salih, N., Ibraheem, H., & Yousif, E., (2010). Synthesis of 2-N-Salicylidene-5-(substituted)-1,3,4-thiadiazole as potential antimicrobial agents. *Asian J. Chem.*, *22*(7), 5289–5296.
26. Yousif, E., Majeed, A., Al-Sammarrae, K., Salih, N., Salimon, J., & Abdullah, B., (2013). Metal complexes of Schiff base: Preparation, characterization and antibacterial activity. *Arabian Journal of Chemistry*, 170–176.
27. Arjmand, F., Mohani, B., & Parveen, S., (2005). Synthesis, antibacterial, antifungal activity and interaction of CT-DNA with a new benzimidazole derived Cu (II) complex. *Bioinorg. Chem. App.*, *2*, 112–119.
28. Bauer, A. W., Perry, D. M., & Kirby, (1959). Single-disk antibiotic-sensitivity testing of staphylococci: An analysis of technique and results. *AMA Arch Intern Med.*, *104*(2), 208–216.
29. Patil, A. M., Dhokte, A. O., Sharma, B. R., Sunil, R., & Mirgane, S. R., (2019). Metal complexes of Schiff base: Synthesis, characterization and antibacterial activity. *J. Biol. Chem. Chron.*, 5(3), 07–12.
30. Nakamoto, K., (2006). Infrared and Raman spectra of inorganic and coordination compounds. *Handbook of Vibrational Spectroscopy*. John Wiley & Sons, Ltd. New York.
31. Temel, H., Ilhan, S., Aslanoglu, M., Kilic, A., & Tas, E., (2006). Synthesis, spectroscopic and electrochemical studied of novel transition metal complexes with quadridentate Schiff base. *J Chin. Chem. Soc.*, *53*, 1027–1031.
32. Shukla, D., Gupta, L. K., & Chandra, S., (2008). Spectroscopic studies on chromium(III), manganese(II), cobalt(II), nickel(II) and copper(II) complexes with hexadentate nitrogen–sulfur donor [N_2S_4] macrocyclic ligand. *Spectrochim Acta.*, *71A*, 746–750.
33. Mohamed, G. G., Omar, M. M., & Hindy, A. M., (2008). Metal complexes of Schiff bases: Preparation, characterization and biological activity. *Turk J. Chem.*, *30*, 361–382.
34. Neelakantan, M. A., Marriappan, S. S., Dharmaraja, J., Jeyakumar, T., & Muthukumaran, K., (2008). Spectral, XRD, SEM, and biological activities of transition metal complexes of polydentate ligands containing thiazole moiety. *Spectrochimica Acta Part A: Molecular and Bimolecular Spectroscopy*, *7*(2), 628–635.
35. Maurya, R. C., Patel, P., & Rajput, S., (2003). Synthesis and characterization of mixed-ligand complexes of Cu(II), Ni(II), Co(II), Zn(II), Sm(III), and U(VI)O2, with a Schiff base derived from the sulfa drug sulfamerazine and 2,2′-bipyridine. *Synthesis and Reactivity in Inorganic and Metal-Organic Chemistry*, *33*(5), 801–816.
36. Rastogi, R. B., Yadav, M., & Singh, K., (2001). Synthesis and characterization of molybdenum and tungsten complexes of 1-Aryl-2,4-dithiobiurets. *Synth. React. Inorg. Met.-Org. Chem.*, *31*(6), 1011–1022.
37. Abd-Elzaher, M. M., Moustafa, S. A., Labib, A. A., Mousa, H. A., Ali, M. M., & Mahmoud, A. E., (2012). Synthesis, characterization, and anticancer studies of ferrocenyl complexes containing thiazole moiety. *Applied Organometallic Chemistry*, *26*(5), 230–236.
38. Sampal, S. N., Thakur, S. V., Rajbhoj, A. S., & Gaikwad, S. T., (2017). Synthesis, characterization and antimicrobial screening of 1,3-Dione with their metal complexes. *Asian J. Chem.*, *30*(2), 398–400.

39. Ucan, S. Y., Ucan, M., & Bedrettin, (2005). Synthesis and characterization of new Schiff bases and their cobalt(II), nickel(II), copper(II), zinc(II), cadmium(II) and mercury(II) complexes. *Synth. React. Inorg. Met. Org. Nano-Metal Chem.*, *35*, 417–421.
40. Turan, N., & Sekerci, M., (2009). Metal complexes of Schiff base derived from terephthalaldehyde and 2-amino-5-ethyl-1,3,4-thiadiazole synthesis, spectral and thermal characterization. *Synthesis and Reactivity in Inorganic, Metal-Organic, and Nano-Metal Chem.*, *39*(10), 651–657.
41. Khedr, A. M., & Marwani, H. M., (2012). Synthesis, spectral, thermal analyses and molecular modeling of bioactive Cu(II)-complexes with 1,3,4-thiadiazole Schiff base derivatives. Their catalytic effect on the cathodic reduction of oxygen. *Int. J. Electrochem. Sci.*, *7*(5), 10074–10093.
42. Turan, N., & Şekerci, M., (2010). Synthesis, characterization and thermal behavior of some Zn(II) complexes with ligands having 1,3,4-thiadiazole moieties. *Heteroatom Chemistry*, *21*, 14–23.
43. Turan, N., & Şekerci, M., (2009). Synthesis and spectral studies of novel Co(II), Ni(II), Cu(II), Cd(II), and Fe(II) metal complexes with N-[5′-Amino-2,2′-bis(1,3,4-thiadiazole)-5-yl]-2-hydroxybenzaldehyde Imine (HL). *Spectroscopy Letters*, *42*(5), 258–267.
44. Chohan, Z. H., Munawar, A., & Supuran, C. T., (2001). Transition metal ion complexes of Schiff bases synthesis, characterization, and antibacterial properties. *Metal Based Drugs,* 8, 137–143.
45. Hanna, W. G., & Moawad, M. M., (2005). Synthesis, characterization and antimicrobial activity cobalt(II), nickel(II) and copper(II) complexes with new asymmetrical Schiff base ligands derived from 7-formalin-substituted diamine-sulphoxine and acetylacetone. *Transit Metal Chem.*, *26*(6), 644–651.
46. Singh, V. P., & Katiyar, A., (2008). Synthesis, characterization of some transition metal(II) complexes of acetone p-amino acetophenone salicyloyl hydrazone and their antimicrobial activity. *Bio. Metals*, *21*(4), 491–501.
47. Azam, F., Singh, S., Khokhra, S. L., & Prakash, O., (2007). Synthesis of Schiff bases of naphtha [1,2-d] thiazol-2-amine and metal complexes of 2-(20-hydroxy)benzylidene amino naphthothiazole as potential antimicrobial agent. *J. Zhejiang. Univ. Sci.*, *8*(6), 446–452.
48. Chohan, Z. H., (1999). Ni(II), Cu(II) and Zn(II) metal chelates with some thiazole derived Schiff-bases: Their synthesis, characterization and bactericidal properties. *Metal Based Drugs*, *6*, 75–79.

CHAPTER 12

Study on the Effect of Global Warming and Greenhouse Gases on Environmental System

ONKAR K. JOGDAND

Department of Environmental Science, Deogiri College, Station Road, Aurangabad – 431005, Maharashtra, India,
E-mail: onkar.jogdand@gmail.com

ABSTRACT

The Greenhouse effect is a prominent influence in observance the Earth warm for the reason that it keeps some of the planet's warmth that would otherwise seepage from the atmosphere out to space. The homework explosion on the Greenhouse gases (GHGs) and their impression on Global warming. Without the greenhouse effect, the Earth's average global temperature would be much cooler and life on Earth as we distinguish it would be intolerable. GHGs include water vapor, CO_2, methane, nitrous oxide (N_2O) and additional gases. Carbon dioxide (CO_2) and other GHGs turn like a comprehensive, gripping Infra-Red radiation. Worldwide warming is a significant environmental issue which is rapidly attractive a part of well-liked civilization. This manuscript provides an explanation of the science linked with this important issue. Past evidence for long-ago climate change is deliberated. The dissimilarity between climate and weather is highlighted. The physics of the greenhouse consequence and the perception of GHGs are presented. The concepts of radioactive forcing of climate change and worldwide warming latent as measures of the complete and relative strengths of GHGs are discussed. Global warming, the improvement of the natural greenhouse effect caused by emissions linked with person behavior of GHGs such as carbon dioxide, nitrous oxide, methane, and halogenated compounds (e.g., CFCs, and SF6), is described. Techniques used to representation history, present, and prospect climate are discussed. The models are based upon essential well recognized technical principals and incorporate the existing thoughtful of the multifaceted comment and couplings among the atmosphere, hydrosphere, and biosphere. Projections of prospect universal climate change from up to date. With our present level of scientific perceptive, we anticipate that over the next century the planet will warm considerably.

12.1 INTRODUCTION

The Earth has an usual surface temperature pleasantly between the boiling point and freezing point of water, consequently appropriate for our kind of life, cannot be illuminated by just propositioning that earth orbits at just the accurate planetary since the sun to absorb just the right amount of solar radiation. The troposphere in planet Venus would goods hellish, Venus-like conditions on planet Earth; the Mars troposphere would leave earth trembling in a Martian-type deep freeze. Moreover, parts of the earth's atmosphere act as protecting blanket of just the correct thickness, getting appropriate solar energy to keep the global average temperature in an entertaining range. The extraterrestrial blanket is too cracked, and the Venusian comprehensive is way too thick. The 'blanket' as specified here, is termed as a gathering of impressive gases called greenhouse gases (GHGs) based on the knowledge that the gases also capture heat similar to the glass walls of a greenhouse. These gases, mostly water vapor, carbon dioxide, methane, and nitrous oxide, all perform as effective global insulators. The discussion of inbound and outward-bound radiation that warms the earth is often referred to as the greenhouse effect because a greenhouse works in much the same way. Inbound ultraviolet (UV) radiation effortlessly passes through the crystal walls of a greenhouse and is immersed by the plants and hard surfaces inside. Weaker Infrared (IR) radiation, however, has struggled passing through the glass walls and is trapped inside, that is, reheating the greenhouse. This consequence lets humid plants embellishment inside a greenhouse, even during a cold winter. The greenhouse inspiration expansions the infection of the Earth by deceiving temperature in our air. This remembers the fever of the Earth higher than it would be if direct boiler by the Sun was the only source of heating. When sunlight influences the surface of the Earth, some of it is fascinated which warms the crushed and some jumps back to planetary as heat. Most GHGs that are in the atmosphere attract and then transmit some of this heat back towards the Earth.

The greenhouse effect is a primary factor in keeping the Earth heartfelt because it retains some of the planet's temperature that would otherwise seepage from the air out to space. In fact, without the conservatory effect, the Earth's usual global temperature would be much remoter and life on Earth as we identify it would not be possible. The difference between the Terrain's actual average temperature 14°C (57.2°F) and the probable effective fever just with the Sun's radiation –19°C (–2.2F) bounces us the strong point of the greenhouse effect, which is 33°C. The greenhouse effect is a usual procedure that is a lot of years old. It plays a life-threatening role in an adjustable the overall temperature of the Earth. The greenhouse consequence was first revealed by Joseph Fourier in 1827, experimentally confirmed by John Tyndall in 1861 [1], and estimated by Svante Arrhenius in 1896 [2] has printed a newspaper on (A outline on the effects of anthropogenic GHGs emissions from power generation and energy consumption). It stretches evidence about notwithstanding the forthcoming difficult energy context in the mainstream of nations in the world, global change in ecological self-possession causing from power group and energy consumption situation is quickly flattering a worldwide alarming phenomenon. The current study intensive on the greenhouse effect and their impacts on global warming.

The international heating is an important green distress which is fast gratifying a portion of respected development. This chapter offers an elucidation of the punishment associated with this important matter. The physics of the conservatory consequence and the theory of GHGs are accessible. The thoughts of weather change and global heating potential as actions of the absolute and comparative assets of GHGs. Worldwide warming up, an enhancement of the usual hot house significance caused by releases connected among human life actions of conservatory gases such as carbon dioxide, methane, nitrous oxide, and halogenated complexes (e.g., CFCs, and SF_6), is described. The fundamental healthy old-fashioned systematic chiefs assimilate the present sensitive of the multifaceted reaction and join between the atmosphere, hydrosphere, and biosphere. Through our present level of methodical understanding, we look advancing to that over the subsequent century the world will moderate considerably. Worldwide heating is an enduring rise in the average high temperature of the earth's weather system an aspect of universal heating shown by temperature capacities and by multiple properties of the heating. Though earlier life periods also experienced chapters of heating the term typically refers to the experimental and ongoing increase in average air and ocean fevers since 1900 caused mainly by releases of GHGs in the modern manufacturing cost-cutting measure.

Climate and temperature have reflective impact on livelihood organisms on the earth. Environmental systems have evolved greater than geographical time scales to suit the general climate. The past 10 to 20 years have brought alarming support that human activities may origin important changes in prospect global climate. Worldwide warming" is a present subject familiar to hundreds of millions of community on or after corner to corner of the world. We supply in this an overview of the recent state of understanding in relation to GHGs and global warming. Weather is the condition of the environment (temperature, humidity, precipitation, wind, cloud cover, etc.) in a particular position at a meticulous time; it fluctuates very much and is infamously complex to forecast. Climate is the time-averaged conditions in a given biological region. Weather is a temporal and spatial common and is automatically much more conventional than weather. Thus, the average temperature during a specified month in particular vicinity (climate) can be predicted with some assurance, however, the temperature at a given moment and position (weather) is much more complex to forecast. Type of weather varies from month to month, season to season, and year to year. Statistically important changes in typical weather taking place more than a time level of decades or longer constitute "climate change."

A GHG is a gas that absorbs and emits radiant energy within the thermal IR range. Energy emitted from the sun ("solar radiation") is concentrated in a region of short wavelengths including visible light. Much of the short wave solar radiation travels down through the Earth's atmosphere to the surface virtually unimpeded. Some of the solar radiation is reflected straight back into space by clouds and by the earth's surface. Much of the solar radiation is absorbed at the earth's surface, causing the surface and the lower parts of the atmosphere to warm that's why the GHGs cause the greenhouse effect. Without GHGs, the average temperature of Earth's surface would be about –18°C (0°F) rather than the present average of 15°C (59°F). Human activities since the beginning of the Industrial Revolution (around 1750) have produced a 45% increase in the atmospheric concentration of carbon dioxide (CO_2), from 280 ppm in 1750 to 406 ppm in early 2017 [3]. This increase has occurred despite the uptake of more than half of the emissions by various natural "sinks" involved in the carbon cycle. The vast majority of anthropogenic carbon dioxide emissions (i.e., emissions produced by human activities) come from combustion of fossil fuels, principally coal, oil, and natural gas, with additional contributions coming from deforestation, changes in land use, soil erosion and agriculture (including livestock). Should GHG emissions continue at their rate in 2017 [4], Earth's surface temperature could exceed historical values as early as 2047, with potentially harmful effects on ecosystem, biodiversity, and human lively hoods. At current emission rates, temperatures could increase by 2°C, which the United Nations' IPCC designated as the upper limit to avoid "dangerous" levels, by 2036. Global warming occurs when carbon dioxide (CO_2) and other air pollutants and GHGs collect in the atmosphere and absorb sunlight and solar radiation that have bounced off the earth's surface. Normally, this radiation would escape into space but these pollutants, which can last for years to centuries in the atmosphere, trap the heat and cause the planet to get hotter. That's what's known as the greenhouse effect. In the United States, the burning of fossil fuels to make electricity is the largest source of heat-trapping pollution, producing about 2 billion tons of CO_2 every year. Coal-burning power plants are by far the biggest polluters. The country's second-largest source of carbon pollution is the transportation sector, which generates about 1.7 billion tons of CO_2 emissions a year. An estimation of GHG emissions from organic and conventional farming systems. It gives information on carbon (C) and nitrogen (N) fluxes in the system soil-plant-animal-environment. The model couples the balancing of C, N, and energy fluxes with the target to estimate the climate-relevant CO_2, CH_4 and N_2O sources and sinks of farming systems.

12.2 GLOBAL WARMING

Global warming is a long-term rise in the average temperature of the Earth's climate system, an aspect of climate change shown by temperature measurements and by multiple effects of the warming. Though earlier geological periods also experienced episodes of warming, the term commonly refers to the observed and continuing increase in average air and ocean temperatures since 1900 caused mainly by emissions of GHGs in the modern industrial economy. In the modern context, the terms global warming and climate change are commonly used interchangeably, but climate change includes both global warming and its effects, such as changes to precipitation and impacts that differ by region. Many of the observed warming changes since the 1950s are unprecedented in the instrumental temperature record and in historical and paleoclimate proxy records of climate change over thousands to millions of years. Imagine you live in a timber shack in Alaska. It's chilly up there, so you build yourself a huge log fire and pile on all the wood you can find. To start with, the fire seems a great idea especially since it's so cold outside. The shack warms up slowly, but predictably, and it's soon pretty cozy. Since the shack is much warmer than the atmosphere and ground that surround it, it loses heat quite quickly. If the fire supplies heat at the same rate as the shack loses it, the shack stays at roughly the same temperature. But if you make the fire too big, the shack will get hotter and hotter. Before long, you'll start feeling uncomfortable. You might wish you'd never made the fire so big in the first place. But once it's burning, there's nothing you can do to stop it. The shack will keep getting hotter long after you stop piling wood on the fire. Global warming is working a bit like this. Thanks to a variety of things that people do, Earth is getting slightly warmer year by year. It's not really warming up noticeably at least not in the short term. In fact, since 1900, the whole planet has warmed up only by around 0.8°C. By the end of the 21st century, however, global warming is likely to cause an increase in Earth's temperature of around 2–5°C. There is a 75% chance of a 2–3° warming and a 50% chance of a 5° warming, and scientists agree that the warming is most likely to be around 3°. Now even a 5° warming might not sound like much to worry about, but 5° is roughly how much difference there is between the worlds as it is today and as it was during the last Ice Age. In other words, when we came out of the Ice Age, the planet warmed by 5° over about 5000 years. Modern climate change threatens to produce the same amount of a warming in as little as a century. Once something as big as a planet starts to warm up, it's very hard to slow down the process and almost impossible to stop it completely. Global warming means Big Trouble. Global warming is caused by a phenomenon known as the greenhouse effect. A greenhouse (or glasshouse) is good for growing things because it traps heat inside and stays hotter than the atmosphere around it.

12.3 GREEN HOUSE GASES

In order, the most abundant GHGs in Earth's atmosphere are:

- Water vapor (H_2O);
- Carbon dioxide (CO_2);
- Methane (CH_4);
- Nitrous oxide (N_2O);
- Ozone (O_3);
- Chlorofluorocarbons (CFCs);
- Hydro fluorocarbons (incl. HCFCs and HFCs).

Global warming potential (GWP) and atmospheric lifetime for major GHGs are summarized in (Table 12.1).

TABLE 12.1 Global Warming Potential and Atmospheric Lifetime for Major Greenhouse Gases

Greenhouse Gas	Chemical Formula	Global Warming Potential, 100-Year Time Horizon	Atmospheric Lifetime (Years)
Carbon Dioxide	CO_2	1	100*
Methane	CH_4	25	12
Nitrous Oxide	N_2O	298	114
Chlorofluorocarbon-12 (CFC-12)	CCl_2F_2	10,900	100
Hydrofluorocarbon-23 (HFC-23)	CHF3	14,800	270
Sulfur Hexafluoride	SF_6	22,800	3,200
Nitrogen Trifluoride	NF_3	17,200	740

12.3.1 WATER VAPOR

The most abundant GHG, but importantly, it acts as a feedback to the climate. Water vapor increases as the Earth's atmosphere warms, but so does the possibility of clouds and precipitation, making these some of the most important feedback mechanisms to the greenhouse effect.

12.3.2 CARBON DIOXIDE (CO_2)

A minor but very important component of the atmosphere, carbon dioxide is released through natural processes such as respiration and volcano eruptions and through human activities such as deforestation, land-use changes, and burning fossil fuels. Humans have increased atmospheric CO_2 concentration by more than a third since the Industrial Revolution began. This is the most important long-lived "forcing" of climate change.

12.3.3 METHANE

A hydrocarbon gas produced both through natural sources and human activities, including the decomposition of wastes in landfills, agriculture, and especially rice cultivation, as well as ruminant digestion and manure management associated with domestic livestock. On a molecule-for-molecule basis, methane is a far more active GHG than carbon dioxide, but also one which is much less abundant in the atmosphere.

12.3.4 NITROUS OXIDE

A powerful GHG produced by soil cultivation practices, especially the use of commercial and organic fertilizers, fossil fuel combustion, nitric acid production, and biomass burning.

12.3.5 CHLOROFLUOROCARBONS (CFCS)

Synthetic compounds entirely of industrial origin used in a number of applications, but now largely regulated in production and release to the atmosphere by international agreement for their ability to contribute to the destruction of the ozone layer. They are also GHGs.

12.3.6 SOURCES OF GREENHOUSE GAS (GHG)

In recent times, one of the major sources of GHG emission is from water resource recovery facilities (wastewater treatment plants (WWTPs). WWTPs are recognized as one of the larger minor sources of GHG emissions. The WWTPs emit gases such as N_2O, CO_2, and CH_4. Increasing emission of GHG from this source pose harm to our climate. Biological mechanisms such as emissions of CO_2 due to microbial respiration, emission of N_2O by nitrification and de-nitrification, and emission of CH_4 from anaerobic digestion processes are direct emissions from WWTPs. Sources that not regulated directly within the WWTP are indirect internal emission sources; consumption of thermal energy and indirect external emission sources; third-party biosolids hauling, chemical productions and their transportation to the plant, etc. The increasing rate of GHG emissions is due to the changes in the economic output, extended energy consumption, increasing emission from landfills, livestock, rice farming, septic processes, and fertilizers as well as other factors. Increase industrialization, use of fertilizers, burning of fossil fuels and other human and natural activities result in a rise above normal average atmospheric temperature; thus posing threat to our environment. Research identifies methane and carbon dioxide as the main GHGs. Therefore, the reduction of methane concentration in the atmosphere, both from natural and anthropogenic sources, is indispensable to tackle the negative outcomes of global warming.

12.4 GREEN HOUSE EFFECTS

Atmospheric scientists first used the word 'greenhouse effect' in the later 1800s [5]. At that time, it was used to designate the naturally happening functions of trace gases in the atmosphere and did not have any negative implications. It was not up until the mid-1950s that the term greenhouse effect was attached to concern over climate alteration. And in contemporary decades, we often hear about the greenhouse effect in somewhat negative terms. The negative concerns are related to the possible impacts of an improved greenhouse effect. It is important to remember that without the greenhouse effect, lifecycle on earth as we know it would not be possible. While the earth's temperature is reliant on upon the greenhouse-like action of the atmosphere, the extent of heating and cooling are toughly influenced by several factors just as greenhouses are pretentious by various factors. In the atmospheric greenhouse effect, the type of surface that sunlight first happenstances are the most important factor. Forests, grasslands, ocean surfaces, ice caps, deserts, and cities all absorb, reflect, and radiate radiation differently. Sunlight falling on a white glacier surface strongly reflects back into space, resulting in minimal heating of the surface and lower atmosphere. Sunlight falling on a dark desert soil is strongly absorbed, on the other hand, and contributes to significant heating of the surface and lower atmosphere. Cloud cover also affects greenhouse warming by both reducing the amount of solar radiation reaching the earth's surface and by reducing the amount of radiation energy emitted into space. Scientists outline the percentage of solar energy reflected back by a surface. Understanding local, regional, and global effects are life-threatening to foretelling global climate change.

In 2013, the Intergovernmental Panel on Climate Change (IPCC) Fifth Assessment Report concluded [6], "It is *extremely likely* that human influence has been the dominant cause of the observed warming since the mid-20th century." The largest human influence has been the emission of GHGs such as carbon dioxide, methane, and nitrous oxide. Climate model projections summarized in the report indicated that during the 21st century, the global surface temperature is likely to rise a further 0.3 to 1.7°C (0.5 to 3.1°F) to 2.6 to 4.8°C (4.7 to 8.6°F) depending on the rate of GHG emissions and on climate feedback effects. These findings have been recognized by the national science academies of the major industrialized nations and are not disputed by any scientific body of national or international standing. Future climate change effects are expected to include rising sea levels, ocean acidification, regional changes in precipitation, and expansion of deserts in the subtropics. Surface temperature increases are greatest in the Arctic, with the continuing retreat of glaciers, permafrost, and sea ice. Predicted regional precipitation effects include more frequent extreme weather events such as heatwaves, droughts, wildfires, heavy rainfall with floods, and heavy snowfall. Effects directly significant to humans are predicted to include the threat to food security from decreasing crop yields, and the abandonment of populated areas due to rising sea levels. Environmental impacts appear likely to include the extinction or relocation of ecosystems as they adapt to climate change, with coral reefs, mountain ecosystems, and Arctic ecosystems most immediately threatened. Because the climate system has a large "inertia" and GHGs will remain in the atmosphere for a long time, climatic changes and their effects will continue to become more pronounced for many centuries even if further increases to GHGs stop. Measurements of temperature taken by instruments all over the world, on land and at sea have revealed that during the 20th century the earth's surface and lowest part of the atmosphere warmed up on average by about 0.6C. During this period, man-made emissions of GHGs, including carbon dioxide, methane, and nitrous oxide have increased, largely as a result of the burning of fossil fuels for energy and transportation, and land-use changes including deforestation for agriculture. In the last 20 years, concern has grown that these two phenomena are, at least in part, associated with each other. That is to say, global warming is now considered most probably to be due to the increases in GHG emissions and concurrent increases in atmospheric GHG concentrations, which have enhanced the Earth's natural greenhouse effect. Whilst other natural causes of climate change can cause the global climate to change over similar periods of time, computer models demonstrate that in all probability there is a real discernible human influence on the global climate. If the climate changes as current computer models have projected, global average surface temperature could be anywhere from 1.4 to 5.8C higher by the end of the 21st century than in 1990. To put this temperature change into context, the increase in global average surface temperature which brought the Earth out of the last major ice age 14,000 years ago was of the order of 4 to 5C. Such a rapid change in climate will probably be too great to allow many eco-systems to suitably adapt, and the rate of species extinction will most likely increase. In addition to impacts on wildlife and species biodiversity, human agriculture, forestry, water resources, and health will all be affected. Such impacts will be related to changes in precipitation (rainfall and snowfall), sea level, and the frequency and intensity of extreme weather events, resulting from global warming. It is expected that the societies currently experiencing existing social, economic, and climatic stresses will be both worst affected and least able to adapt. These will include many in the developing world, low-lying islands and coastal regions, and the urban poor. The Framework Convention on Climate Change (1992) [7] and the Kyoto Protocol (1997) [8] represent the first steps taken by the international community to protect the Earth's climate from dangerous man-made interference. Currently, nations have agreed to reduce GHG emissions by an average of about 5% from 1990 levels by the period 2008 to 2012. The UK, through its Climate Change Programme, has committed itself to a 12.5% cut in GHG emissions. Additional commitments for further GHG emission reduction will need to be negotiated during the early part of the 21st century, if levels of GHG concentrations in the atmosphere are to be stabilized at reasonable levels. Existing and future targets can be achieved by embracing the concept of sustainable development-development today that does not compromise the development needs of future generations. In practical terms, this means using resources, particularly fossil-fuel-derived energy, more efficiently, re-using, and recycling products where possible, and developing renewable forms of energy which are inexhaustible and do not pollute the atmosphere.

The greenhouse effect is mostly caused by the interaction of the sun's energy with GHGs such as carbon dioxide, methane, nitrous oxide and fluorinated gases in the Earth's atmosphere. The ability of these gases to capture heat is what causes the greenhouse effect. GHGs consist of three or more atoms. This molecular structure makes it possible for these gases to trap heat in the atmosphere and then transfer it to the surface which further warms the Earth. This uninterrupted cycle of trapping heat clues to an overall increase in global temperatures. The procedure, which is very similar to the way a greenhouse works, is the main reason why the gases that can produce this outcome are collectively called as GHGs. The prime forcing gases of the greenhouse effect are: carbon dioxide (CO_2), methane (C_{H4}), nitrous oxide (N_2O), and fluorinated gases. The greenhouse effect of solar radiation on the Earth's surface caused by GHGs (Figure 12.1).

The greenhouse effect is the procedure by which radioactivity from a globe's atmosphere warms the planet's surface to a temperature above what it would be without its atmosphere. Uncertainty a planet's atmosphere contains radioactively active gases (i.e., Greenhouse gases) they will radiate energy in all directions. Part of this radiation is directed towards the surface, warming it. The intensity of the downward radiation that is, the strength of the greenhouse effect will depend on the atmosphere's temperature and on the amount of GHGs that the atmosphere contains. Earth's natural greenhouse effect is critical to supporting life. Human activities, mainly the burning of fossil fuels and clearing of forests, have strengthened the greenhouse effect and caused global warming. The term "Greenhouse Effect" is an inaccuracy that ascended from a faulty analogy with the effect of sunlight passing through glass and warming a greenhouse. The way a greenhouse retains heat is fundamentally different, as a greenhouse works mostly by reducing airflow so that warm air is kept inside. GHGs such as carbon dioxide, methane, nitrous oxide, and halogenated compounds emissions are caused by human activities and some do occur naturally. The GHGs absorb IR radiation and trap heat in the atmosphere, thereby enhancing the natural greenhouse effect defined as global warming. This natural occurrence warms the atmosphere and makes life on earth possible, without which the low temperature will make life impossible to live on earth. "Gas molecules that captivate thermal IR radiation, and are in a substantial amount, can force the climate system. These type of gas molecules are called GHGs," Michael Daley, an associate professor of Environmental Science at Lasell College told Live Science. Carbon dioxide (CO_2) and other GHGs turn like a blanket, gripping IR radiation and preventing it from evading into outer space. The net effect is the steady heating of Earth's atmosphere and surface, and this process is called global warming.

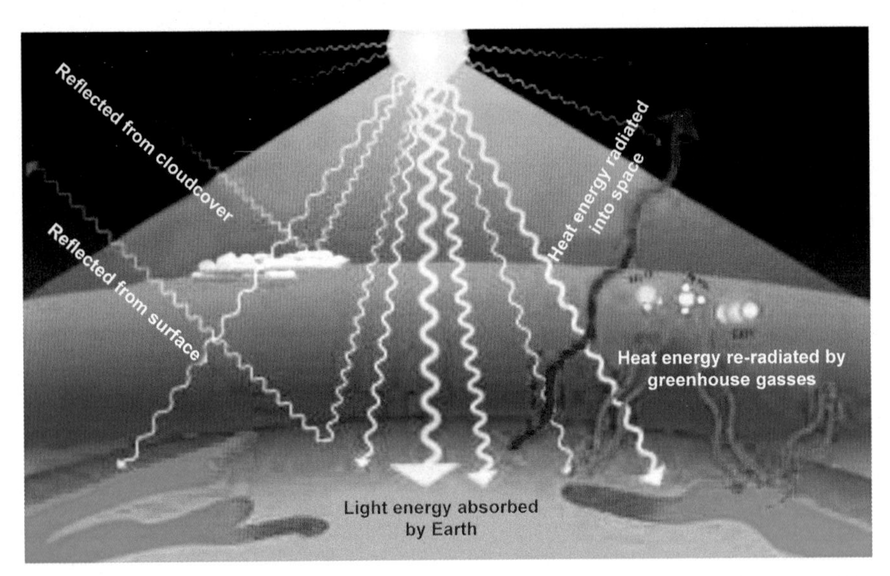

FIGURE 12.1 The greenhouse effect of solar radiation on the earth's surface caused by greenhouse gases.

These GHGs include water vapor, CO_2, methane, nitrous oxide (N_2O), and other gases. Since the dawn of the Industrial Revolution in the early 1800s, the scorching of fossil fuels like coal, oil, and gasoline have greatly increased the concentration of GHGs in the atmosphere, specifically CO_2, National Oceanic and Atmospheric Administration (NOAA). "Deforestation is the second largest anthropogenic basis of carbon dioxide to the atmosphere ranging between 6% and 17%," said Daley. Some human activities like the production and consumption of fossil fuels, use of various chemicals agriculture, burning bush, waste from incineration processes and other industrial activities have increased the concentration of GHGs, particularly CO_2, CH_4, and N_2O in the atmosphere making them harmful. This increase in atmospheric GHG concentration has led to climate change and global warming effect, which is motivating international efforts such as the Kyoto Protocol, signing of Paris Agreement on climate change and other initiatives to control negative outcomes of the greenhouse effect. The contribution of a GHG to global warming is commonly expressed by its GWP which enables the comparison of global warming impact of the gas and that of a reference gas, typically carbon dioxide. Atmospheric CO2 intensities have increased by more than 40% since the beginning of the Industrial Revolution, from about 280 parts per million (ppm) in the 1800s to 400 ppm today. The last time Earth's atmospheric levels of CO_2 reached 400 ppm was during the Pliocene Epoch, between 5 million and 3 million years ago, according to the University of California, San Diego's Scripps Institutions of Oceanography. The greenhouse effect, collective with growing levels of GHGs and the resultant global warming, is expected to have profound consequences, according to the near-universal consensus of scientists. If global warming undergoes unimpeded, it will cause noteworthy climate change, a rise in sea levels, increasing ocean acidification, life-threatening weather events and other severe natural and societal impacts, according to NASA, the Environmental Protection Agency (EPA) [9] and other scientific and governmental bodies.

12.5 GAS (WATER VAPOR) OF THE GREENHOUSE EFFECTS

Carbon dioxide is to some extent one of the GHGs. It involves one carbon atom with an oxygen atom bonded to each side. As soon as its atoms are bonded tightly together, the carbon dioxide molecule can absorb IR radiation and the molecule starts to vibrate. Eventually, the vibrating molecule will emit the radiation again, and it will likely be absorbed by yet another GHG molecule.

This absorption-emission-absorption cycle serves to keep the heat near the surface, effectively insulating the surface from the cold of space. Carbon dioxide, water vapor (H_2O), methane (CH_4), nitrous oxide (N_2O), and some limited other gases are GHGs. They all are molecules made up of more than two constituents atoms, bound loosely enough together to be able to vibrate with the absorption of heat. The foremost mechanisms of the atmosphere (N_2 and O_2) are two-atom molecules too closely bound together to vibrate and consequently, they do not absorb heat and subsidize to the greenhouse effect. Carbon dioxide, methane, nitrous oxide, and the fluorinated gases are all well-mixed gases in the atmosphere that do not react to changes in temperature and air pressure, so the levels of these gases are not affected by condensation effect. Water vapor also is a highly active component of the climate system that retorts briskly to fluctuations in conditions by either dwindling into rain or snow or evaporating to return to the atmosphere. Consequently, the imprint of the greenhouse effect is principally circulated through water vapor, and it turns as a fast reaction effect. Carbon dioxide and the other non-condensing GHGs are the vital gases within the Earth's atmosphere that tolerate the greenhouse effect and rheostat its strength. Water vapor is a fast-acting feedback but its atmospheric concentration is controlled by the radiative forcing supplied by the non-condensing GHGs. In fact, the greenhouse effect would collapse were it not for the presence of carbon dioxide and the other non-condensing GHGs. Together the feedback by the condensing and the forcing by the non-condensing gases within the atmosphere both play an important role in the greenhouse effect.

12.6 EFFECTS OF GLOBAL WARMING

The effects of global warming are the environmental and social changes caused (directly or indirectly) by human emissions of GHGs. There is a scientific consensus that climate change is occurring, and that human activities are the chief driver. Many influences of climate change have already been practical, including glacier retreat, changes in the timing of cyclical events (e.g., earlier flowering of plants), and changes in agricultural efficiency. Anthropogenic imposing has expected underwritten to some of the observed changes, comprising sea level rise, changes in climate extreme, declines in Arctic sea ice extent and glacier departure. Future effects of climate change will vary liable on climate change policies and social development. The two main policies to address climate change are dropping human GHG discharges climate change justification and adapting to the impacts of climate change. Near-term climate change rules could expressively affect long-term climate change effects. Severe mitigation rules might be able to limit global warming (in 2100) to around 2°C or below, comparative to pre-industrial levels. Without modification, increased energy demand and extensive use of fossil fuels might lead to global warming of around 4°C. Higher degrees of global warming would be more problematic to adapt to, and would increase the risk of undesirable impacts.

This article doesn't protection ocean acidification, which is straight caused by atmospheric carbon dioxide, not global warming. In this object, "climate change" means a change in climate that perseveres over a sustained period of time. The World Meteorological Organization describes this time period as 30 years. Examples of climate change include surges in global surface temperature (global warming), changes in rainfall forms, and changes in the occurrence of extreme weather events. Changes in climate may be due to natural causes, e.g., changes in the sun's output, or due to human actions, e.g., changing the composition of the atmosphere. Any human-induced changes in climate will occur against a related of natural climatic variations and of variations in human activity such as population growth on shores or in arid areas which increase or decrease climate susceptibility. Also, the term "anthropogenic imposing" refers to the influence exerted on a habitat or chemical environment by persons, as opposed to a natural process. "Discovery" is the process of representing that climate has changed in some defined numerical sense, without providing a reason for that change. Detection does not suggest attribution of the detected change to a specific cause. Acknowledgment of causes of climate change is the process of creating the most likely causes for detected change with some defined level of self-reliance. Detection and provenance may also be applied to observed changes in physical, ecological, and social systems.

12.6.1 TEMPERATURE CHANGES

The graph above shows the average of a usual of temperature replications for the 20th century (black line), trailed by projected temperatures for the 21st century based on three GHG discharges scenarios (colored lines) (Figure 12.2). This thing breaks down some of the impacts of weather change according to changed levels of future global warming. This technique of relating impacts has, for instance, been used in the IPCC Assessment

Reports on weather change. The instrumental temperature record displays global warming of around 0.6°C throughout the 20th century. In a revision carried out by David R. Easter ling et al. trends were observed finished a period of time. "It is clear from the detected record that there has been an increase in the global mean temperature of about 0.6°C. Since the start of the 20th century and that, this increase is related with a stronger heating in daily lowest temperatures than in maximums leading to a decrease in the daytime temperature range. This object is primarily about effects during the 21st century. For longer-term properties, see the long-term effects of global warming. See also the effects of global warming on humans (Figure 12.3).

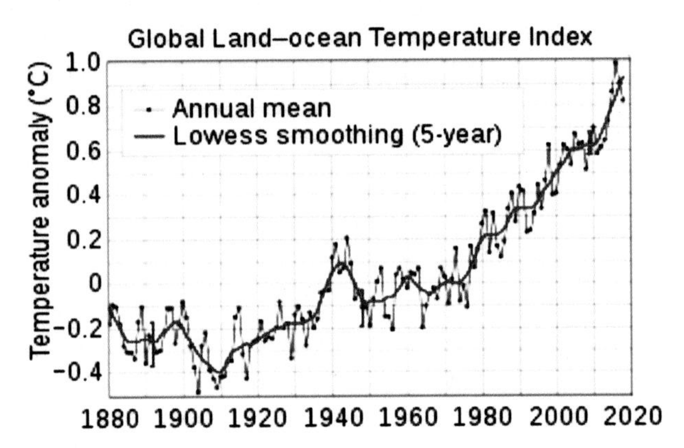

FIGURE 12.2 Global unpleasant surface temperature change since 1880, relative to the 1951–1980 mean.

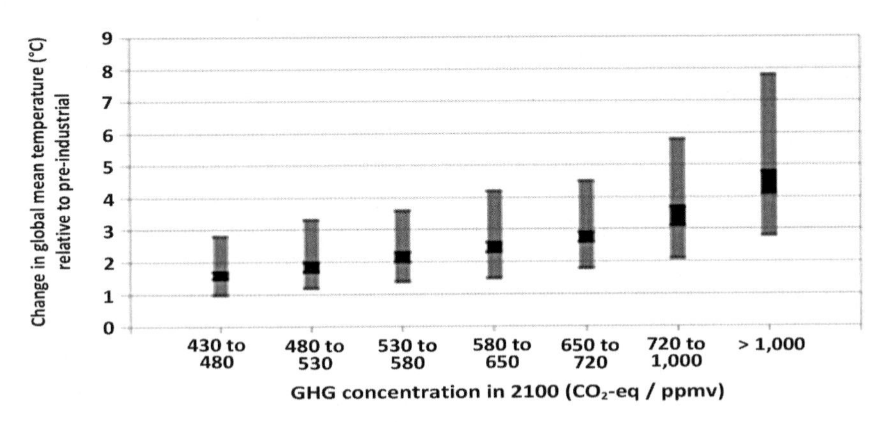

FIGURE 12.3 Projected global warming in 2100 for a range of emission scenarios.

The effects of global warming are the environmental and social changes produced (directly or indirectly) by human releases of GHGs. There is a scientific agreement that climate change is occurring, and that human activities are the primary driver. Many impacts of climate change have already been observed, including glacier retreat, changes in the judgment of seasonal events (e.g., earlier flowering of plants), and changes in agricultural output. Anthropogenic driving has likely contributed to certain of the observed changes, including sea-level rise, changes in climate extremes, and failures in Arctic sea ice extent and glacier retreat. Future properties of climate change will vary depending on climate change strategies and public development. The two main rules to statement climate change are reducing human GHG discharges (climate change mitigation) and adapting to the impacts of climate change. Near-term climate change policies could knowingly affect long-term climate change effects. Stringent mitigation policies might be able to limit global warming (in 2100) to around 2°C or below, relative to pre-industrial levels. Without mitigation, increased energy demand and extensive use of fossil fuels might lead to global warming of around 4°C. Higher magnitudes of global warming would be more difficult to adapt to, and would increase the risk of negative impacts. This article doesn't cover ocean acidification, which is directly caused by atmospheric carbon dioxide, not global warming. Each year, scientists learn more about the consequences of global warming, and many agree that environmental, economic, and health consequences are likely to occur if current trends continue. Melting glaciers, early snowmelt, and severe droughts will cause more dramatic water shortages and increase the risk of wildfires in the American West. Rising sea levels will lead to coastal flooding on the Eastern Seaboard, especially in Florida, and in other areas such as the Gulf of Mexico. Forests, farms, and cities will face troublesome new pests, heat waves, heavy downpours, and increased flooding. All those factors will damage or destroy agriculture and fisheries. Disruption of habitats such as coral reefs and Alpine meadows could drive many plant and animal species to extinction. Allergies, asthma, and infectious disease outbreaks will become more common due to increased growth of pollen-producing ragweed, higher levels of air pollution, and the spread of conditions favorable to pathogens and mosquitoes. Historical sea-level reconstruction and projections up to 2100 published in January 2017 by the U.S. Global Change Research Program [10] for the Fourth National Climate Assessment (Figure 12.4) and Map of the Earth with a six-meter sea level rise represented in red (Figure 12.5).

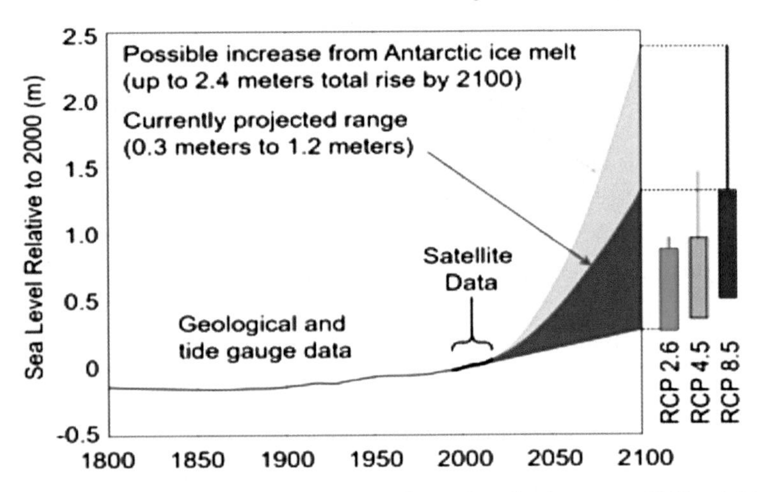

FIGURE 12.4 Historical sea-level reconstruction and projections up to 2100 published in January 2017 by the U.S. Global Change Research Program for the Fourth National Climate Assessment.

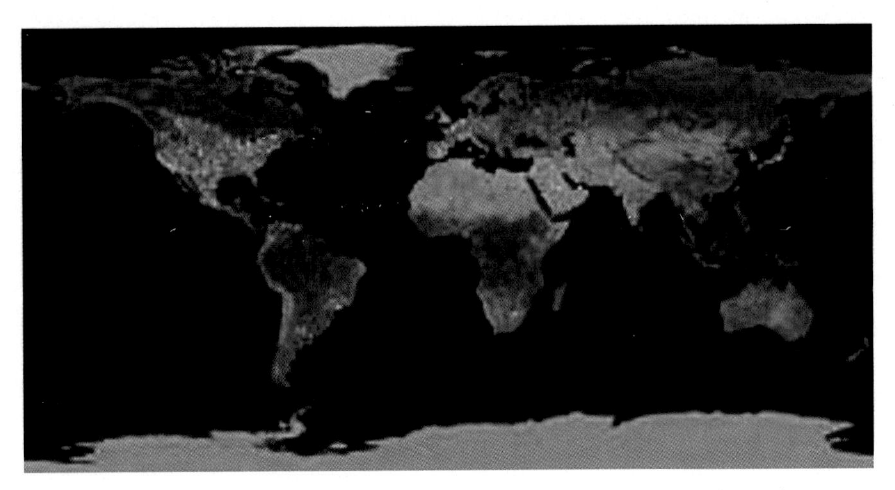

FIGURE 12.5 Map of the Earth with a six-meter sea level rise represented in red.

The effects of global warming are as follows (Figure 12.6):

1. **Physical Environmental:** The climate change melts sea ice, Geological Survey projects that two-thirds of polar bears will evaporate by 2050. Physical effects of climate change. The environmental properties

of global warming are comprehensive and extensive. They include the subsequent diverse effects: Arctic sea ice decline, sea-level rise, retreat of glaciers: global warming has controlled to periods of decrease and retreating of the Arctic sea ice, construction it susceptible to distinctive anomalies. Estimates of declines in Arctic sea ice vary. Current forecasts suggest that Arctic summers could be ice-free (defined as an ice extent of less than 1 million square km) as primary as 2025–2030. Since 1993, sea level has on normal risen with 3.1 ± 0.3 mm per year. Moreover, sea-level rise has enhanced from 1993 to 2017. Over the 21st century, the IPCC projects (for a high discharges scenario) that global mean sea level could increase by 52–98 cm. The rate of ice loss from glaciers and ice sheets in the Antarctic is a key area of uncertainty since Antarctica contains 90% of potential sea-level rise. Polar amplification and amplified ocean warmth are discouragement and threatening to clear Antarctic glacier outlets, potentially subsequent in more rapid sea-level rise.

FIGURE 12.6 Physical environmental.

2. **Extreme Weather, Extreme Events, and Tropical Cyclones:** Data exploration of extreme events from 1960 until 2010 recommends that droughts and heat waves appear simultaneously with increased

existence. Tremendously wet or dry events inside the monsoon period have increased since 1980. Predictions suggest a credible growth in the frequency and severity of some extreme weather actions, such as heatwaves, studies have also linked the quickly warming Arctic to extreme weather in mid-latitudes as the stream develops more erratic.

3. **Changes in Ocean Properties:** Increases in atmospheric CO_2 concentrations have led to an increase in dissolved CO_2 and as moment ocean acidity. Additionally, oxygen levels decrease because oxygen is less soluble in warmer water, an effect known as ocean deoxygenating.
4. The amount of global warming will be strong-minded primarily by anthropogenic CO_2 discharges. This is due to carbon dioxide's much extended lifetime in the atmosphere. Long-term effects also include a reply from the Earth's crust, due to ice melting and deglaciation, in a progression called post-glacial rebound, once land masses are no longer down by the weight of ice. This could lead to landslides and increased seismic and volcanic actions. Tsunamis could be produced by marine landslides produced by warmer ocean aquatic thawing ocean-floor permafrost or releasing gas hydrates. Sea level rise will remain over many centuries. A climate change, tipping points in the climate system: Climate change could consequence in global, large-scale changes. Some large-scale changes could happen abruptly, i.e., over a short time period, and might also be irreversible. Examples of unexpected climate change are the rapid release of methane and carbon dioxide from permafrost, which would lead to amplified global warming. Another instance is the possibility for the Atlantic Meridional Overturning Circulation to slow or to shut down. This could activate cooling in the North Atlantic, Europe, and North America. It would predominantly affect areas such as the British Isles, France, and the Nordic countries, which are warmed by the North Atlantic drift.
5. **Biosphere:** In terrestrial ecosystems, the previous timing of spring events, as well as opposite ward and upward shifts in plant and animal varieties, have been linked with high confidence to recent warming It is probable that most ecosystems will be artificial by higher atmospheric CO_2 levels, combined with higher global temperatures. Expansion in the subtropics is probably linked to global warming. Ocean acidification threatens impairment to coral reefs, fisheries, protected species, and other natural resources of value to society.

Without substantial actions to reduce the rate of global warming, land-based ecosystems are at risk of major ecological shifts, transforming composition and structure. Overall, it is expected that climate change will result in the extinction of many species and compact diversity of ecosystems. Rising temperatures have been found to push bees to their physiological limits, and could cause the destruction of bee populations continued ocean uptake of CO_2 may affect the intelligences and central nervous system of certain fish species, and that this impacts their capability to hear, smell, and avoid predators.

6. **Social Systems:** The effects of global warming on humans, Effects of global warming on human health, Climate security. The effects of climate change on human systems, frequently due to warming or shifts in precipitation designs, or both, have been noticed worldwide. The future social impacts of climate change will be rough through the world. Several risks are expected to increase with higher magnitudes of global warming. All regions are at risk of facing negative impacts, with low-latitude, less developed areas facing the greatest risk. Economic growth (measured in GDP growth) of lesser countries is much more impaired with predictable future climate warming than more developed countries. In minor islands and mega deltas, inundation as a result of sea-level rise is anticipated to threaten a vital organization and human disbursements. This could lead to concerns of homelessness in nations with low-lying areas such as Bangladesh, as well as statelessness for populations in countries such as the Maldives and Tuvalu. Climate change can be an important driver of migration, both within and between countries.' In 2014 [11], a meta-analysis concluded rise violence increases by up to 20% for each degree of warming, which includes fist fights, violent crimes, civil unrest, or wars. The violent herder-farmer conflicts in Nigeria, Sudan, and other countries in the Sahel region have been impaired by climate change. The crop production will probably be destructively affected in low latitude nations, while effects at northern latitudes may be positive or negative. Global warming of about 4.6°C relative to pre-industrial levels could attitude a large risk to global and regional food security. The impact of climate change on crop productivity for the four major yields was negative for wheat and maize, and neutral for soy and rice, in the years 1960–2013. Realize also Climate change and agriculture. Usually, impacts will be more negative than positive.

Impacts include: the effects of dangerous weather, leading to injury and loss of life; and indirect effects, such as undernutrition brought on by crop failures. There has been a shift from cold- to heat-related humanity in some regions as a result of warming. Temperature rise has been associated with increased numbers of suicides.

7. **Regional:** Regional impacts of climate change are now noticeable at more locations than before, on all landmasses, and across ocean regions. The Arctic, Africa, small islands, and Asian mega deltas are areas that are likely to be especially affected by future climate change. Africa is one of the most vulnerable continents to climate unevenness and change because of multiple existing stresses and low adaptive capacity. Present stresses include poverty, political conflicts, and ecosystem degradation. By 2050, between 350 million and 600 million people are projected to experience increased water stress due to climate change (see Climate change in Africa). Climate unevenness and change are projected to severely compromise agricultural production, including access to food, across Africa. Research projects that districts may even become run-down, with humidity and temperature reaching levels too high for humans to survive. Livelihoods of indigenous peoples of the Arctic have been altered by climate change, and there is emerging evidence of climate change impacts on livelihoods of indigenous peoples in other regions. Polar bears enter occupied areas more than in the past, owing to climate change. Global warming decreases sea-ice and forces bears to visit land in search of food.
8. **Responses:** The mitigation and adaptation to climate change are two paired responses to global warming. Effective version is easier in the case of considerable emission reduction. Many of the nations that contributed least to global GHG productions are most vulnerable to climate change, which increases about justice and equality with regard to mitigation and adaptation.
9. **Mitigation:** Each pathway shows how numerous measures (e.g., enhanced liveliness efficiency, increased use of renewable energy) could subsidize to releases reductions. Mitigation of climate change is the reduction of GHG discharges, or the enrichment of the capacity of carbon descends to absorb GHGs from the atmosphere. There is a large potential for future decreases in emissions by a mixture of activities, including energy conservation and increased energy efficiency; the use of energy skills, such as renewable energy, nuclear

energy, and carbon capture and storage; decarbonizing buildings and transport; and ornamental carbon sinks through, for example, reforestation, and foiling deforestation. A 2015 [12] report by Citibank determined that transitioning to a low carbon economy would yield a confident return on savings.

Near and long-term trends in the global energy system are unpredictable with limiting global warming at below 1.5 or 2°C, comparative to pre-industrial levels. Current pledges through as part of the Paris Agreement would lead to about 3.0°C of warming at the end of the 21st century, comparative to pre-industrial levels. In limiting warming at below 2°C, more severe emission decreases in the near-term would allow for less rapid decreases after 2030, and be inexpensive complete. Many combined replicas are unable to meet the 2°C target if pessimistic assumptions are made about the availability of mitigation technologies. CO may help society and individuals more quickly. The biking reduces GHG emissions while dropping the effects of a sedentary lifestyle at the similar time The development and scaling-up of clean technology, such as cement that produces less CO_2 is critical to achieve sufficient announcement decreases for the Paris settlement goals. Mitigation at an individual level is also possible. The most significant action individuals could make to mitigate their own carbon footprint is to have fewer children, followed by living car-free, forgoing air travel, and adopting a plant-based diet.

10. **Adaptation:** Climate change adaptation is the procedure of adjusting to actual or predictable climate change and its effects. Humans can struggle to moderate or avoid harm due to climate change and activity chances. Examples of adaptation are improved coast-line guard, better disaster management and the development of crops that are more impervious. The adaptation may be planned, either in reaction to or anticipation of global warming, or spontaneous, i.e., without government interference. The public section, private sector, and communities are all gaining experience with adaptation and adaptation is becoming embedded within certain planning procedures. While some adaptation responses call for trade-offs, others bring synergies and co-benefits. Environmental governments and public figures have highlighted changes in the climate and the risks they entail, while promoting adaptation to changes in infrastructural needs and releases reductions. Adaptation is especially significant

in developing countries since those countries are predicted to bear the brunt of the effects of global warming. The potential for humans to adapt (called adaptive capacity) is unequally distributed across different regions and populations, and developing countries generally have less capability to adapt.

11. **Climate Engineering:** (occasionally called geoengineering or climate interference) is the cautious modification of the climate. It has been examined as a likely response to global warming. Procedures under research fall normally into the classes' solar radiation management and carbon dioxide removal, although several other schemes have been suggested. A study from 2014 [13] investigated the most common climate engineering methods and concluded they are either ineffective or have hypothetically severe side effects and cannot be immobile without causing rapid climate change.
12. **Society and Culture:** The atmospheric attentiveness of CO_2, emissions worldwide would need to be intensely reduced from their present level. Most countries in the world are parties to the United Nations Framework Convention on Climate Change (UNFCCC). The final objective of the Convention is to prevent hazardous human interference of the climate system. As stated in the Convention, this requires that GHG absorptions are stabilized in the atmosphere at a level where ecosystems can adapt logically to climate change, food production is not threatened, and economic development can proceed in a sustainable fashion. The Framework Convention was agreed on in 1992, but global discharges have risen since then. During discussions, the G77 (a lobbying group in the United Nations representing developing countries) pushed for a mandate requiring developed countries to "[take] the lead" in reducing their emissions. This was justified on the basis that the developed countries' emissions had contributed most to the accumulation of GHGs in the atmosphere, per-capita emissions (i.e., emissions per head of population) were still relatively low in developing countries, and the emissions of developing countries would grow to meet their development needs. These first-round promises expired in 2012. The United States President Bush prohibited the treaty on the basis that "it exempts 80% of the world, including major population centers such as China and India, from compliance, and would cause serious harm to the US economy. In 2009 [14], several UNFCCC Parties produced widely depicted as dissatisfied because of its low goals, leading poor nation to reject it.

Parties associated with the Accord aim to limit the future increase in global mean temperature to below 2°C. In 2015, a binding agreement was negotiated in Paris with all UN countries with the aim to keep climate change well below 2°C. The agreement replaces the Kyoto protocol. Unlike Kyoto, no obligatory emission targets are set in the Paris agreement. Instead, the procedure of uniformly setting ever more ambitious goals and reevaluating these goals every five years has been made obligatory. The Paris agreement repeated that developing countries must be commercially supported.

13. **Scientific Discussion:** Scientific discussion takes place in objects that are reviewed and evaluated by scientists who work in the significant fields and participate in the IPCC. The scientific consensus as of 2013 [15] stated in the IPCC Fifth Assessment Report is that it "is extremely likely that human effect has been the dominant cause of the observed warming since the mid-20th century." A 2008 report by the U.S. National Academy of Sciences stated that most scientists by then agreed that experiential warming in recent decades was chiefly caused by human activities increasing the amount of GHGs in the atmosphere. In 2005 [16], the Society quantified that while the irresistible majority of scientists were in agreement on the main points, some individuals and organizations opposed to the agreement on urgent action needed to reduce GHG emissions had tried to undermine the science and work of the IPCC. National science conservatoires have called on world leaders for policies to cut global emissions. In 2018, the IPCC published a Special Report on Global Warming of 1.5°C which cautioned that, if the current rate of GHG emissions is not lessened, global warming is likely to reach 1.5°C (2.7°F) between 2030 and 2052 causing major disasters. The report said that avoiding such crises will require a swift alteration of the global economy that has "no documented historic pattern." In the scientific literature, there is a strong consensus that global surface temperatures have increased in current decades and that the trend is caused mostly by human-induced releases of GHGs. No scientific body of countrywide or global standing disagrees. In November 2017 [17], a second warning to humanity signed by 15,364 scientists from 184 countries stated that "the current trajectory of hypothetically catastrophic climate change due to rising conservatory gases from burning fossil fuels, deforestation, and agronomic production-particularly from farming ruminants for meat consumption" is "especially disconcerting."

14. **Public Opinion and Disputes:** The global warming controversy refers to a variety of disagreements, substantially more noticeable in the popular media than in the scientific literature, about the nature, causes, and consequences of global warming. The uncertain issues include the causes of increased global average air temperature, particularly since the mid-20th century, whether this warming trend is extraordinary or within normal climatic variations, whether civilization has contributed significantly to it, and whether the increase is completely or partially an artifact of poor measurements. Further disputes concern estimates of climate sensitivity, predictions of additional warming, and what the consequences of global warming will be. From about 1990, onward, conservative think had begun challenging the legitimacy of global warming as a social problem. They challenged the scientific evidence, argued that global warming would have benefits, and asserted that proposed solutions would do more harm than good. Organizations such as the libertarian Competitive, conservative commentators, and some corporations have challenged IPCC climate change scenarios, funded scientists who affect with the scientific consensus, and provided their own projections of the economic cost of stricter controls. On the other hand, some fossil fuel companies have ascended back their efforts in current years, or even called for strategies to reduce global warming. Global oil companies have begun to recognize climate change happens and is caused by human events and the burning of fossil fuels.

The global warming problem came to worldwide public consideration in the late 1980s and polling collections began to track opinions on the subject. The longest reliable polling, by Gallup in the US, found relatively small deviations of 10% or so from 1998 to 2015 in opinion on the seriousness of global warming, but with cumulative polarization between those concerned and those unconcerned. By 2010 in the US, just a little over half the population (53%) observed it as a thoughtful concern for either themselves or their families. Latin America and developed Asia saw themselves most at risk at 73% and 74%. In the assessed 111 countries, people were more likely to characteristic global warming to human activities than to usual causes, except in the US where nearly half (47%) of the population attributed global warming to usual causes. A 2013 survey by Center sampled 39 countries about global threats. According to 54% of those questioned, global warming featured top of the perceived global threats. Due to unclear media coverage in the early 1990s, issues such as ozone depletion and climate change were often mixed up, affecting public understanding of these issues. According to a 2010 survey of Americans, a common thought that the ozone layer and spray cans contribute to global warming. Although there are a few areas of linkage, the association between the two is not strong. Reduced stratospheric ozone has had a slight cooling influence on surface temperatures, while increased troposphere ozone has had a slightly larger warming effect. However, chemicals causing ozone depletion are also powerful GHGs, and as such, the Montreal protocol against their emissions may have done more than any other measure to mitigate climate change. In response to seeming inaction on climate change, a climate movement is protesting in various ways, such as fossil fuel divestment, worldwide protests.

12.7 CONCLUSIONS

The proportions of positive scheme gases to be moderately see-through to in coming observable lively from the sun, however impervious to the potency produced from the sphere is one of the best quiet events in the characteristic disciplines. This quantity, the greenhouse effect, is whatever varieties the earth a comfortable place for human being's activities. I acclamation future work to be done on GHGs and global warming. The present rate of different increase of $CO_2$3 ppm per year [1] is an order of magnitude or more greater than the increase in atmospheric CO_2 during the last deglaciation, when rapid departure of northern hemisphere ice expanses led to rates of sea-level rise of up to 5 m per century [18]. The last time the troposphere contained CO_2 levels comparable to today's values, during the Pliocene, surface heats were on average ~3°C warmer, the Greenland ice sheet collapsed, and sea level rose by up to 30 m^2 [19]. With the grouping of constant scorching of fossil fuels and the additional influence of GHGs to the troposphere through positive feedbacks in the climate system, future atmospheric CO_2 levels could exceed 1,000 ppm [20] levels well above the constancy verge values for mainland ice on Earth [21]. In fact, it is essential to look back at least 34 million years prior to the present icehouse to examine climate change under such CO_2 levels. In this context, the greatness and rate of the present conservatory gas increase place the climate system in what could be one of the most severe increases in radiative limiting of the world wide weather organization in ground anti quality. To entirely assess temperature helpful responses and listing arguments that may exemplify Earth's upcoming, and to better understand climate change influences and retrieval, it is necessary to examine the records from past warm periods when there were similar sizes and rates of GHG making. Exciting research occasions to help achieve this task exist in the untapped potential of the deep-time geological record. This study recognizes a six-element research agenda calculated to describe past climate unevenness and to better oblige how. Earth's climate system has replied to episodes of altering GHG levels. The knowledge increased by this scientific agenda will be significant for addressing questions regarding the projected rise in atmospheric CO_2 and the societal implications of this rise. The boom also describes the research organization essential for successful implementation of the deep-time pale climatology agenda, as well as an education and outreach strategy designed to broaden our collective understanding of the unique perspective that the full range of the geological record provides for future climate change. Improved Understanding of Climate Sensitivity and CO_2-Climate Coupling. Responsible for the sensitivity of Earth's mean surface temperature to increased GHG levels in the atmosphere is a key obligation for estimating the likely magnitude and effects of future climate change. The current understanding of climate compassion, defined on the basis of modern data and relatively recent paleoclimate records ($\leq$ 20,000 years), is connected with large uncertainty (1.5 to $\geq$6°C). Positive responses typically considered to have been active on longer timescales, but that may become increasingly relevant with continued warming, are not considered in these estimates. A better definition of long-term equilibrium climate sympathy including more refined limits on its lower boundary—over the full range and timescales of past radiative forcing is a major research priority. A connected focus is on gaining an improved understanding of how climate feedbacks and their role in amplifying climate change have varied with changes in GHG forcing. Accomplishing this objective will require the development of more accurate and precise paleo-CO_2 and paleotemperature proxies, as well as the growth of new proxies for the full range of GHGs. A complementary requirement is for high-resolution and high-precision time-series accounts, based on assimilating multiproxy methods. Earth's climate system would reply to cumulative levels of distinctive CO_2. Climate dynamics of hot tropics and warm poles.

Present climate demonstrating and deep-time paleo-climatology studies have confirmed that the long-standing example that the temperatures of tropical climates do not rise knowingly during warm periods since a particular type of temperature safeguarding mechanism is probably improper. Therefore, the devices and feedbacks in the modern climate system those have controlled humid and polar surface temperatures eventually leading to the existing comparatively high pole-to-equator thermal gradient may not operate in heater biospheres. A decreased latitudinal grade in the future, which would almost certainly be related with polar sea ice and mainland ice sheet losses, would change impressive wind patterns and, in turn, ocean circulation all having potential detrimental effects through teleconnections [22]. To improve information of the procedures and climate responses that may affect surface high temperature under complex atmospheric CO_2, it is important that high-temporal-resolution, higher-precision proxy time series be developed across latitudinal transects, with a focus on recreating terrestrial-marine connections. This will require a greatly increased effort in high-precision geochronological dating, coupled with considerably more spatially determined substitution archives. A more inclusive sympathetic of the limits of humid climate constancy, the origin of irregular polar heat, and an sympathetic of how a pathetic current pitch is established and preserved in warmer temperature organizations will require further climate model development and deep-time assessments. These assessments would also make available a much desirable test of the effectiveness of model prognoses of future climate.

Study of the present icehouse climate state has providing better restraints on CO_2 and surface temperature threshold levels for ice sheet stability [23]. Large gaps, however, remain in the understanding of ice sheet undercurrents, with subsequent confines on the applicability of current attached climate-ice sheet replicas. These issues climax the reservations that still exist in forecasts regarding the periods at which ice sheets would respond to continued warming and in thoughtful the influence of responses not exposed by recent paleoclimate records or considered by future climate model forecasts (e.g., the projections used in IPPC, 2007) [24]. Therefore, the magnitude of sea-level rise, once climate stability is reached, remains elusive despite deep-time paleoclimate evidence that it could be substantially higher than model projections [25]. To markedly improve the sympathetic of climate-ice sheet-sea level dynamics applicable to a warming Earth, it will be essential to probe deeper into Earth's history to the periods of truly disastrous ice sheet collapse that accompanied past icehouse-to-greenhouse transitions. To fully exploit such deep-time archives will require radiometrically constrained and spatially resolved marine, paralic, and terrestrial records for both high and low latitudes. In addition, improved methods for de-convolving temperature and seawater $\delta^{18}O$ from proxy records are required, as well as targeted efforts to couple land-ice component replicas with complex global climate simulations that are capable of assimilating the atmospheric hydrological cycle.

There are broad technical agreements that one of the largest influences of continued CO_2 forcing would be major regional climate modifications, with the probability of substantial common impacts (e.g., water shortages, flooding). The understandings gained from recreating the processes and climate feedbacks that influence surface temperatures below higher atmospheric CO_2 levels are an important component of this research agenda, chiefly because of the sensitivity of climate to small changes in high-latitude and tropical surface temperatures as an importance of teleconnections. Such studies should lead to an improved sympathetic of how various components of the climate system replied to abrupt changes, in particular during times when the rates of change were adequately large to chance diversity. Furthermore, directing such intervals for more detailed study is a critical requirement for obliging how long any unexpected climate change might persist. Thoughtful Ecosystem Thresholds and Flexibility in a heating-up world. Mutually ecosystems and human civilization are highly sensitive to abrupt shifts in climate, because such shifts may exceed the acceptance of organisms and, consequently, have major properties on biotic diversity as well as human investments and societal stability. Representative future biodiversity fatalities and biosphere-climate responses, however, is fundamentally challenging because of the complex, nonlinear relatives with stimulating effects that consequence in an unknown net response to climatic pressuring. Normal and physical schemes can adjust to unexpected climate change is a very important question additional present-day global heating. An important tool to address is to describe and appreciate the consequence of equal "natural trials" in the deep-time geological record, particularly where the magnitude and/or rates of change in the global climate system were sufficiently large to threaten the sustainability and multiplicity of species, which at times led to mass extinctions.

KEYWORDS

- **atmospheric lifetime**
- **carbon dioxide**
- **chlorofluorocarbons**
- **climate change**
- **climate models**

- **global warming**
- **global warming potential**
- **greenhouse effect**
- **greenhouse gas**
- **nitrous oxide**
- **ozone**
- **solar constant**
- **solar radiation**

REFERENCES

1. In 2013, the Intergovernmental Panel on Climate Change (IPCC) Fifth Assessment Report.
2. IPCC AR4 WG1, (2007). In: Solomon, S., Qin, D., Manning, M., Chen, Z., Marquis, M., Averyt, K. B., Tignor, M., & Miller, H. L., (eds.), *Climate Change 2007: The Physical Science Basis, Contribution of Working Group I to the Fourth Assessment Report of the Intergovernmental Panel on Climate Change.* Cambridge University Press, ISBN: 978-0-521-88009-1.
3. Pagani et al., (2010). *The Atmosphere Contained CO_2 Levels Comparable to Today's Values, During the Pliocene, Surface Temperatures Were on Average ~3°C Warmer, the Greenland ice Sheet Collapsed, and Sea Level Rose by up to 30 m.*
4. *Enhanced Greenhouse Effect*, (2010). Ace.mmu.ac.uk. Archived from the original on 2010-10-24. Retrieved 2010-10-15.
5. *NASA Earth Fact Sheet*. Nssdc.gsfc.nasa.gov. Retrieved 2010-10-15.
6. NASA: Climate Forcings and Global Warming, (2009). "*Enhanced Greenhouse Effect-Glossary*." Nova. Australian Academy of Scihuman impact on the environment-2006.
7. Synthesis Report: Summary for Policymakers (PDF). *IPCC Fifth Assessment Report* (p. 4).
8. Allaby, A., & Allaby, M., (1999). *A Dictionary of Earth Sciences* (p. 244). Oxford University Press. ISBN: 978-0-19-280079-4.
9. Bowen, M., (2006). *Thin Ice: Unlocking the Secrets of Climate in the World's Highest Mountains.* Owl Books. ISBN: 978-1429932707.
10. Temperature change and carbon dioxide change, U.S. National Oceanic and Atmospheric Administration.
11. Brian, S., (2004). *Favorite Demonstrations for College Science: An NSTA Press Journals Collection* (p. 57). NSTA Press. ISBN 978-0-87355-242-4.
12. Claussen, E., Cochran, V. A., & Davis, D. P., (2001). Global climate data. *Climate Change: Science, Strategies, and Solutions* (p. 373). University of Michigan. ISBN: 978-9004120242.
13. Easterbrook, S., (2015). "*Who First Coined the Term "Greenhouse Effect."*" Serendipity.
14. Hansen, J., (2005). "A slippery slope: How much global warming constitutes dangerous anthropogenic interference"? *Climatic Change, 68*(333), 269–279.

15. Held, I. M., & Soden, B. J., (2000). "Water vapor feedback and global warming." *Annual Review of Energy and the Environment, 25,* 441–475.
16. Hileman, B., (2005). "Ice core record extended." *Chemical and Engineering News, 83*(48), 7.
17. *IPCC Fourth Assessment Report.* Working Group I Report "The Physical Science Basis" Chapter 7" Atmospheric Carbon Dioxide-Mauna Loa." NOAA. "Climate Milestone: Earth's CO_2 Level Passes 400 ppm." National Geographic. 2013-05-12. Retrieved 2017-12-10.
18. Jacob, D. J., (1999). "*7. The Greenhouse Effect*" (p. 97). Introduction to atmospheric chemistry. Princeton University Press. ISBN 978-1400841547. "Solar Radiation and the Earth's Energy Balance." *Eesc.columbia.edu.* Retrieved 2010-10-1 Intergovernmental Panel on Climate Change Fourth Assessment Report. Chapter 1: Historical overview of climate change science. The elusive "absolute surface air temperature," see GISS discussion.
19. Kasting, J. F., (1991). "Runaway and moist greenhouse atmospheres and the evolution of earth and Venus." *Planetary Sciences: American and Soviet Research/Proceedings from the U.S.-U.S.S.R. Workshop on Planetary Sciences* (pp. 234–245). Commission on Engineering and Technical Systems (CETS).
20. Kiehl, J. T., & Trenberth, K. E., (1997). "Earth's annual global mean energy budget" (PDF). *Bulletin of the American Meteorological Society, 78*(2), 197–208.
21. Kurpaska, S., (2014). "*Energy Effects During Using the Glass with Different Properties in a Heated Greenhouse" (PDF): Technical Sciences,* (Vol. 17, No. 4, pp. 351–360). "Titan: Greenhouse and Anti-greenhouse." Astrobiology Magazine-earth science-evolution distribution origin of life universe-life beyond: Astrobiology is study of earth.
22. McKay, C., Pollack, J., & Courtin, R., (1991). "The greenhouse and ant greenhouse effects on Titan." *Science, 253*(5024), 1118–1121.
23. McNeill, L., (2019). "*This Lady Scientist Defined the Greenhouse Effect But Didn't Get the Credit, Because Sexi.*" Smithsonian.
24. Mitchell, J. F. B., (1989). "The "greenhouse" effect and climate change" (PDF). *Reviews of Geophysics, 27*(1), 115–139.
25. Oort, A. H., & Peixoto, J. P., (1992). *Physics of Climate.* New York: American Institute of Physics. ISBN978-0-88318-711-1. The name water vapor-greenhouse effect is actually a misnomer since heating in the usual greenhouse is due to the reduction of convection.

Index

Printed in the United States
by Baker & Taylor Publisher Services